Basic Steel Design with LRFD

Theodore V. Galambos

James L. Record Professor of Structural Engineering
University of Minnesota

F. J. Lin

Chairman and Chief Executive Officer,
Science, Engineering, Management, Inc.
South Pasadena

Bruce G. Johnston

The Late Professor Emeritus of Structural Engineering
University of Michigan

Prentice Hall, Upper Saddle River, New Jersey 07458

Library of Congress Cataloging-in-Publication Data
Galambos, Theodore V.

    Basic steel design with LRFD / Theodore V. Galambos, F. J. Lin,
Bruce G. Johnston
     p. cm.
    Includes bibliographical references and index.
    ISBN 0-13-059577-2
    1. Building, Iron and steel. 2. Load factor design. 3. Steel,
Structural.
TA684.J6 1996
624.1'821--dc20
                                            95-33165
                                              CIP

Editorial production: *bookworks*
Acquisitions editor: *Bill Stenquist*
Manufacturing manager: *Donna Sullivan*
Cover designer: *Bruce Kenselaar*

© 1996 by Prentice Hall
Prentice-Hall, Inc.
A Pearson Education Company
Upper Saddle River, NJ 07458

Printed in the United States of America

10 9 8 7 6 5 4 3 2

ISBN 0-13-059577-2

Prentice-Hall International (UK) Limited, London
Prentice-Hall of Australia Pty. Limited, Sydney
Prentice-Hall Canada Inc., Toronto
Prentice-Hall Hispanoamericana, S.A., Mexico
Prentice-Hall of India Private Limited, New Delhi
Prentice-Hall of Japan, Inc., Tokyo
Pearson Education Asia Pte. Ltd., Singapore
Editoria Prentice-Hall do Brasil, Ltda., Rio De Janeiro

This book is dedicated to the memory of
Bruce G. Johnston
who had the idea for this type of a steel design book
and who shepherded
BASIC STEEL DESIGN
through three successful editions

Contents

FOREWORD   ix
PREFACE   xi
ABBREVIATIONS   xv

1   THE STEEL STRUCTURE   1

  1.1   Introduction   1
  1.2   The Structure and Its Parts   2
  1.3   Structural Steel   4
  1.4   Loads on Structures   7
  1.5   Historical Development   7
  1.6   Structural Design Economy   10
  1.7   Structural Safety   12
  1.8   Planning and Site Exploration
       for the Specific Structure   17
  1.9   Structural Layout, Details, and Drawings   17
  1.10   Fabrication Methods   19
  1.11   Construction Methods   20
  1.12   Service and Maintenance Requirements   20

2   TENSION MEMBERS   23

2.1   Introduction   23
2.2   Types of Tension Members   25
2.3   Design Limit States and Effective Net Area   30
2.4   Design for Repeated Load   32
2.5   Flowchart   33
2.6   Illustrative Examples   36

3   BEAMS   45

3.1   Introduction   45
3.2   Elastic Bending of Steel Beams   50
3.3   Inelastic Behavior of Steel Beams   53
3.4   Flexural Design of Beams   55
3.5   Lateral Support Requirements   63
3.6   Beam Deflection Limitations   64
3.7   Beams Under Repeated Load   65
3.8   Biaxial Bending of Beams   65
3.9   Load and Support Details   66
3.10   Load Tables for Beams   68
3.11   Illustrative Examples   68

4   COLUMNS UNDER AXIAL LOAD   86

4.1   Introduction   86
4.2   Basic Column Strength   87
4.3   Effective Length of Columns   92
4.4   Types of Steel Columns   94
4.5   Width/Thickness Ratios   99
4.6   Column Base Plates and Splices   101
4.7   Design Compressive Strength   101
4.8   Illustrative Examples   102

5   MEMBERS UNDER COMBINED FORCES   119

5.1   Introduction   119
5.2   Design by Use of Interaction Formulas   122
5.3   Equivalent Axial Compression Load   125
5.4   Illustrative Examples   126

6   CONNECTIONS   136

6.1   Introduction   136
6.2   Riveted and Bolted Connections   139
6.3   Pinned Connections   148
6.4   Welded Connections   150
6.5   Eccentrically Loaded Connections   159

6.6   Shear Connections for Building Frames   166
6.7   Moment-Resisting Connections   173
6.8   Bolted End-Plate Connections   179
6.9   Concluding Remarks Concerning Connections   180

7   PLATE GIRDERS   183

7.1   Introduction   183
7.2   Selection of Girder Web Plate   185
7.3   Selection of Girder Flanges   189
7.4   Intermediate Stiffeners   194
7.5   Bearing Stiffeners   197
7.6   Connections of Girder Elements   198
7.7   Illustrative Examples   202

8   CONTINUOUS BEAMS AND FRAMES   215

8.1   Introduction   215
8.2   Moment-Distribution Analysis: A Summary   216
8.3   Elastic Design of Continuous Beams   218
8.4   Elastic Design of Continuous Frames   225
8.5   Introduction to Plastic Design   234
8.6   Plastic Design of Frames   238

9   COMPUTER-AIDED TECHNOLOGY   249

9.1   Introduction   249
9.2   Basic Flowchart Programming   250
9.3   Computer-Aided Design   251
9.4   Computer-Aided Optimization   257

10   COMPOSITE CONSTRUCTION   259

10.1   Introduction   259
10.2   Flexural Strength of Composite Cross Section   261
10.3   Design of Composite Beams   270
10.4   Composite Columns   273
10.5   Composite Beam Design Examples   275

11   SPECIAL TOPICS IN BEAM DESIGN   282

11.1   Introduction   282
11.2   Torsion   282
11.3   Combined Bending and Torsion   286
11.4   Biaxial Bending and Lateral-Torsional Buckling   300
11.5   Shear Center   307

INDEX   317

Foreword

This is the first edition of a book which was originally published under the name *Basic Steel Design* in 1974. The first edition of that book had several features which were unique for the time it appeared. 1. It was a book which was primarily directed toward beginning structural engineering students as the text for their first course in design, and therefore it was not burdened down with all the knowledge necessary for being a full fledged structural engineer; it contained just enough material that could be learned and retained in one term of instruction. 2. It was a book that presented the basic steps of structural design in the form of flowcharts which then could be used by the student to program the procedures on the computer. 3. It was meant to go in tandem with the *Specification for the Design, Fabrication and Erection of Structural Steel for Buildings* of the American Institute of Steel Construction (AISC) because this design standard was prototypical of all other codes which the student would later encounter in the real design world.

This new edition attempts to retain all of the features of the earlier versions. Much has happened in the time since the publication of the third edition of *Basic Steel Design* in 1986: The world of steel design specifications has moved from *Allowable Stress Design* to *Load and Resistance Factor Design (LRFD)*. This latter method is also known in many parts of the world as *Limit States Design*. We retain the acronym LRFD in the title to signify that the total context of the book was changed to this method, following the new AISC *Load and Resistance Factor Design Specification for Structural Steel Buildings*. The other big change is the ubiquitous availability of personal computers with spreadsheet or mathematical software. The example problems are tailored for these types of programs. Finally, the next change taking place is the move toward the adoption of the SI units in design offices. Most chapters of the book have problems in SI units also. While the book is intended primarily as a first text for students in universities and technical schools, it will be invaluable help to the practicing engineer who has been away from the classroom and who wants to, or needs to, relearn steel design by the new LRFD method.

Preface

This book is concerned with the basics of structural steel design. It is suitable for reference use or as a text and is unique in at least two respects: (1) reference is made primarily to a single design specification—that of the American Institute of Steel Construction, and (2) the chapters on individual member design include flow diagrams, similar to those used to guide the development of a computer program. These diagrams have been found to be excellent teaching aids.

At a time when the electronic digital computer has entered into every phase of structural design, from initial planning to the production of finalized detail drawings, it is increasingly important to be able to understand and visualize every phase of structural behavior. The computer is a robot and it must be guided with intelligence by the engineer. In this book the use of complex analyses is minimized. In the initial study of steel design, the acquisition of a basic understanding of structural behavior and the meaning of specification requirements can best be attained by simplicity of approach and emphasis on the development of sound structural judgment.

Chapter 1 is a broad and descriptive introduction to the steel structure, covering the properties of steel, the history of the development of steel structures, and touching on the topics of economy, safety, planning, fabrication, construction, and maintenance.

Chapters 2 through 7 are devoted to the various types of structural members in common use: the tension member, the beam, the column, and so on. Each of these chapters takes up the structural behavior problem, explains pertinent AISC Specification clauses, and summarizes the logical application of the specification by means of flow diagrams. Although the flow diagram was developed primarily as an aid to the development of a computer program, it also serves admirably as a summary and as a guide to the logical sequence of steps that must be taken in the design selection of a particular structural member.

Chapter 8 goes beyond the treatment of the individual member to provide a study of both the elastic and plastic design of continuous beams and frames. It provides a review and summary of the moment-distribution method as applied to the design of such structures.

Chapter 9, on computer-aided technology, does not attempt to teach computer programming. It presents a broad survey of the current scope of computer technology and design.

Chapter 10, on composite construction, covers design procedures that take advantage of the economy resulting from the joint action of steel and concrete when structural shapes are encased in concrete.

Chapter 11 covers special topics that are of occasional importance, but less common than those treated in the earlier chapters. These topics include the torsion of both open and box members, combined bending and torsion, biaxial bending, lateral-torsional buckling, and the shear center.

Steel design specifications are essentially similar, but the diversity of formulas pertaining to identical problems (such as the column design formulas) is confusing to the beginning student. But after learning to design with a particular specification the student can readily adapt his basic knowledge to a different one.

The purchase of the *Steel Construction Manual* of the American Institute of Steel Construction is essential to the complete use and understanding of this book. The AISC *Manual* also includes the AISC Specification, which will be referred to throughout the book. Moreover, practice in use of the AISC *Manual* is essential as a secondary educational objective toward structural design practice. The nomenclature that will be used herein is nearly identical with that found in the AISC Specification and will not be repeated herein, except in individual references to the presentation of equations or formulas.

The flowchart symbols have the following significance:

Decision requirement — A flow chart location where a specification criterion is either met, or not met, and the answer determines which of two alternative paths must be followed in making an exit from that particular location.

Process requirement or statement — A flow chart location where an opening or closing statement is made, or where an operation, as stated, is to be performed.

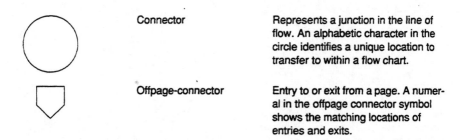

| | Connector | Represents a junction in the line of flow. An alphabetic character in the circle identifies a unique location to transfer to within a flow chart. |
| | Offpage-connector | Entry to or exit from a page. A numeral in the offpage connector symbol shows the matching locations of entries and exits. |

Each chapter includes a number of example problems which are presented with more complete details than would be required by an experienced designer. Problems for assignment are also included, with initial emphasis on variations of the example problems, thus adding incentive to the careful study of the examples in the text. In each chapter one or several problems are worked with SI units.

T. V. GALAMBOS
F. J. LIN
BRUCE G. JOHNSTON

Abbreviations

AISC	American Institute of Steel Construction
AISCM	AISC *Manual of Steel Construction*
AISCS	AISC *Load and Resistance Factor Design Specification for Structural Steel Buildings*, Dec. 1, 1993, edition
AISI	American Iron and Steel Institute
ASCE	American Society of Civil Engineers
ASD	Allowable stress design
ASTM	American Society for Testing and Materials
C	Channel shape
CAT	Computer-aided technology
CRC	Column Research Council
FLB	Limit-state moment for flange local buckling
FS	Factor of safety
L	Angle shape
LF	Load factor
LRFD	Load and resistance factor design
LTB	Limit-state moment for torsional buckling
PD	Plastic design
PL	Plate

S	Standard beam shape
SF	Shape factor
SSRC	Structural Stability Research Council
ST	Structural tee cut from S shape
TS	Structural tubing
W	Wide-flange shape
WLB	Limit-state moment for web local buckling
WT	Structural tee cut from W shape
WW	Wide-flange (W) shape made by welding three plates
ϕ	Resistance factor

1

The Steel Structure

1.1 Introduction

It is only by means of *structure* that the observable external details of our planet's surface are altered. Structures are the earmarks of our civilization, and the structural engineer—through the practice of construction within the framework of civil engineering—helps to create them: the buildings, dams, bridges, power plants, and towers that make possible our shelter, power, transportation, and communication. Thus the civil engineer has a responsible role in determining whether or not the structures that he or she builds enhance or detract from their environment.

After the prospective owner of a structure has considered alternatives and selected the site and has made subsurface exploration of soil conditions, the structural design is initiated by a consideration of various structural systems, alternative types and disposition of members, and the preparation of preliminary design drawings. Subsequently, the structural designer determines the required sizes of members and their connections, describing these in detail through drawings and written notes, so as to facilitate the fabrication and construction of the structural frame. One must first learn to design the parts before he or she can plan the whole. Hence the emphasis herein is on the design and selection of steel tension members, beams, compression members (columns), beam-columns, plate girders, and connections that join these members to form a bridge, building, tower, or other steel structure. In addition, attention is given to the design of simple frames that involve an assemblage of members into a structure.

The adequacy of a structural member is in part determined by a set of design rules, called *specifications*, which include formulas that guide the designer in checking the strength, stiffness, proportions, and other criteria that may govern the acceptability of the member. There are a variety of specifications that have been developed for both materials and structures. Each is based on years of prior experience gained through actual structural usage. The diversity of specification formulas and rules pertaining to essentially similar problems is a source of confusion in the study of structural design. In this book reference will be made primarily to a single specification—the widely used American Institute of Steel Construction (AISC) *Load and Resistance Factor Design Specification for Structural Steel Buildings* (adopted Dec. 1, 1993) (Ref. 1.1). Those who master the use of this specification, and understand the structural meaning and significance of its requirements, can readily turn to some other specification pertaining to the design of steel structures and understand the parallel set of design rules that it will contain. Further explanations of the design criteria are contained in a variety of Handbooks and design Guides, such as Ref. 1.6 which explains the rules for stability design.

The 1993 AISC Specification will be found in the second (1993) edition of the AISC *Manual of Steel Construction* (Ref. 1.2) along with much additional design information and tabular data in its two volumes. The AISC *Manual* must be considered to be an essential companion volume to this book and frequent reference is made to it. To abbreviate the repeated references that are made to the manual and the specification, they will be referred to herein as AISCM and AISCS, respectively. At this point one should read the Foreword and Preface in AISCM and thumb briefly through the entire two volumes to get a preliminary idea of the contents.

1.2 The Structure and Its Parts

The basic framework that gives strength and form to a structure does so in the same way that the human skeleton gives strength and form to the human body, and in accordance with the same principles. The creation of the complete structure calls for the combined services of the architect, civil engineer, environmentalist, city planner, and other specialists in engineering fields that may include acoustics, machine design, illumination, heat, ventilation, and other facilities. The overall design, processing, and scheduling of these inputs and the consideration of the involved interrelations while a structure is being planned and built have become known as *systems engineering*.

A steel design text must be concerned initially with the structural members that are the component parts of the overall structure. In a steel structure these are beams, which carry loads transverse to their long axis, columns or compression members, which transmit compression force along their long axis (the trunk of a tree is a most efficient column), and tension members, exemplified by a wire rope, most effectively capable of transmitting tension force, or pull, and built up of many individual wires that have been cold drawn to greatly increase their strength. Compression members usually must also carry some transverse loads and as such are called *beam-columns*. The way a structure is made up of these component parts is illustrated in Fig. 1.1, in which the upper part of a building frame is carried over an auditorium by means of a truss. In this figure the columns, beams, beam-columns, and tension members are labeled by the letters C, B, BC, and T, respectively. At every juncture point or joint between the ends of the members, connections must be provided, often offering the most challenging design problems, as they are the least standardized—yet essential to the continuity of the structure and its resistance to collapse.

Qualified structural designers *think* about the actual structure as much or more than the mathematical model that they use to check the internal forces for which they will determine the required material and type, size, and location of the members that carry the loads. The "structural engineering mind" is one that can visualize the real structure, the loads upon it, and in a sense "feel" how these loads are transmitted through the various members down into the foundations. The best designers are gifted with what sometimes has been called *structural intuition*. To develop intuition and feel, the engineer is a keen observer of other structures. He or she may even contemplate the behavior of a tree, designed by nature to withstand violent storms, flexible where it is

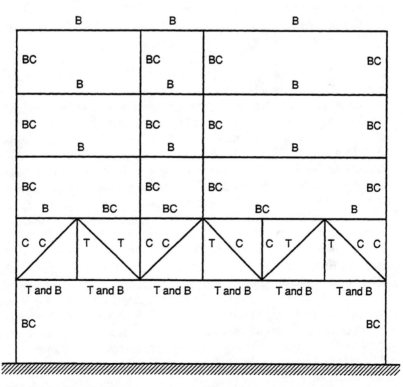

Figure 1.1 Structural frames are composed of beams (B), columns (C), tension members (T), and beam columns (BC).

frail in leaves and small branches, but growing in strength and never abandoning continuity as the branches merge with the trunk, which in turn spreads below its base into the root system, which provides its foundations and connection with the earth.

1.3 Structural Steel[*]

A knowledge of the elastic, inelastic, fracture, and fatigue characteristics of a metal is needed for the evaluation of its suitability for the making of a structural member for a particular structural application. *Elasticity* is the ability of a metal to return to its original shape after loading and subsequent unloading. *Fatigue* of a metal occurs when it is repeatedly stressed above its *endurance limit* through many cycles of loading and unloading. *Ductility* is the ability to be deformed without fracture in the inelastic range—that is, beyond the *elastic limit*. In steel, loaded in a simple tension state of stress, there occurs a sharp *yield point* at a stress only slightly greater than the elastic limit. When loaded beyond the yield point, the ductility of structural steel permits it to experience large *inelastic* elongation. Finally, the ultimate breaking strength is reached and the specimen fractures. The tensile load at fracture, divided by the original area of the unloaded specimen, is termed the *ultimate tensile strength*. Minimum specification values of the yield point, ultimate tensile strength, ductility indices, and chemistry have been established by the American Society for Testing and Materials (ASTM) to control the acceptance of structural steels.

Section A3.1a of AISCS lists 15 steels that are used in the manufacture of structural steel products. Of these, those listed in Table 1.1 are used in structural steel shapes and plates and will be designated in design examples and problems to follow.

The mechanical properties of structural steel, that describe its strength, ductility, and so on, are given in terms of the behavior in a simple tension test. The initial portion of a typical tension stress–strain curve for structural steel is shown in Fig. 1.2(b). To a greatly different horizontal scale, the complete curve is given in Fig. 1.2(a), but the load-carrying capacity of beams and columns is very largely determined within the range of Fig. 1.2(b). The slope of the stress–strain curve in the elastic range is termed E, the modulus of elasticity, and is taken as 29,000 kips per square inch (ksi), or 200,000 N/mm^2 (MPa), for the structural steels. The yield point, F_y, is the most significant property that differentiates the structural steels.

The yield point of steel will vary somewhat with temperature, speed of test, and the characteristics (size, shape, and surface finish) of the test specimen. After initial yield, the specimen elongates in the *plastic* range without appreciable change in stress. Actually, yield occurs at very localized regions, which *strain-harden*, that is, strengthen, so as to force yielding into a new location. After all the elastic regions have been exhausted, at strains of from 4 to 15 times the elastic strain, the stress starts to increase and a more general strain hardening or strengthening commences. The sharp yield point and flat yield stress level shown in Fig. 1.2(a) are peculiar to the non-heat-treated structural steels.

---

[*] The reader may wish to supplement the study of this section by reference to Chapter 1, "The Structural Steels and Their Mechanical Properties," of Ref. 1.7.

Table 1.1   Steels Used in Structural Steel Shapes and Plates

Designation	Minimum Yield Point [ksi (MPa)][a]	Ultimate Strength [ksi (MPa)][a]	
Structural steel, ASTM A36	36 (248)	58–80 (400–552)	
Hot-formed welded and seamless carbon steel structural tubing, ASTM A501	36 (248)	58 (400)	min.
High-strength low-alloy structural steel, ASTM A242	42 (290)	63 (434)	
High-strength low-alloy structural steel with 50,000 psi minimum yield point to 4 in. thick, ASTM A588	46 (317) 50 (345)	67 (462) 70 (483)	min.
Hot-formed welded and seamless high-strength low-alloy structural tubing, ASTM A618	50 (345) 50 (345)	65 (448) 70 (483)	min.
High-strength low-alloy columbium–vanadium steels of structural quality, ASTM A572	42 (290) 50 (345) 60 (414) 65 (448)	60 (414) 65 (448) 75 (517) 80 (552)	min.
High-yield-strength quenched and tempered alloy steel plate, suitable for welding, ASTM A514	90 (621) 100 (689)	110–130 (758–896) 110–130 (758–896)	

[a] ksi, kips per square inch; 1 kip = 1000 lb. One megapascal (MPa) equals 1 newton per millimeter squared (N/mm^2). U.S. structural engineering practice has not yet (1996) decided which notation is preferable.

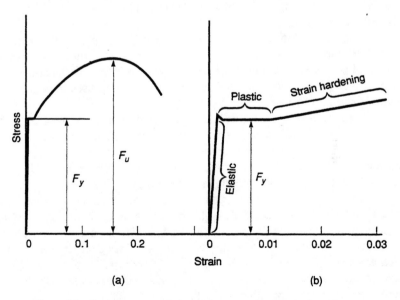

Figure 1.2   Typical stress–strain curves from a tension test of structural steel.

Structural steels are unique in that they are tough. *Toughness* may be defined as a combination of strength and ductility. After the general strain-hardening range commences in the tension test, the stress continues to increase, and inelastic extension of the test specimen continues uniformly (without local reduction in cross-sectional area) until the maximum load is reached. The specimen then experiences a local constriction and is said to *neck down*. The nominal stress based on the original area is termed the *ultimate tensile strength, F_u,* of the material. The ability of steel to withstand inelastic deformation without fracture also permits it to sustain local yielding during fabrication and construction, thus allowing it to be sheared, punched, bent, and hammered without visible damage.

Under certain combinations of circumstances, steel structures may develop cracks without appreciable prior ductile deformation. The designer should avoid sharp reentry corners that cause stress concentrations, especially in large boxlike or tank structures. Sheared edges and punched holes also cause minute stress concentrations and damage the edge material where cracks are apt to start. Operation in extremely low temperature is another factor conducive to brittle fracture. Thus careful attention to smooth edge preparation, avoidance of stress raisers, and quality control of material and fabrication processes will reduce the likelihood of brittle fracture to a minimum.

Most frequently the designer will use a standard steel shape as a structural member. These are hot rolled from billets, and their standardized dimensions for detailing and properties for design are completely tabulated in AISCM. These include wide-flange and miscellaneous beams, channels, and angles. The beams range from a depth of 3 inches (in.) (76 mm) up to 44 in. (303 mm) at 335 pounds per foot (lb/ft) or 498 kg/m, and include (primarily for use in tall buildings) a series of very wide column sections having a nominal depth of 14 in. (1118 mm) with weights of 90 to 808 lb/ft (134 to 1202 kg/m).

The reader should turn to AISCM, where the availability and selection of the appropriate grade of steel as well as the availability of shapes, plates, and bars are discussed and tabulated. The abbreviations and symbols used in designating hot-rolled structural steel shapes also are listed in AISCM and will be used throughout the text and in assigned problems.

In some parts of the country, less accessible than others to those mills that roll heavy shapes, equivalent shapes are made up by welding three plates together. Of course, when either beam or column section requirements exceed those available in standard rolled shapes, such sections are tailor made, as it were, by welding together plates to form heavier girder or column sections.

In addition to hot-rolled shapes, standard sizes of plates, bar, pipe, and hot-formed tubing, either square, rectangular, or circular in cross section, are available. Supplementing the available range of hot-rolled sections is a wide variety of both standard and special cold-formed shapes. Their use in design is covered by the *Specification for the Design of Cold-Formed Steel Structural Members* (Ref. 1.8) of the American Iron and Steel Institute and its accompanying manual (Ref. 1.5), which is coordinated as much as possible with AISCS.

To quote from the commentary on the cold-formed member specification (Ref. 1.8):

> Cold-formed members, as distinct from heavier, hot-rolled sections, are used essentially in three situations: (1) where moderate loads and spans render the thicker, hot-rolled shapes uneconomical, (2) where, regardless of thickness, members are wanted of cross-sectional configurations which cannot economically be produced by hot-rolling or by welding of flat plates, and (3) where it is desired that load-carrying members also provide useful surfaces, such as in floor and wall panels, roof decks, and the like.

Another popular type of flexural member is the manufactured truss, such as the steel joist and the joist-girder. These products are governed by the *Standard Specifications and Load Tables* of the Steel Joist Institute (Ref. 1.9).

1.4 Loads on Structures

The structural designer is required to determine the member shape, size, and arrangement that will safely carry the loads that the structure is expected to experience. The weight of the structure, which must be estimated in advance, together with all permanently attached equipment, is termed the *dead load*. Especially in very tall buildings or long-span structures it is important to check the final design with regard to the adequacy of the initial dead-load estimates. *Live load* consists of stored material, people, vehicles, snow, ice, wind, explosive blast, water in motion, earth pressure, impact, earthquake effects, and so on—all dependent on the type of structure, its intended use, and its geographic location. Loads can occur in combination, and the probability of such combinations as well as the magnitude of the loads must be considered. Live loads on standard-type structures are generally specified by the various building codes. Structures of unusual shape may require wind-tunnel-model tests to determine the magnitude and distribution of the load. Similarly, loads induced by earthquake may be based on building code requirements for conventional structures but may require dynamic analyses and/or dynamic tests of models in the case of unusual structures.

In the design office practice of designing building structures, the loads are obtained from local building codes. These local codes derive from regional model codes such as the *Uniform Building Code* or the *Southern Building Code,* or from the national load code maintained by the American Society of Civil Engineers (Ref. 1.10).

1.5 Historical Development

Structural design in ancient times was simply a matter of repeating what had been done in the past, with little knowledge of material behavior or structural theory. Success or failure was determined simply on the basis of whether the building or bridge supported the actual load or collapsed under it. Experience then was the only teacher; it is still today a most important element of good design. Gradually, through centuries of experience, the art of proportioning members evolved. Empirical rules were established. The columns in Grecian temples were said to be proportioned with the slenderness ratio of a woman's leg. The great builders of the Renaissance had no knowledge of stress analyses, yet achieved structures that required more than empiricism. They were artist, architect, engineer, and builder combined, and their cathedral domes stand now as evidence that they were able to intuitively design magnificent structures that today would not be attempted without the use of sophisticated procedures based on mathematical analyses.

Structures of the past and present, and predictions regarding structures of the future, are directly conditioned by the development and commercial availability of structural engineering materials. Certain of these materials, such as stone, brick, timber, and rope, have been used since the beginning of recorded history. Columns of stone blocks, hewn with precision, are dominant features of Egyptian, Greek, and Roman temples. The aqueducts and bridges of Rome were stone arches, which, like columns, transmit primarily compressive stresses. The Stone Age of structures persisted into the early part of the nineteenth century, when most arches and domes were still built of stone masonry and held in place by stone buttresses.

The commercial development of iron provided the first of the structural metals that were to open up an entirely new world to the structural engineer. The first bridge to be constructed completely of cast iron in 1779 still stands at Coalbrookdale in England. But (in bridges) the use of cast iron, which failed with a brittle fracture in tension, was short-lived. The commercial production of wrought-iron shapes in 1783 brought rapid changes, as it made available a product with that added quality of toughness exemplified by an ability to take large tensile deformation in the inelastic range without fracture. Moreover, wrought iron could be formed into flat plates that could be bent and joined by rivets, making possible the steam locomotive, which, in turn, created a demand for long-span metal bridges. Noteworthy among the early wrought-iron bridges was the Britannia Bridge across the Menai Straits of the Irish Sea. It consists of twin parallel box girders continuous over four spans, with two center spans of 460 ft each, flanked by 230-ft end spans. It was completed in 1850 and is the prototype of a current trend in bridge construction that may be called "the rebirth of the box girder bridge."

The development of the Bessemer converter in 1856 and the open-hearth furnace in 1867 introduced structural steel, and this is the material that has been used most in bridges, as well as in many buildings, for the past 100 years. The first major bridge to be constructed entirely of structural steel was the famous Eads Bridge across the Mississippi at St. Louis. Completed in 1874, it incorporates tubular steel arches with a central span of 520 ft, flanked by 502-ft side spans.

Paralleling the development of iron and steel as engineering materials were advances in material-testing techniques and in structural analysis that made possible the transition of structural design from an art to an applied science. Hooke (1660) demonstrated that load and deformation were proportional, and Bernoulli (1705) introduced the concept that the resistance of a beam in bending is proportional to the curvature of the beam. Bernoulli passed this concept on to Euler, who in 1744 determined the elastic curve of a slender column under compressive load. Important developments in the late 1800s included (1) the manufacture of mechanical strain-measuring instruments that made possible the determination of the elastic moduli that related stress to strain, (2) correct theories for the analysis of stress and deformation resulting from either the bending or twisting of a structural member, and (3) the extension of column-buckling theory to the buckling of plates and the lateral-torsional buckling of beams.

The foregoing advances made possible the development of engineering specifications built around the *allowable-stress method* of selecting structural members. The first general specification for steel railway bridges was developed in 1905, and the first highway bridge specification in 1931. In 1923 the AISC brought out its first general specification for building construction. Under each of these specifications, the criterion for acceptable design strength is as follows: the calculated maximum stress, assuming elastic behavior up to anticipated maximum loads, is kept lower than a specified allowable stress. The allowable stress is intended to be less than the calculated stress at failure by a *factor of safety*. Typical values of this factor are 1.65 to 2.00 in AISC's *Allowable Stress Design* specification (Ref. 1.11). Unfortunately, the calculated maximum elastic stress at failure load varies widely. A slender column or laterally unsupported beam may fail at a fraction of the yield point stress, but a very short column will reach the yield point before it fails. A statically loaded tension member may develop the ultimate tensile strength of the material, or nearly twice the yield point; but the same member, loaded and unloaded repetitively for thousands of cycles, may fail due to fatigue at a fraction of the yield point. A connection, because it yields locally, may not fail until the calculated *elastic* stress is several times the yield point; but it, too, is susceptible to fatigue failure at much lower stresses. It is evident that the true criterion of acceptability is strength—not stress—and thus, on the basis of experience and strength analyses, specified allowable stresses have had to be adjusted upward and downward over a wide range to provide a reasonably uniform index of structural strength.

The allowable stress design (ASD) method is still a frequently used method of designing steel building structures in the 1990s and is governed by the AISC *Specification for Structural Steel Building—Allowable Stress Design and Plastic Design* (Ref. 1.11). This method will not be discussed further in this book. Instead, the load and resistance factor design (LRFD) method is considered. Since both specifications are based on the behavior of the same type of structure, the students can easily pick up the procedures of ASD if they have a thorough grounding in the LRFD method.

During the past 50 years increasing attention has been given to the evaluation of the inelastic properties of materials and to the direct calculation of the ultimate strength of a member. This information is useful in improving the allowable-stress procedure, but it also permits bypassing stress calculation by using the calculated member strength as a direct basis for the design. *Load-factor* design has resulted. The maximum anticipated service loads are multiplied by load factors to yield a required strength, which must be less than the directly calculated strength. Philosophically, this is a more realistic, direct, and natural procedure. The load-factor approach has been used for many years in aircraft design, and AISCS, starting in 1961, permits it as an acceptable alternative for the design of continuous frames in building structures.

Plastic design with only one load factor is also governed by Ref. 1.11. Plastic design will be considered in this book in the context of LRFD (Ref. 1.1). Although the current trend in design is to deemphasize the calculation of stress, such calculations are still essential in the design of machine parts and structural elements that must endure many load repetitions. Total resultant stresses must also be calculated in truss analysis and design.

Structural design methods have undergone rapid change in the decade of the 1980s as more and more specifications put increasing emphasis on load-factor design (also called *load and resistance factor design* in the United States and *limit states design* in Canada). This procedure uses different load factors for dead, live, wind, and snow loads and resistance factors by which the computed strength of beams, columns, connectors, and so on, are multiplied to account for the various uncertainties inherent in predicting loads and strengths. Furthermore, the load and resistance factors are determined by probabilistic methods from the statistical data on loads and strengths. Details of the LRFD method are discussed in the following sections of this chapter.

The 1980s and 1990s also witnessed the ever-increasing use of the computer in every aspect of design, including layout, analysis, and in the production of detailed design drawings.

1.6 Structural Design Economy

In a competitive world, with increasing costs for materials and labor, the search for greatest design economy consistent with safety and the desired life of the structure is of major importance. Members must be shaped, arranged, and connected in ways that will provide an efficient and economical solution to the design problem, having in mind not only the cost per pound of the material itself, but also the labor costs of shop fabrication and field erection. Minimum weight is often a design goal. However, if simplicity of fabrication is sacrificed to achieve minimum weight, the overall cost may be increased. In Fig. 1.3(a) a steel beam under uniform load will be adequate in strength if it is fabricated as shown out of three segments—two end pieces, labeled (1), which weigh less per foot of length than the center section, labeled (2). But the

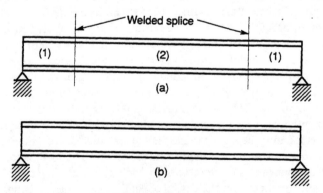

Figure 1.3   Reducing weight may increase cost.

cost of welding the three segments together may (or may not) exceed the cost of the added weight if the beam is made of a single nonwelded member having the same size as the center segment, as shown in Fig 1.3(b).

A similar situation may occur in plate girder design (see Chapter 7). The use of very thin webs is made possible by vertical and (in some cases) horizontal stiffeners welded to the web. In borderline instances, the use of a thicker web, which eliminates the need for stiffeners, can result in a saving in fabrication cost even though the overall girder weight is increased.

Figure 1.4 illustrates how member arrangement can affect economy. In each trussed rectangular frame, load H is applied horizontally at the top, acting as shown, and only at A is there horizontal reaction support. In arrangement (a), only two of the five members are directly stressed by the load. These are called *load-bearing* members and are indicated by the heavily weighted lines. But in arrangement (b), four of the five members are stressed by the load. The unstressed members would be called *secondary* members. Moreover, in arrangement (b) the compression (C) member is longer than in arrangement (a), thus using more material because of the lesser efficiency of compression members in comparison with tension (T) members. The foregoing illustrates the general principle that greatest economy results by providing the most direct path possible for the transmission of force from load point to footing.

Method of shipment can have an important bearing on economy. Connections can be made in the fabricating shop at a fraction of the cost for the same connections made during field erection. A fabricating plant built on a navigable waterway has a great advantage in building a bridge over a river accessible to the same waterway. Girders several hundred feet in length can be shop-fabricated with no field splices and shipped direct to the site on barges. The same girders, if shipped by rail or truck, would need several field-splice connections, and if their overall height exceeded rail or highway clearance limitations, horizontal field splices would also be necessary. In the case of major bridges, a temporary fabrication shop may be built adjacent to the site to avoid shipment of bridge segments.

In short-span structures, dead weight contributes but little to the stress. But as the span increases, so does the proportion of the dead-load stress to the total combined

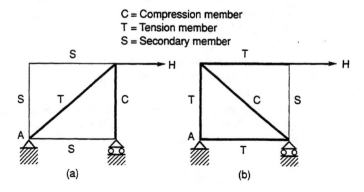

Figure 1.4   Economy as affected by member arrangement.

stress. Finally, when the span is so great that most of the stress is due to dead load, the upper limit of span has been reached for that material and that type of structure. Thus in long-span bridges or in tall buildings, careful attention to weight reduction and accuracy in dead-weight computations take on increasing importance. In such structures the use of high-strength steels for load-carrying members and lightweight metals for non-load-carrying elements is advantageous.

1.7 Structural Safety

Structural safety can be assured by a combination of good design, good workmanship in fabrication, and good construction methods. The avoidance of any possibility of structural failure should be a primary concern for the designer. In design, the choice of a proper load factor requires experience and sound engineering judgment. Questions of deterioration due to corrosion during the planned life of the structure, variation in material properties, and many other factors also need to be considered. The most rational approach to the problem of structural safety requires a statistical evaluation of the random nature of all the variables that determine the strength of the structure, on the one hand, and those that may cause it to fail (primarily, the loads) on the other hand. Then, by elementary probability theory, the risk of failure may be evaluated and the probability of its occurrence kept at an acceptable level, dependent on the importance of the structure, risk to human life, and other factors. Increasing attention is being given to this approach to safety evaluation, and statistical studies are being made of material properties, variation in strength of various types of members, and loads. Particular attention also has to be given to uncertain loads, such as those due to wind and earthquake.

Structural safety in design is accomplished by assuring, by means of design calculations, that the limits of structural usefulness given by the applicable structural specification, such as AISCS, are not violated. In its Preface, AISCS cautions that "independent professional judgment must be exercised" in applying the specification and that "it is not intended to cover the infrequently encountered problems within the full range of structural design practice . . . ." AISCS is regularly updated to include new developments in research and practice and its writers are especially concerned with safety. In 1978, AISCS was composed of two parts: one segment provided rules for allowable stress design (ASD), and the other one defined the criteria for plastic design (PD). In 1986, AISC published the LRFD Specification (Ref. 1.1). This new procedure defines the criteria for LRFD, a procedure that aims to make full use of available test information, design experience, and engineering judgment, applied by use of probabilistic analyses.

In ASD (Ref. 1.11) the limits of structural usefulness are allowable stresses that must not be exceeded when the forces in the steel structure are determined by an elastic analysis. The allowable stresses F_{all} are defined by the relationship

$$F_{all} = \frac{F_{lim}}{FS} \tag{1.1}$$

where FS is the factor of safety and $F_{\lim}$ is a stress that denotes a limit of usefulness such as the yield stress F_y, a critical (buckling) stress F_{cr} (column stability, beam stability, or plate stability), the tensile stress F_u at which the member fractures, or the stress range F_{sr} in fatigue. For example, the allowable stresses in tension are the smaller of $0.6F_y$ (limit state of yielding, FS $= 1/0.6 = 1.667$) and $0.5F_u$ (limit state of fracture FS $= 1/0.5 = 2.0$). The actual stresses that must not exceed the allowable stresses are determined by elastic analysis for the working loads on the structure.

In Part 2 of Ref. 1.11 the limit of structural usefulness is a load P_u which will cause a plastic mechanism to form. This limiting load is then compared with the factored working loads:

$$(LF)\,P_w \leq P_u \qquad (1.2)$$

where P_w represents the working loads and LF is a load factor. (LF $= 1.7$ for gravity loads and LF $= 1.3$ for combined gravity and wind or earthquake loads.)

In the LRFD criteria (Ref. 1.1) the designer is not expected to manipulate statistical data. Instead, he or she proceeds by following prescribed rules in determining resistances and using multiple load factors. The design check to be made is as follows:

$$\sum_{i=1}^{w} \gamma_i Q_i \leq \phi R_n \qquad (1.3)$$

In this formula the γ_i's are load factors by which the individual load effects Q_i are multiplied to account for the uncertainties of the loads, R_n is a nominal resistance, and ϕ is a resistance factor that accounts for the uncertainties inherent in the determination of the resistance. The γ's are larger than unity and ϕ is less than unity. The γ and ϕ factors are given in the various sections of AISCS for different load combinations and types of members. These criteria are discussed in subsequent chapters for tension members, beams, columns, and so on.

The general design criteria are illustrated in Fig. 1.5, where R represents the resistance (strength) of a structural element and Q is the load effect (calculated force due to the maximum loads expected during the life of the structure). The symbols R_n and Q_n represent, respectively, the load effect due to the specified working loads and the minimum specified resistance. The factored load effects $\Sigma \gamma Q_n$ (e.g., $\gamma_D Q_D + \gamma_L Q_L$, where Q_D and Q_L are the dead- and live-load effects, respectively, and γ_D and γ_L are the corresponding load factors equal to, respectively, 1.2 and 1.6) and the resistance factor ϕ (e.g., $\phi = 0.9$ for flexural members) serve the purpose of providing a margin of safety between R_n and Q_n to take care of the unforeseen but possible eventuality that the actual load might exceed the specified value and/or that the actual resistance is smaller than the specified value. These uncertainties are in the nature of loads and resistances. In fact, we can easily visualize that both load effects and resistances have a form of a probabilistic distribution (Fig. 1.5), characterized by a bell-shaped curve that has a mean value (R_m or Q_m) and a standard deviation. Exceedance of a limit state is, then, the condition that $R < Q$, and this is always possible. Structural safety is thus defined as the acceptably small probability of Q exceeding R, and the true role of the load factors and the resistance factors is to assure that this probability is negligibly small.

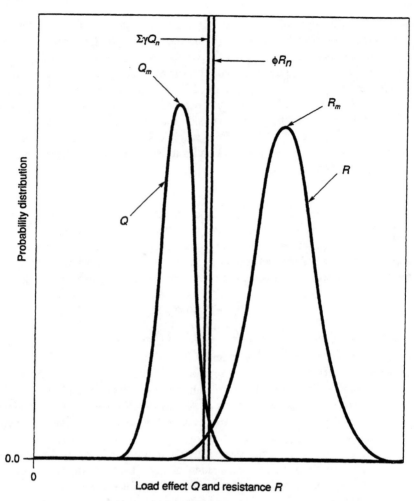

Figure 1.5 Definition of structural safety.

One could start the determination of the required margin of safety by stipulating an acceptably small probability of exceeding a limit of structural usefulness (*limit state*), and then from the known probabilistic distributions of R and Q one could, by the calculus of probability theory, arrive at the appropriate margin of safety. This is how the various factors were arrived at for the LRFD method (Ref. 1.1).

The method of arriving at a probabilistic safety margin is as follows. A structure is safe (i.e., a limit state is not violated) if $R - Q \geq 0$, or $R/Q \geq 1$, or $\ln(R/Q) \geq 0$. The distribution of $\ln(R/Q)$ is shown in Fig. 1.6. The limit state is violated if $\ln(R/Q)$ is negative, and the probability of this happening is represented as the shaded area in Fig. 1.6. The smaller this area is, the more reliable is the structural element. The shaded area varies in size as the distance of the mean value of $\ln(R/Q)$ from the origin (Fig. 1.6). This distance depends on two factors: the width of the distribution curve, as

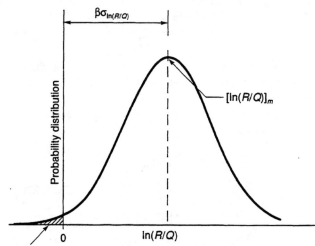

Figure 1.6   Definition of reliability index.

characterized by its standard deviation $\sigma_{\ln(R/Q)}$ (i.e., the scatter of the data which make up the distribution), and a factor β, which is called the *reliability index*. For any given distribution $\ln(R/Q)$, the larger β is, the smaller is the probability of exceeding a limit state. In fact, if we know the exact distribution curve, we can directly relate this probability to the reliability index β.

Unfortunately, the probabilistic distributions of R and Q are known only for very few resistance and load components, but at least we know the mean values and the standard deviation from the analysis of data on loads and material properties. From this knowledge, and from some approximations described in Refs. 1.12 and 1.13, we can obtain the following simple formula for the reliability index β:

$$\beta = \frac{\ln(R_m/Q_m)}{\sqrt{V_R^2 + V_Q^2}} \tag{1.4}$$

In this equation the terms R_m and Q_m are the mean values of the resistance R and the load effect Q, respectively, and V_R and V_Q are the corresponding coefficients of variation (= standard deviation/mean).

We can now find out the value of β inherent in, say, Part 1 of the 1978 AISCS by designing an element, then obtaining the appropriate statistical data, and so computing the value of β. This process is called *calibration* (Ref. 1.12).

Calibrations were performed for many types of structural elements (Refs. 1.12–1.16) and on the basis of these calibrations, $\beta = 2.6$ for members and $\beta = 4.0$ for connections were selected for the LRFD criteria. Since the factors of safety in the previous AISCS resulted from an evolutionary process of many years of experience, there was some scatter in the β's. The new LRFD method removes this scatter, thereby ensuring uniform reliability at least equal to the accepted designs in the current AISCS. One advantage of probability-based design is therefore that a more uniform reliability

results. The other advantage is that the variability of the various design elements can
be properly accounted for by selecting the appropriate load factors γ and resistance
factors ϕ determined from the specified index β and the appropriate statistical data
(Ref. 1.13).

The resistance factors ϕ are given in Ref. 1.1 for the various limit states that
must be considered: for example, $\phi = 0.9$ for beams and $\phi = 0.85$ for columns. The
load factors γ are given in Ref. 1.10 and in Sec. A4.1 of AISCS, and following are the
basic combinations of loading which must be investigated:

$$1.4D_n \tag{1.5}$$

$$1.2D_n + 1.6L_n + 0.5(L_{rn} \text{ or } S_n \text{ or } R_n) \tag{1.6}$$

$$1.2D_n + 1.6(L_{rn} \text{ or } S_n \text{ or } R_n) + (0.5L_n \text{ or } 0.8W_n) \tag{1.7}$$

$$1.2D_n + 1.3W_n + 0.5L_n + 0.5(L_{rn} \text{ or } S_n \text{ or } R_n) \tag{1.8}$$

$$1.2D_n \pm 1.0E_n + 0.5L_n + 0.2S_n \tag{1.9}$$

$$0.9D_n \pm (1.3W_n \text{ or } 1.0E_n) \tag{1.10}$$

where $D_n, L_n, L_{rn}, S_n, R_n, W_n$, and E_n are the nominal dead, live, roof live, snow, rain,
wind, and earthquake load, respectively.

For example, for beams under dead and live loads,

$$\phi = 0.9 \qquad \gamma_D = 1.2 \qquad \gamma_L = 1.6$$

and the design criterion then becomes

$$0.9M_u \geq 1.2M_D + 1.6M_L \tag{1.11}$$

where M_D and M_L are the moments due to dead and live loads, respectively, and M_u
is the nominal ultimate moment of the beam. In Eq. (1.11) we can see the advantage
of LRFD over ASD or PD: the smaller load factor for dead loads reflects the fact that
the determination of dead loads is more certain than that of live loads. Thus where
dead load dominates, this fact is taken into account in design.

For example, a 30-ft-long simply supported beam is subjected to a uniformly distributed dead load of 0.8 kip/ft (11.7 kN/m) and a live load of 1.2 kips/ft (17.5 kN/m).
The required distributed design load is then

$$w_u = 1.2w_D + 1.6w_L = 1.2 \times 0.8 + 1.6 \times 1.2 = 2.88 \text{ kips/ft (42 kN/m)}$$

The corresponding required moment is

$$M_u = \frac{w_u L^2}{8} = \frac{2.88 \times 30^2}{8} = 324 \text{ kip/ft (439 kN·m)}$$

The designer must designate a beam with a nominal moment capacity M_n larger than
or equal to

$$\frac{M_u}{\phi} = \frac{324}{0.9} = 360 \text{ kip/ft (488 kN·m)}$$

Years of design experience, conditioned by both unsuccessful and successful behavior, have produced criteria that aid the choice of safe design criteria. These may not always produce the most economical structure; nevertheless, the overall cumulative experience in engineering design has provided a background from which the engineer derives confidence in many particular design applications. Obviously, great skill, care, and more detailed analyses, possibly supplemented by laboratory tests of models or portions of a prototype structure, are needed when the designer attempts a new and venturesome type of structure.

1.8 Planning and Site Exploration for the Specific Structure

After the decision is made to build a steel structure to fill particular service functions, consideration is given to those factors that may influence the overall economy. If alternative sites are available, in the case of large and heavy structures, preliminary explorations are required of the various locations with a mapping of the terrain and partial preliminary study of subsurface foundation conditions by means of borings and/or open-pit excavations. Load-bearing tests may be required. If the terrain is uneven, certain functions of a building may make advantageous use of the changing elevation of the ground, and this will of course affect the overall structural layout. Other influencing factors are transportation facilities, availability of water, gas, and other utilities, drainage features, orientation with respect to prevalent winds, consideration for daylight lighting, and the general type of foundation to be required. The minimization of energy requirements is of steadily increasing importance. After all these considerations and after selection of the exact site, more test borings should be made if there is any question at all about the foundation conditions and their uniformity. It may very well happen that the preliminary borings have straddled some subterranean stream location, hard strata lenses, or rock faults with associated local poor foundation conditions. In such a case, misleading estimates resulting in costly changes in design may be avoided by complete subsurface exploration. In recent years subsurface seismic surveys have proved remarkably accurate in locating bedrock and other hard layers. Such surveys are much cheaper than borings and may be used as a preliminary step covering wide areas, followed up by dry sample borings covering smaller areas selected by the results of the seismic survey. In regions subject to settlement of foundation or questionable soil capacity, undisturbed soil samples should be taken for laboratory tests of unconfined and confined compressive strength, shear strength, degree of consolidation, permeability, and so on. If pile foundations are used, test piles may be required.

1.9 Structural Layout, Details, and Drawings

After preliminary drawings have been made of the space and area requirements, in plan and in elevation, and general decisions made as to materials, type of structure, and so on, the designer may proceed with a preliminary trial location of columns and footings.

With respect to both design and fabrication, *economy results from simplicity and duplication*. Duplication also leads to the use of standardized mass-produced elements, such as windows and doors, and has led to what is termed *modular construction*. The module is a basic space dimension that repeats itself throughout the structure and may apply to the column spacing or to smaller details. The module in building construction is frequently about 5 ft, usually a multiple of 4 in. This module is used throughout the building and applies to partitions, ceilings, lighting, windows, and so on. Columns may be four or five modules on center, as part of the modular scheme. Thus standardized spacing results in an increase in duplication and standardization of details. Duplication in floor and roof construction will also result from the constancy of column spacing chosen in the basic modular design concept. Such duplication leads in the fabrication shop to fewer different sizes and lengths of beams in the overall steel order, and the duplication of beam and column details reduces the number of design detail drawings that are required. Repetition speeds up the work in the shop with corresponding reduction in cost.

The choice of roof and exterior wall construction involves the possibility of selecting some commercially developed standardized roof or wall product that will determine within reasonable limits the spacing of purlins and girts. However, within the range of feasible variations in purlin and girt spacing consistent with the modular layout, preliminary designs and cost estimates of various spacings should be made to determine the least weight of steel, and this will usually be of least cost as well.

Procedures for the design of main members, such as beams, columns, and tension members, are fairly simple and precisely laid down by specifications. It is in the design of the connecting details between members and their supports that the structural engineer is called upon for the greatest judgment and design skill. Poorly designed connections may lead indirectly to the failure of main members or even the entire structure. In any structure the load must be transmitted through successive connections from points of application down to the footings. The designer must follow the same sequence—for each succeeding component of a structure must carry the accumulated dead weight of tributary components, and in preliminary design studies these weights can only be approximated.

Important in the design of details is the elimination of bending or eccentricity in local elements. As one specific example, if a column is supported by a beam (or acts as a local beam support), the webs of the column and beam should be in alignment; but, since the major load in the column is carried in the flanges, bearing stiffeners may be needed to transfer the load from the column flanges to the beam web. Thus the load is transmitted from point to point throughout the structure in the most efficient manner without possibility of local failure.

Careful attention must be given in the design of structural details to the method of fabrication and erection, with proper clearances for bolting or riveting, welding electrodes and holders, or whatever else may be required for the fabrication process in question. The designer of details must visualize the complete construction operation. Care must be taken to provide drainage holes in pockets of exposed steel construction, because moisture and dirt should not be permitted to collect at these points of greatest incipient corrosion.

Design drawings should be complete and easy to follow. The AISC textbooks *Detailing for Steel Construction* (Ref. 1.3) and *Engineering for Steel Construction* (Ref. 1.4) provide an excellent guide. As in many other separate aspects making up the whole of an engineering design, the *saving* of money in preparation of design drawings may result in greatly *increasing* the overall costs of the structure. If the structure is to be bid upon by private organizations, lack of sufficient detail in the design drawings may cause the bidder to add an appreciable amount for the contingencies that he or she must be prepared to face later when the details of connections and other framing problems are fully brought to light. If these are completely shown in the initial design drawings, the bidder will be able to give the best possible price.

1.10 Fabrication Methods

For greatest economy the design must be made in the light of a preliminary decision as to material and mode of fabrication. For example, the economy due to welding largely results from the introduction of continuity and from the elimination of the connecting pieces that would be needed in riveted or bolted construction. Although it is usually uneconomical to use both bolting and welding in the shop fabrication of a particular member, because of the double handling, the designer should consider the use of shop welding with bolted field connections. This is especially appropriate in the case of truss bridges, with high-strength bolts used for the field connections. Such connections have an excellent record for resistance to repeated load.

The use of welding requires careful and competent inspection both with regard to procedure and finished product. Both shop and field inspection of welding are important, as the quality of welds depends to a large extent on the skill, character, and endurance of the welder. Standards for quality of materials, procedures, and inspection of welds and welding processes as established by the American Welding Society are quite generally accepted by AISCS.

When bolting or riveting is to be used, the question arises as to whether punching of holes, subpunching with reaming, or drilling should be employed. Punching with automatic spacing equipment and repetition of members having the same punch pattern is a very economical means of preparation for bolts and rivets. However, punching damages the material locally at the edges of the holes, and such members are not as good under repeated load as are members with drilled holes. Of course, only such members as will receive large fluctuations in applied load require consideration of fatigue strength. There would be no point in subpunching and reaming—or drilling—holes for the connection of roof purlins to their truss supports, because the maximum loads are repeated a relatively few times and the stresses are minimal. In the case of shop assemblies joining several different plates or members, economy may be achieved by clamping the pieces into a single "pack" for single or multiple drilling through all pieces in one operation. Drilling provides smooth edges of holes and the best possible resistance to repeated load.

1.11 Construction Methods

Structural designs should be prepared with ample consideration of the manner and
facility with which field erection can proceed. The arrangement, number, type, and
location of field splices and connections should be planned so as to avoid unnecessary
duplication of construction equipment and provide the simplest possible erection plan
with a minimum of field work. Connections should be arranged to facilitate field
assembly. Careful design planning in relation to construction will minimize the total
cost of the project. In important large projects a definite erection plan should be presented, but the contractor should have freedom to exercise ingenuity through alternative schemes that meet the approval of the owner.

In one particular sense, proper construction methods have a special relation to
overall economy, since it is during construction that failures in engineering structures
most often occur. During lifting operations, members of trusses that normally are in
tension, or the lower flange of a plate girder, which is normally in tension, may be
placed in compression with consequent possible buckling failures. In the case of very
long plate girders used in bridge construction, special horizontal temporary truss systems fixed to the plate girders may be used during erection. Although erection is normally the responsibility of the steel contractor, the design engineer can help in
complex cases by scheduling the bracing that must be supplied as the construction is
in progress. Alternatively, the contractor may be required to submit erection procedure plans to the engineer for approval.

Even after the main frames and members are successfully placed in the structure, failures have occasionally occurred because of the haste with which construction
of main framing has proceeded without attention to the cross bracing that may be
planned for the final structure in the planes of the walls and roof. After permanent
bracing, roof, and walls are in place, the wind-load resistance of the structure will be
greatly increased.

In summary, it may be said that construction failures are usually caused by lack
of three-dimensional or *space frame* stability and that many more failures occur during erection than during service of the finished structures.

1.12 Service and Maintenance Requirements

The engineer, together with the architect and special consultants on such matters as
heating, lighting, and ventilation, should give careful attention to the way in which the
utility of the structure may be affected by the engineering design. Especially in an
industrial building, structural design must be conditioned by the service functions of
the structure.

Inadequate initial planning regarding the service which the structure is to perform will inevitably result in revisions in layout and corresponding costly design and
material order changes before the structure is complete. It is obvious also that the
locations of the electric wiring, heating ducts, and other service ducts for water, gas,
chemicals, and so forth, as well as the locations of all special items of equipment,
must be carefully predetermined, as all affect and are affected by the structural design.

Another service requirement of concern to the engineer is the desired life of the structure, together with consideration as to any special problems of corrosion that may exist due to atmospheric conditions, humidity, and so on. By proper design the engineer should avoid pockets where dirt and water may collect and should provide access to all parts of the structure that will require repeated painting and inspection during its life. Under adverse conditions, when maintenance cannot be assured, an extra thickness of metal may be furnished to allow for corrosion. Special corrosion-resistant steels are available, and another alternative is the use of *weathering steels*, which require no paint and develop a surface oxide that resists corrosion and presents a pleasing burnt-brown color.

During the useful life of an industrial plant, changes in processing procedures or even a complete change in use may come about. Thus, along with all necessary attention to special service requirements, an effort should be made in incorporate flexibility with respect to possible future alterations. The use of temporary interior partition walls is an example of such flexibility with respect to future change.

A structure should be designed to provide a life consistent with the buyer's wishes. A structure that is to last 100 years will be of quite different construction than one designed to survive for only a few years. The choice of materials used in construction may be affected; but even if both structures were of steel, there would need to be different consideration given to the problems of permissible stress, load evaluation, corrosion, painting, and other matters of upkeep for the two different life expectancies. The use of closed tube or box sections may materially reduce painting maintenance costs and be more justified in a long-lived structure than in a temporary one. Consideration under this category should also be given to fireproofing and fire protection. The difference in cost of fire insurance over the life of the structure must be weighted against the difference in initial cost between various degrees of fireproofing, assuming, however, that safety to human life is not an overriding consideration.

REFERENCES

1.1. *Load and Resistance Factor Design Specification for Structural Steel Buildings,* American Institute of Steel Construction, Chicago, 1993.*

1.2. *Manual of Steel Construction: Load and Resistance Factor Design,* Vols. I and II, American Institute of Steel Construction, Chicago, 1993.*

1.3. *Detailing for Steel Construction,* American Institute of Steel Construction, Chicago, 1983.

1.4. *Engineering for Steel Construction,* American Institute of Steel Construction, Chicago, 1984.

1.5. *Cold-Formed Design Manual,* American Iron and Steel Institute, Washington, DC, 1990.

1.6. T. V. GALAMBOS, ed., *Guide to Stability Design Criteria for Metal Structures,* 4th ed., Wiley, New York, 1988.

* A mandatory supplement for the complete use of this book. Reference 1.1 will be referred to herein simply as AISCS; Ref. 1.2 will be referred to as AISCM.

1.7. R. L. BROCKENBROUGH AND B. G. JOHNSTON, *The USS Steel Design Manual*, rev. ed., The United States Steel Corporation, Pittsburgh, PA, 1981.

1.8. *Specification for the Design of Cold-Formed Steel Structural Members*, American Iron and Steel Institute, Washington, DC, 1994.

1.9. *Standard Specifications and Load Tables*, Steel Joist Institute, Myrtle Beach, SC, 1994.

1.10. *Minimum Design Loads for Buildings and Other Structures*, ASCE7-93, American Society of Civil Engineers, Washington, DC, 1993.

1.11. *Specification for Structural Steel Buildings: Allowable Stress Design and Plastic Design*, American Institute of Steel Construction, Chicago, 1989.

1.12. T. V. GALAMBOS, B. ELLINGWOOD, J. G. MACGREGOR, AND A. C. CORNELL, "Probability-Based Load Criteria: Assessment of Current Design Practice," *ASCE Journal of the Structural Division*, Vol. 108, No. ST5, May 1982.

1.13. T. V. GALAMBOS, B. ELLINGWOOD, J. G. MACGREGOR, AND A. C. CORNELL, "Probability-Based Load Criteria: Load Factors and Load Combinations," *ASCE Journal of the Structural Division*, Vol. 108, No. ST5, May 1982.

1.14. *Building Code Requirements for Minimum Design Loads in Buildings and Other Structures*, ANSI A58.1, American National Standards Institute, New York, 1972.

1.15. T. V. GALAMBOS AND M. K. RAVINDRA, "Load and Resistance Factor Design for Steel," *ASCE Journal of the Structural Division*, Vol. 104, No. ST9, Sept. 1978.

1.16. J. A. YURA, T. V. GALAMBOS, AND M. K. RAVINDRA, "The Bending Resistance of Steel Beams," *ASCE Journal of the Structural Division*, Vol. 104, No. ST9, Sept. 1978.

1.17. P. B. COOPER, T. V. GALAMBOS, AND M. K. RAVINDRA, "LRFD Criteria for Plate Girders," *ASCE Journal of the Structural Division*, Vol. 104, No. ST9, Sept. 1978.

2

Tension Members

2.1 Introduction

The most efficient way to use structural steel is in a tension member, that is, one that transmits "pull" between two points in a structure. Of course, if under certain load conditions the stress reverses in the member and becomes compression, the member must be designed both as a tension member and a column, and the efficiency is lost.

To make all the material in the tension member fully effective, the end connections must be designed to be stronger than the body of the member. If overloaded to failure, such a tension member will not only reach the yield stress but go above this level up to the ultimate strength of the material. In so doing it can absorb a great deal more energy per unit weight of material than any other type of member. This is an important consideration if impact or dynamic loads are a possibility. Beams and columns do not utilize material at full efficiency for two reasons: (1) metal failure is localized at highly stressed locations, and (2) some type of buckling failure always occurs at or below the yield stress, and the ultimate tensile strength of the material can never be reached.

Four types of tension members that can achieve high efficiency are illustrated in Fig. 2.1, showing (a) the wire rope or cable with socketed ends in which the use of cold-drawn steel wires having tensile strengths up to 150 ksi (or more, in special

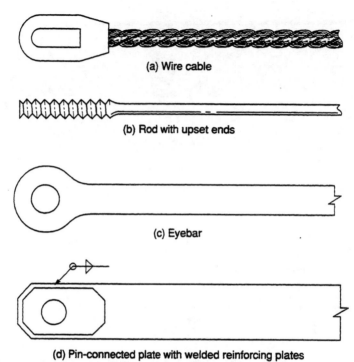

(a) Wire cable

(b) Rod with upset ends

(c) Eyebar

(d) Pin-connected plate with welded reinforcing plates

Figure 2.1 High-efficiency tension members.

applications) provide the greatest strength/weight ratio available in the use of steel; (b) the simple round rod with threaded upset ends; (c) the eyebar, with forged ends for pin connections that are stronger than the body of the bar; and (d) the pin-connected plate with welded reinforcing plates at the ends.

In contrast to the foregoing, a tension member that can fail within its end connection before yielding in the body of the member will absorb little energy before failure—possibly less than 1% of the capacity it would have with uniform yielding throughout its length. Regardless of where failure under static load might occur, the tension member and its end connections should be designed to guard against fatigue failure if alternate loading and unloading is to be expected for a large number of repetitions. Because of their efficiency, and because buckling is not a problem, tension members make more advantageous use of the higher-strength steels than any other type member.

No structural member is perfectly straight, and an intended axial force will never act precisely along the longitudinal axis. As a result, there are always "accidental" bending moments in such a member. In a column, as illustrated in Fig. 2.2(a), these bending moments cause added deflection, which further increases or "amplifies" both the deflection and the bending moment caused thereby, equal to the product of the axial load and the deflection.

In an initially curved and eccentrically loaded tension member [Fig. 2.2(b)] the member tends to straighten, and the bending moments are reduced everywhere except

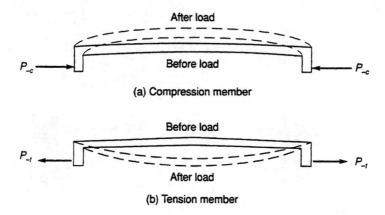

(a) Compression member

(b) Tension member

Figure 2.2 Deflection of eccentrically loaded compression and tension members.

at the end. Thus for very small accidental curvatures and end eccentricities, the additional tension stress induced by bending can usually be neglected unless design for repeated load is required.

2.2 Types of Tension Members

Four efficient tension-member types have been illustrated in Fig. 2.1. In addition, structural shapes and built-up members may be used, especially in trusses where tension and compression members must frame into a common joint, as shown in Fig. 2.3.

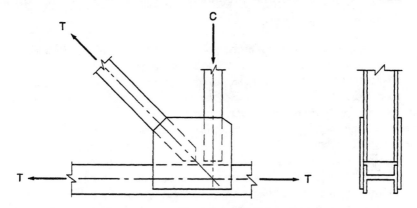

Figure 2.3   Tension members (T) and compression member (C) entering lower chord joint of a truss.

2.2.1 Wire Ropes and Cables

A cable is defined as a flexible tension member consisting of one or more groups of wires, strands, or ropes. A strand is formed by wires laid helically about a

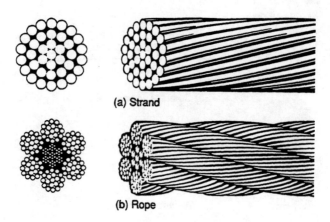

(a) Strand

(b) Rope

Figure 2.4 Wire strand and wire rope (From U. S. Steel Corporation wire rope catalog, by permission.)

center wire so as to produce a symmetrical section; a wire rope is a plurality of strands laid helically around a central core that is composed of a fabric core, or another wire strand, as illustrated in Fig. 2.4.

Wire cables are finding increasing use in structural steel design, and have been used both as primary and secondary supporting members in a wide variety of structures, including suspension bridges, prestressed concrete members, guyed towers, and wide-span roof structures. In roof construction the cables may radiate outward from a central tower, or may run radially inward from an outer compression ring, as illustrated in Fig. 2.5. The major U. S. steel producers issue catalogues that provide extensive design information and illustrations of the use of cables in roof structures.

2.2.2 Rods and Bars

The simplest tension member is the square or round rod. Round bars with threaded ends are less costly than bars with upset ends[*] [Fig. 2.1(b)], but have certain disadvantages. Failure under impact overload or repeated load is apt to occur in the threaded portion. Bars with upset ends yield throughout their entire length and are recommended for the design of diagonal bracing for simple tower structures in earthquake regions. Large-diameter bars with threaded ends should be used with caution, because the lateral contraction in the diameter of the bar as yielding commences in the threaded portion may result in sufficient loss of thread-bearing area to result in failure by stripping of the threads before fully developing the maximum desired strength. To guard against rods loosening after overloading, provision should be made for tightening at the ends of the member or by means of a turnbuckle between the ends of a two-piece member.

Round bars are frequently grouted into holes in rock formations to stabilize tunnel liners or retaining walls. They are also useful in reducing and restraining movement, such as a cracked machinery pedestal or spreading walls in old masonry structures.

---

[*] Upset ends were originally formed by forging. Currently, the threaded end segments may be made from a larger-diameter rod than the center segment and the three parts are then butt-welded together.

Figure 2.5   Wire strand roof support system. (Courtesy of American Institute of Steel Construction.)

In designing a rod with upset ends, the average stress on the area at the root of the thread should be less than the stress in the body of the bar. This will ensure yielding in the body of the bar if there is severe overload, as in an earthquake. If there is a misalignment or bending in a tie rod, the use of upset ends offers the additional advantage that any added stress due to bending is greatest in the main body of the bar, which is most flexible and able to adjust to such a condition.

Areas of rods and bars are tabulated in AISCM as are dimensions of threads, turnbuckles, clevises, and sleeve nuts for a wide variety of rod diameters.

2.2.3 Eyebars and Pin-Connected Plates

Eyebars and pin-connected plates [Fig. 2.1(c) and (d)] are used in a variety of special situations. Examples include the transfer of tensile load from a wire rope or cable to a structural steel assemblage or to an anchorage, as in the case of a suspension bridge. The use of eyebars as tension members in a modern long-span bridge is illustrated in Fig. 2.6.

If failure were to occur in the head of the eyebar at the connection, tests have demonstrated that it would be one of the following types:

Figure 2.6 Eyebar tension members in the Brent Spence Bridge over the Ohio River.

1. Fracture behind the pin in a direction parallel to the axis of the bar. This type of failure will occur if insufficient edge distance behind the pin is provided [Fig. 2.7(a)].

2. Failure in the net section through the pin transverse to the axis of the bar. This type of failure will occur if the gross area of the main section of the bar is equal to or greater than the net section through the pin hold [Fig. 2.7(b)].

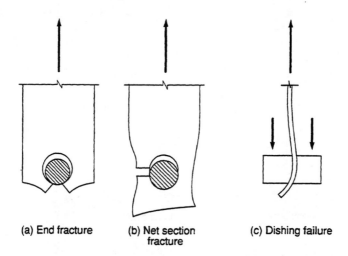

(a) End fracture   (b) Net section   (c) Dishing failure
                          fracture

Figure 2.7   Various failure modes of a pin-connected plate.

3. Failure by *dishing*. This is an inelastic lateral stability type of failure, which will occur if the width/thickness ratio behind the pin is too great. Dishing failure is akin to the lateral instability of a short deep beam [Fig. 2.7(c)].

Since it is desirable to ensure that general yielding and ultimate failure occur in the main body of the bar rather than at the end, all specifications provide dimensional requirements that prevent failure of the types enumerated above. When several eye-bars come together or connect at the same pin with packing and external nuts to prevent spreading of the package, there will be no need to restrict the width/thickness ratio of the eyebar, since dishing will be prevented by the lateral constraints provided. The required proportion for eyebars is similar in the various specifications.

In general, connection details are treated in Chapter 6; but in the case of eyebars or pin-connected plates, the design of end-connecting details, except for the proportioning of the pin itself, is related to the design of the whole member and will be covered in this chapter. Section D of AISCS specifies proportions that provide for a balanced design with respect to the various possible modes of failure. It should be studied in detail by reference to Example 2.1.

2.2.4 Structural Shapes and Built-up Members

Structural shapes and built-up members are used when rigidity is required in a tension member, to resist small lateral loads, or when reversal of load may subject the member to alternate compression and tension, as in a truss diagonal near the center of a span. The most commonly used shapes are the angle, tee, and W, S, or M shapes, as shown in Fig. 2.8. For exposed use, to minimize wind load, or for esthetic reasons, the pipe section may be favored. Built-up members are formed by connecting two or more structural shapes with separators, battens, lacing, or continuous plates, so that they will work together as a unit, as shown in Fig. 2.8. The angle and channel members, as shown, may be used in single-plane truss construction connected to end gusset plates

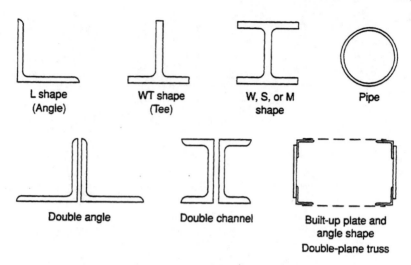

Figure 2.8 Structural shapes as tension members.

with rivets, bolts, or welds. The built-up open-box shape, as shown (the dashed lines indicate lacing or battens), is suitable for double-plane truss construction and is conveniently bolted, riveted, or welded between two gusset plates at each end connection. W, M, or S shapes are especially suited to double-plane welded truss construction.

Although end connections are of paramount importance in the design of a complete tension member, this topic, except for the eyebar, is covered in Chapter 6.

A nonmandatory clause, Sec. B7 of AISCS, recommends a maximum slenderness ratio (l/r) of 300. Rods and cables are excepted from these limitations.

2.3 Design Limit States and Effective Net Area

Axially loaded tension members are proportioned so that the required design force P_u is less than or equal to the limit state strength P_n multiplied by the resistance factor ϕ:

$$\phi_t P_n \geq P_u$$

Two limit-state conditions must be checked according to Sec. D1 of AISCS:

1. The limit state of yielding of the gross cross-sectional area A_g:

$$\phi_t = 0.9 \qquad P_n = F_y A_g$$

   where F_y is the specified minimum yield stress of the steel.

2. The limit state of fracture of the effective net area A_e at the ends of the tension member:

$$\phi_t = 0.75 \qquad P_n = F_u A_e$$

   where F_u is the specified minimum tensile strength of the steel. The smaller resistance factor of 0.75 for fracture is used because of the serious consequence of this type of failure of the connection.

The effective net area of a tension member is defined in Sec. B3 of AISCS: $A_e = AU$. A is the area equal to the gross cross-sectional area when the tension load is transmitted by longitudinal welds only or in combination with transverse welds, and it is equal to the net area if it is transmitted by bolts or rivets. U is the *shear-lag reduction coefficient* to be used when only part of the cross section is connected, as, for example, one leg of an angle tension member. The determination of U is discussed in Sec. B3 of the AISCS Commentary for many different types of welded and bolted or riveted joints. Application of the effective net section principle is illustrated in Examples 2.4 and 2.5.

The actual net area A_n, in the case of a chain of holes extending across an element, is determined by an empirical rule as defined in Sec. B2 of AISCS: "the net width of the part shall be obtained by deducting from the gross width the sum of the diameters of all the holes in the chain, and adding, for each gage space in the chain, the quantity

$$\frac{s^2}{4g}$$

where s is the longitudinal spacing (pitch) of any two consecutive holes, in., and g is the transverse spacing (gage) of the same two holes, in. The critical net area, A_n, of the part is obtained from that chain which gives the least net width."

The following example illustrates the application of the foregoing rule for net width calculation. In the example, a plate, $\frac{3}{4}$ by 10, is in tension and is attached to another plate by means of 14 high-strength $\frac{3}{4}$ ϕ bolts. The connection strength of the bolts will not be evaluated, that being a topic considered in Chapter 6. Example 2.1 considers only the strength as determined by net section, either through line *abde*, deducting two holes, or through *abcde*, deducting three holes and adding, by the rule above, the value of $s^2/4g$ as determined by *bc* and *cd*.

Example 2.1

Determine the allowable tensile force as determined by the following hole pattern:

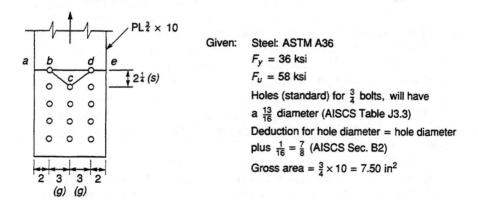

Given: Steel: ASTM A36
F_y = 36 ksi
F_u = 58 ksi
Holes (standard) for $\frac{3}{4}$ bolts, will have
a $\frac{13}{16}$ diameter (AISCS Table J3.3)
Deduction for hole diameter = hole diameter
plus $\frac{1}{16}$ = $\frac{7}{8}$ (AISCS Sec. B2)
Gross area = $\frac{3}{4} \times 10 = 7.50$ in²

Solution   The effective net area of this plate is equal to the actual net area (Sec. B3 of AISCS). The net area is the lesser of

$$0.75 \times \left(10 - 2 \times \frac{7}{8}\right) = 6.19 \text{ in}^2 \quad (\text{line } abde)$$

$$0.75 \times \left(10 - 3 \times \frac{7}{8} + \frac{2 \times 2.25^2}{4 \times 3}\right) = 6.16 \text{ in}^2 \quad (\text{line } abcde)$$

The design strength in tension is the lesser of

$$\phi_t P_n = 0.9 \times 36 \times 0.75 \times 10 = 243 \text{ kips} \quad (\text{yield})$$

$$\phi_t P_n = 0.75 \times 58 \times 6.16 = 268 \text{ kips} \quad (\text{fracture})$$

Assuming a capacity of 243 kips, the possibility exists that the full net section with three holes deducted might control the strength after the first two fasteners have taken their share of the load, the remaining load being equal to

$$\frac{12}{24} \times 243 = 208.3 \text{ kips}$$

The net area, three holes deducted, is

$$0.75 \times \left(10 - 3 \times \frac{7}{8}\right) = 5.53 \text{ in}^2$$

The capacity at the line through c is

$$0.75 \times 58 \times 5.53 = 240.6 \text{ kips} \qquad > 208.3 \text{ kips} \quad \text{OK}$$

Eyebars and pin-connected members are designed for the limit states of (1) tension on the net effective area, (2) shear on the effective area, and (3) bearing on the projected area of the pin. Example 2.3 illustrates the design of an eyebar.

2.4 Design for Repeated Load

When load is repeatedly applied and removed, with the number of repetitions running into the many thousands or up into the millions, metal may develop cracks that eventually may spread to the point where they cause *fatigue* or *failure* of the member. Fatigue cracks are most apt to occur when the repeated load is primarily tension. Local stress concentrations increase the susceptibility to fatigue failure. Such concentrations may be due to poorly made welds, small holes, and rough or damaged edges resulting from the fabrication process of shearing, punching, or poor-quality oxygen cutting. The fatigue strength of the higher-strength steels has not been shown to be appreciably greater than the commonly used A36 grade structural steel with a yield stress of 36 ksi.

In 1969, AISCS introduced a simplified approach to design for repeated load, as presented in Appendix K3 of AISCS. The feature of the AISCS approach to repeated load design is its use of the expected *stress range* as the controlling design criterion. Stress range is the algebraic difference between maximum and minimum stress to be expected in any one cycle of loading. Thus the following two cases each have the same stress range of 16 ksi:

Maximum Stress	Minimum Stress
20 ksi tension	4 ksi tension
4 ksi tension	−12 ksi compression

The permissible stress range is a function of (1) loading condition, and (2) stress category.

Loading conditions are defined and tabulated in Table A-K3.1 of AISCS and are determined according to the anticipated number of loading cycles to be used as a basis for design. If less than 20,000 cycles, no consideration need be given to repeated load, but at successive lower limits of 20,000, 100,000, 500,000, and 2,000,000 cycles, respectively, load conditions 1, 2, 3, and 4 are established.

The stress categories, ranging from A to F with increasing severity of the local stress raiser, are tabulated and defined in Table A-K3.2 of AISCS. After establishing the loading condition and the stress category, the allowable stress range F_{sr} is read from Table A-K3.3 of AISCS.

Members in conventional buildings usually do not need to be designed for repeated load because the number of repetitions of maximum load is usually less than 20,000. However, crane runway girders and supporting structures for machinery and equipment do require the consideration of fatigue. Fatigue is a major design consideration for bridges.

2.5 Flowchart

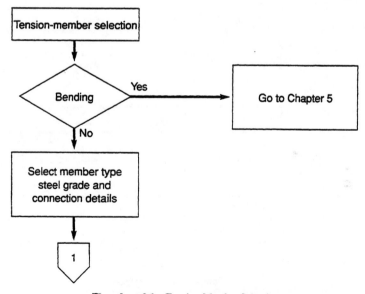

Flowchart 2.1   Tension-Member Selection

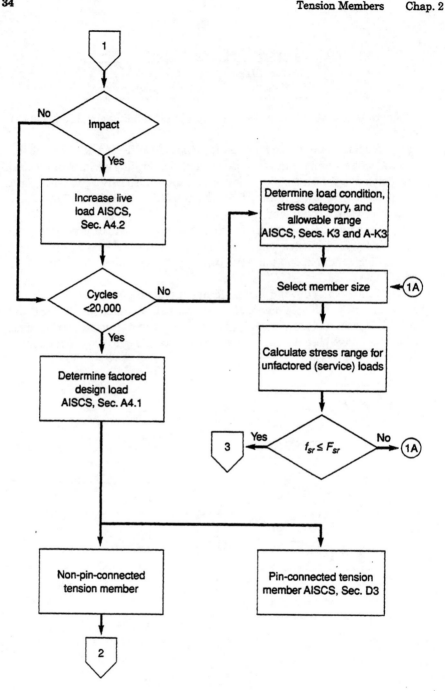

Flowchart 2.1   (continued)

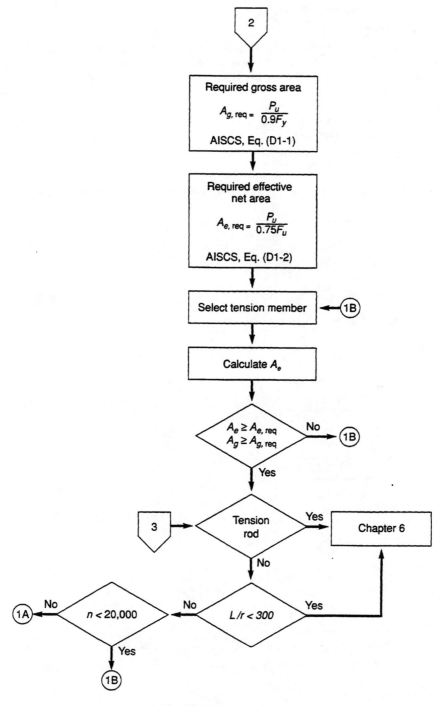

Flowchart 2.1   (continued)

2.6 Illustrative Examples

Example 2.2

A floor-beam suspender is stressed in tension by a dead load of 30 kips and a live load of 40 kips. The full live load will be repeated less than 20,000 times. Select a round steel rod, using upset ends, to satisfy AISCS. Use quenched and tempered alloy steel with $F_y = 100$ ksi.

Solution

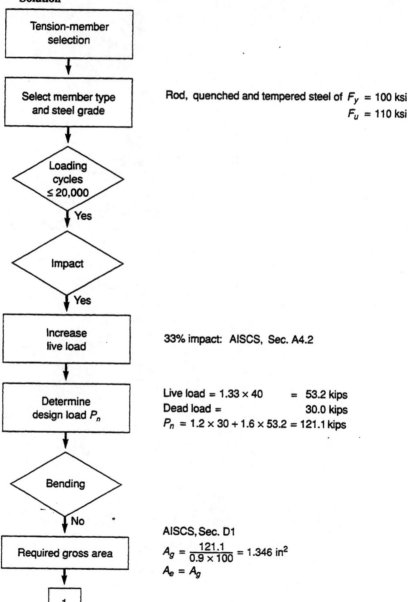

Rod, quenched and tempered steel of $F_y = 100$ ksi

$F_u = 110$ ksi

33% impact: AISCS, Sec. A4.2

Live load $= 1.33 \times 40$ $= 53.2$ kips
Dead load $=$ $\quad$ 30.0 kips
$P_n = 1.2 \times 30 + 1.6 \times 53.2 = 121.1$ kips

AISCS, Sec. D1

$$A_g = \frac{121.1}{0.9 \times 100} = 1.346 \text{ in}^2$$

$$A_e = A_g$$

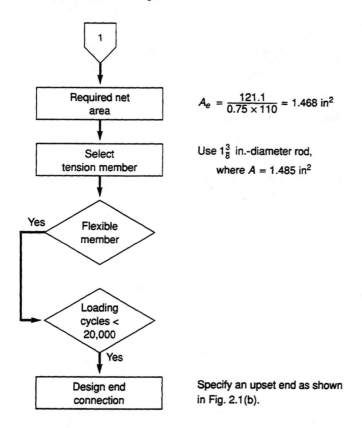

$$A_e = \frac{121.1}{0.75 \times 110} = 1.468 \text{ in}^2$$

Use $1\frac{3}{8}$ in.-diameter rod,
where $A = 1.485$ in^2

Specify an upset end as shown
in Fig. 2.1(b).

Example 2.3
Design an eyebar to carry a tensile dead load of 200 kips and a tensile live load of 400 kips (less than 20,000 repetition of load).

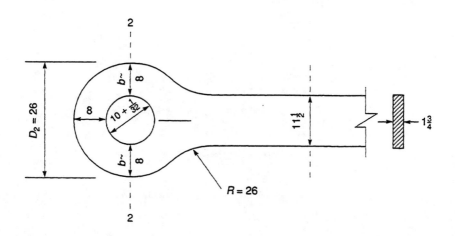

Select ASTM A572 steel with

$$F_y = 50 \text{ ksi} \qquad P_D = 200 \text{ kips}$$
$$F_u = 65 \text{ ksi} \qquad P_L = 400 \text{ kips}$$

Given: pin diameter $d = 10$ in.

Solution   Required design strength:

$$P_u = 1.2P_D + 1.6P_L = 880 \text{ kips}$$

(a) *Required area of eyebar body* (AISCS, Sec. D1a):
   $\phi_t = 0.9$. Select $b_b = 11.5$ in. and $t = 1.75$ in.

Gross area:

$$A_g = b_b t$$

Nominal capacity:

$$P_n = A_g F_y$$
$$\phi_t P_n = 905.625 \text{ kips} \qquad > 880 \text{ kips}\quad \text{OK}$$

(b) *Check geometry* (AISCS, Sec. D3): Eyebar has uniform thickness; it has no reinforcement at the hole; the hole is circular.   OK

$$R = 26 \text{ in.} \qquad > d = 10 \text{ in.}\quad \text{OK}$$
$$8t = 14 \text{ in.} \qquad > b_b = 11.5 \text{ in.}\quad \text{OK}$$
$$t = 1.75 \text{ in.} \qquad > 0.75 \text{ in.}\quad \text{OK}$$

Select $a = 8$ in. (*a* is the distance from edge pinhole to edge of member, measured parallel to the direction of force).

$$\frac{2d}{3} = 6.667 \text{ in.} \qquad < a\quad \text{OK}$$

$$\frac{3b_b}{4} = 8.625 \text{ in.} \qquad > a\quad \text{OK}$$

(c) *Tension on the net effective area:*

$$\phi_t = 0.75$$
$$b_{\text{eff}} = 2t + 0.63 \text{ in.} = 4.13 \text{ in.} \qquad < b = 8 \text{ in.}\quad \text{OK}$$

Nominal capacity:

$$P_n = 2b_{\text{eff}} F_u t$$
$$\phi_t P_n = 704.681 \text{ kips} \qquad < 880 \text{ kips}\quad \text{NG}$$

Redesign: try $t = 2$ in.

$$b_{\text{eff}} = 2t + 0.63 \text{ in.} = 4.63 \text{ in.}$$
$$P_n = 2t b_{\text{eff}} F_u$$
$$\phi_t P_n = 902.85 \text{ kips} \qquad > 880 \text{ kips}\quad \text{OK}$$

Use $t = 2$ in.; all other dimensions are OK.

(d) *Shear on effective area:*

$$\phi_{sf} = 0.75$$

$$A_{sf} = 2t\left(a + \frac{d}{2}\right)$$

Nominal capacity:

$$P_n = A_{sf}F_y$$

$$\phi_{sf}P_n = 1.95 \times 10^3 \text{ kips} \qquad > 880 \text{ kips   OK}$$

(e) *Bearing on projected area of pin:*

$$\phi = 1.0$$

$$A_{pb} = td \qquad \text{projected area of pin}$$

Nominal capacity:

$$P_n = A_{pb}F_y$$

$$\phi P_n = 1000 \text{ kips} \qquad > 880 \text{ kips   OK}$$

Following is the same problem but solved with SI units. Select ASTM A572 steel with

$$F_y = 345 \text{ MPa} \qquad P_D = 890 \text{ kN}$$

$$F_u = 448 \text{ MPa} \qquad P_L = 1780 \text{ kN}$$

Given: pin diameter $d = 254$ mm.

Solution   Required design strength:

$$P_u = 1.2P_D + 1.6P_L = 3.916 \times 10^3 \text{ kN}$$

(a) *Required area of eyebar body* (AISCS, Sec. D1a):
$\phi_t = 0.9$. Select $b_b = 292$ mm and $t = 45$ mm.

Gross area:

$$A_g = b_b t$$

Nominal capacity:

$$P_n = A_g F_y$$

$$\phi_t P_n = 4.08 \times 10^3 \text{ kN} \qquad > 3916 \text{ kN   OK}$$

(b) *Check geometry* (AISCS, Sec. D3): Eyebar has uniform thickness; it has no reinforcement at the hole; the hole is circular. OK

$$R = 660 \text{ mm} > d = 254 \text{ mm   OK}$$

$$8t = 360 \text{ mm} \qquad > b_b = 292 \text{ mm   OK}$$

$$t = 45 \text{ mm} \qquad > 0.75 \text{ in.   OK}$$

Select $a = 203$ mm (a is the distance from edge of pinhole to edge of member, measured parallel to the direction of force).

$$\frac{2d}{3} = 169.333 \text{ mm} \qquad < a \text{   OK}$$

$$\frac{3b_b}{4} = 219 \text{ mm} \qquad > a \text{   OK}$$

(c) *Tension on the net effective area:*

$$\phi_t = 0.75$$

$$b_{eff} = 2t + 16 \text{ mm} = 106 \text{ mm} \qquad < b = 203 \text{ mm} \quad \text{OK}$$

Nominal capacity:

$$P_n = 2b_{eff}F_ut$$

$$\phi_t P_n = 3.205 \times 10^3 \text{ kN} \qquad < 3916 \text{ kN} \quad \text{NG}$$

Redesign: try $t = 51$ mm

$$b_{eff} = 2t + 16 \text{ mm} = 117.6 \text{ mm}$$

$$P_n = 2tb_{eff}F_u$$

$$\phi_t P_n = 4.015 \times 10^3 \text{ kN} \qquad > 3916 \text{ kN} \quad \text{OK}$$

Use $t = 51$ mm; all other dimensions are OK.

(d) *Shear on effective area:*

$$\phi_{sf} = 0.75$$

$$A_{sf} = 2t\left(a + \frac{d}{2}\right)$$

Nominal capacity:

$$P_n = A_{sf}F_y$$

$$\phi_{sf}P_n = 8.675 \times 10^3 \text{ kN} \qquad > 3916 \text{ kN} \quad \text{OK}$$

(e) *Bearing on projected area of pin:*

$$\phi = 1.0$$

$$A_{pb} = td \qquad \text{projected area of pin}$$

Nominal capacity:

$$P_n = A_{pb}F_y$$

$$\phi P_n = 4.452 \times 10^3 \text{ kN} \qquad > 3916 \text{ kN} \quad \text{OK}$$

Example 2.4

A tension member of a roof truss has a length of 25 ft and is stressed in tension by a dead load of 40 kips and a live load of 60 kips. The tension member is a main member and needs some amount of rigidity. Select a single structural tee to satisfy AISCS. Use A36 steel. Connection is made by welding to the flange.

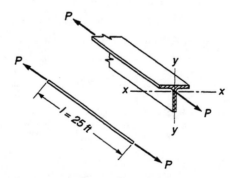

Solution

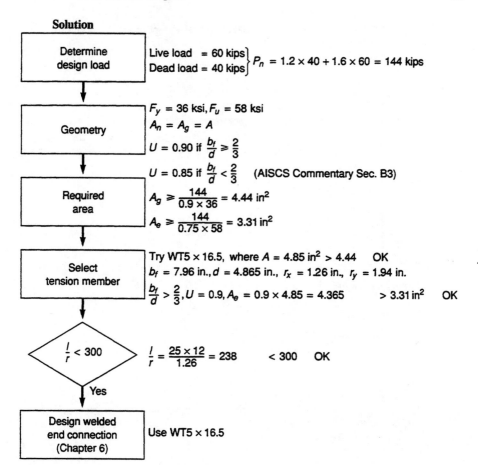

| Determine design load | Live load = 60 kips
Dead load = 40 kips $\Big\}$ $P_n = 1.2 \times 40 + 1.6 \times 60 = 144$ kips |

| Geometry | $F_y = 36$ ksi, $F_u = 58$ ksi
$A_n = A_g = A$
$U = 0.90$ if $\frac{b_f}{d} \geq \frac{2}{3}$
$U = 0.85$ if $\frac{b_f}{d} < \frac{2}{3}$ (AISCS Commentary Sec. B3) |

| Required area | $A_g \geq \dfrac{144}{0.9 \times 36} = 4.44$ in^2
$A_e \geq \dfrac{144}{0.75 \times 58} = 3.31$ in^2 |

| Select tension member | Try WT5 × 16.5, where $A = 4.85$ in^2 > 4.44   OK
$b_f = 7.96$ in., $d = 4.865$ in., $r_x = 1.26$ in., $r_y = 1.94$ in.
$\frac{b_f}{d} > \frac{2}{3}$, $U = 0.9$, $A_e = 0.9 \times 4.85 = 4.365$ > 3.31 in^2 OK |

| $\frac{l}{r} < 300$ | $\frac{l}{r} = \dfrac{25 \times 12}{1.26} = 238$ < 300   OK |

Yes

| Design welded end connection (Chapter 6) | Use WT5 × 16.5 |

Example 2.5

Rework Example 2.4 with an additional axial tension of 45 kips produced by wind. Connection is to the flanges by welding.

Solution

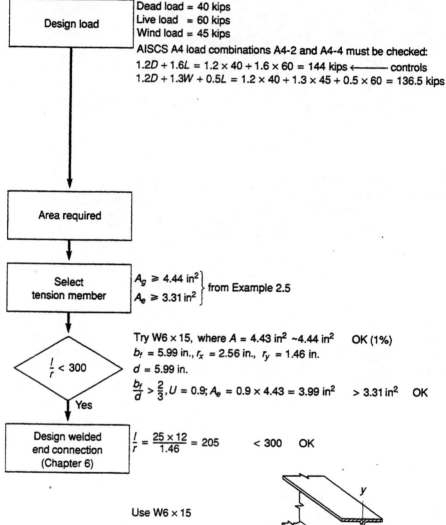

Design load	Dead load = 40 kips
	Live load = 60 kips
	Wind load = 45 kips

AISCS A4 load combinations A4-2 and A4-4 must be checked:

$1.2D + 1.6L = 1.2 \times 40 + 1.6 \times 60 = 144$ kips ←——— controls

$1.2D + 1.3W + 0.5L = 1.2 \times 40 + 1.3 \times 45 + 0.5 \times 60 = 136.5$ kips

Area required

Select tension member

$\left. \begin{array}{l} A_g \geq 4.44 \text{ in}^2 \\ A_e \geq 3.31 \text{ in}^2 \end{array} \right\}$ from Example 2.5

$\dfrac{l}{r} < 300$

Try W6 × 15, where $A = 4.43 \text{ in}^2 \sim 4.44 \text{ in}^2$ OK (1%)

$b_f = 5.99$ in., $r_x = 2.56$ in., $r_y = 1.46$ in.

$d = 5.99$ in.

$\dfrac{b_f}{d} > \dfrac{2}{3}, U = 0.9; A_e = 0.9 \times 4.43 = 3.99 \text{ in}^2$ $> 3.31 \text{ in}^2$ OK

Yes

Design welded end connection (Chapter 6)

$\dfrac{l}{r} = \dfrac{25 \times 12}{1.46} = 205$ < 300 OK

Use W6 × 15

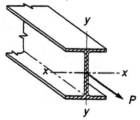

Example 2.6

Rework Example 2.4 but now the live load of 60 kips may be repeated 300,000 times and during each cycle the member is subjected to 10 kips compression. Welded end connection with fillet welds is similar to Case 17 in Appendix K3 of AISCS.

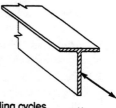

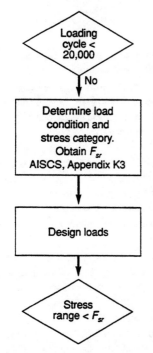

Loading condition 2 for 300,000 loading cycles. Illustrative Example 17 for fillet-welded connection.
Stress category E.

Therefore, allowable stress range
F_{sr} = 13 ksi

Tension:
    Dead load + 40
    Live load + 60 P_t = +100 kips (max.)

Compression:
    Dead load + 40
    Live load − 60 P_t = +30 kips (min.)

WT5 × 16.5 is the selected tension member as in Example 2.4 for the design dead and live loads of 100 kips.

Then check stress range for this fatigue loading condition:

Actual stresses:

$$\text{Max. } f_t = \frac{P_t}{A} = \frac{100}{4.85} = 20.6 \text{ ksi}$$

$$\text{Min. } f_t = \frac{P_t}{A} = \frac{30}{4.85} = 6.18 \text{ ksi}$$

Actual stress range = 20.6 − 6.18
                        = 14.42 ksi > F_{sr} NG

Try WT5 × 19.5 (next larger size) A = 5.73 in²
New stress range, by proportion of areas
$= \frac{4.85}{5.75} \times 14.42 = 12.20$ < F_{sr} OK

PROBLEMS*

2.1. Similar to Example 2.1, with the following changes:
Plate is $\frac{3}{4}$ by 12.
Steel is ASTM 242 with yield strength of 42 ksi and ultimate tensile strength of 63 ksi.
Lateral spacing of bolts is 4 instead of 3.
Longitudinal spacing is $2\frac{1}{2}$ instead of $2\frac{1}{4}$.
Bolts are $\frac{7}{8}$ instead of $\frac{3}{4}$.

2.2. Design an eyebar to carry a factored total load of 600 kips, using ASTM A242 steel with $F_u = 63$ ksi and $F_y = 42$ ksi. Assume fewer than 20,000 load repetitions.

2.3. Design a tension member to carry a dead load of 50 kips and live load of 75 kips. Use two angles with long legs back to back. Long legs are to be separated $\frac{3}{8}$ in. for end connections to gusset plates. In determining effective net section, assume a single line of holes for $\frac{7}{8}$-in.-diameter bolts through long sides of angles and refer to Sec. B3 of AISCS. Use Example 2.4 as a guide.

2.4. Rework Problem 2.3 but now the live load of 75 kips may be repeated 600,000 times, and during each cycle the member may be subjected to a compression of 30 kips. The high-strength bolted end connection may be assumed to be equivalent, in fatigue resistance, to Case 8 in Appendix K4 of AISCS. Example 2.6 should be studied as a guide.

2.5. Rework Problem 2.3 with the addition of a 50-kip tension force due to wind. Refer to Example 2.5.

2.6. The $1\frac{3}{8}$-in.-diameter bar was chosen in Example 2.2 without consideration of repeated load; that is, the repetitions of load were assumed to be less than 20,000. Ruling out the possibility of failure in the end threads, show that the bar is capable of withstanding 300,000 repetitions of live load (without impact). For repeated load, the round bar may be assumed as equivalent to Case 2 in Appendix K4 of AISCS.

2.7. Referring again to Example 2.2, assume that the thread root area in the upset end of the rod is 1.90 in². Assume that this puts the end in repeated load Category F of Table K4 of AISCS. Is the threaded end adequate for 300,000 repetitions of live load alone, excluding impact? If not, what change would you recommend? Such a change could involve a different material, different type of member, or a different end connection.

3

Beams

3.1 Introduction

Beams support loads that are applied at right angles (transverse) to the longitudinal axis of the member. Such loads are usually downward, as illustrated in Fig. 3.1(a). The beam carries the loads to its supports, which may consist of the bearing walls, columns, or other beams into which it frames. At the supports the upward "reactions" have a total magnitude equal to the weight of the beam plus the applied loads P. Since the weight of the beam is not known until after it is designed, the design starts with a preliminary weight estimate that is subject to later revision.

Imagine a free-body diagram of the left portion of the beam [Fig. 3.1(b)] with bending moment (M) and the shear (V) necessary at the cut section to provide static equilibrium. The problem of beam design consists mainly in providing enough bending strength and enough shear strength at every location in the span. For short spans, it is most economical to use a single-beam cross section throughout the span, in which case only the maximum values of bending moment and shear need to be determined.

A *simple beam* [Fig. 3.1(a)] is supported vertically at each end with little or no rotational restraint, and downward loads cause positive bending moment throughout the span. The top part of the beam shortens, due to compression, and the bottom part of the beam lengthens, due to tension [Fig. 3.1(d)]. The most common rolled-steel

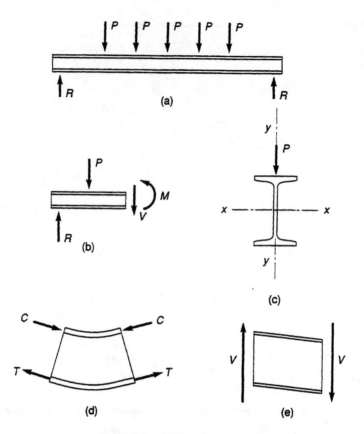

Figure 3.1 Simple beam behavior.

beam cross section, shown in Fig. 3.1(c), is called the W shape, with much of the material in the top and bottom flange, where it is most effective in resisting bending moment. The web of the beam supplies most of the shear resistance and in so doing is slightly distorted, as shown in Fig. 3.1(e). The contribution of this distortion to beam deflection is usually neglected. The bending moment causes curvature of the beam axis, concave upward, as shown in Fig. 3.1(d) for positive moment, concave downward for negative moment. The deflection of beams is usually calculated on the assumption that it is entirely caused by the curvature due to bending moment. Standard AISC nomenclature pertaining to the W (wide-flange) hot-rolled steel beams is illustrated in Fig. 3.2.

The reader should gain a familiarity with information in AISCM relative to rolled shapes, reading the descriptive material and scanning the tabular material, which includes:

1. Discussion and tabular material relative to shape selection, designation, dimensions, availability, size groupings, principal producers, and proper manner of shape designation

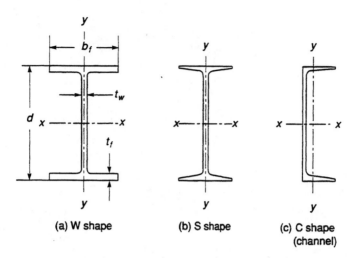

Figure 3.2   Nomenclature pertaining to a beam cross section, where $x - x$ = principal major axis, strong axis of bending, weak plane; $y - y$ = principal minor axis, weak axis of bending, strong plane.

2. Dimensions of shape cross sections for detailing
3. Properties of shape cross sections for use in design calculations
4. Standard mill practice in the rolling, cambering, and cutting of shapes, with corresponding dimensional tolerances.

A *plate girder* (see Chapter 7) is of such large depth and span that a rolled beam is not economically suitable—it is tailor made (built up out of plate material by use of welds, rivets, or bolts) to suit the particular span, clearance, and load requirements.

It is assumed that the reader is familiar with the analysis of shears and moments, with the drawing of corresponding shear and moment diagrams, and with the usual designation of support conditions. Various cases are illustrated in Fig. 3.3. At the top the loads and supports are shown for (a) a cantilever beam, (b) a simple beam with a cantilever overhang at the right end, and (c) a beam fixed at the left end and the same as (b) at the right end. In (c) the shears and moments between the fixed end and the simple support are *statically indeterminate*; that is, they cannot be determined by simple statics. Shear diagrams are shown on the second line, moment diagrams on the third. Although the calculation of shears and moments will be included in many of the illustrative examples, reference should be made to a text on strength of materials or elementary structural theory for additional information on these topics. For uniform or distributed loads the shear and moment diagrams are similar to those shown in Fig. 3.3, but the shear, since it changes with load, is a sloping line instead of a horizontal one, and the moment diagram is a continuous curve between concentrated loads and/or reactions. The reader should review the mathematical relationships between load, shear, and bending moment as found in texts on strength of materials or structural theory (Refs. 3.2 or 3.3).

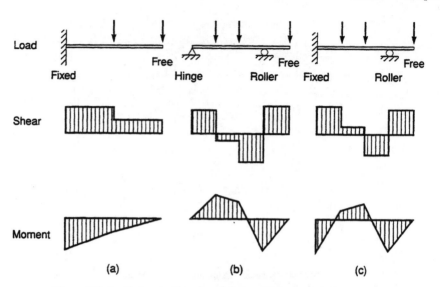

Figure 3.3   Load, shear, and moment diagrams for various beam situations.

Beams usually are framed with other beams, or attached to a floor slab, as shown in Fig. 3.4 , so that the beam cannot move sideways and the beam is forced to deflect vertically in the strong (y–y) plane (see Fig. 3.2). Whenever a beam deflects in

Figure 3.4   Beams supporting permanent metal forms for floor slab. (Courtesy of Bethlehem Steel Corp.)

the plane in which it is loaded, the simple theory of bending[*] may be used. The condition may be forced, as mentioned previously, or it can occur naturally if the plane of the loads contains a principal axis of the cross section. However, if the load is in the strong (y–y) plane (see Fig. 3.2), the beam may need lateral support to prevent it from buckling sideways; alternatively, the specifications provide for reduced allowable loads if lateral support does not meet certain minimal requirements. If loaded in the weak (x–x) plane (see Fig. 3.2), lateral buckling is no problem. Sections that lack two axes of symmetry usually require more positive lateral supports than does the W shape. For example, the laterally unsupported channel member will twist if loaded through the centroidal axis, as shown in Fig. 3.5(b), and requires restraint against both twist and lateral buckling. The Z section does not twist but deflects at an angle to the plane of the loads unless supported as shown in Fig. 3.5(c). An angle loaded as shown in Fig. 3.5(d) must be supported against both twist and lateral deflection. It is also important to recognize that if a Z or angle section is used without lateral support, the stress due to bending cannot be calculated by the simple beam formula. Where lateral support is needed only to prevent lateral buckling [Fig. 3.5(a)] there is no calculable stress in the lateral supports. In cases (b), (c), and (d), however, there is a calculable stress in the lateral supporting members, and thus there is a more clearly defined design problem with regard to lateral support.

Most beams are designed by simple bending theory. The design process involves the calculation of the maximum bending moment and the selection of a beam having an equal or greater bending moment resistance. The selection is then checked for maximum shear capacity, and the end connections or bearing support details are designed. A deflection check may also be required.

Some of the more complex beam design problems, such as general biaxial bending and combined bending and torsion, are treated in Chapter 10. An introduction to plastic design of continuous beams and frames is presented in Chapter 8.

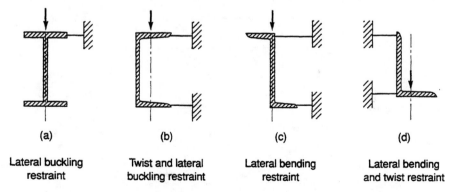

(a)	(b)	(c)	(d)
Lateral buckling restraint	Twist and lateral buckling restraint	Lateral bending restraint	Lateral bending and twist restraint

Figure 3.5 Type of lateral restraint required to permit beam selection by simple bending theory.

[*] The simple theory of bending is reviewed briefly in Section 3.2.

In the AISC LRFD specification the required cross-sectional moment M_u and shear V_u are calculated by structural analysis for the factored forces acting on the beam, and the following design criteria are checked:

$$\phi M_n > M_u \quad \text{and} \quad \phi V_n > V_u$$

The moment ϕM_n and the shear ϕV_n are the factored flexural and shear design strengths, respectively, as prescribed in Sec. F of AISCS. The resistance factor $\phi = 0.90$ for beams; M_n and V_n are the limiting moment and shear capacities of the member.

The values of V_u and M_u are determined by statics if the beam is statically determinate. In the case of statically indeterminate structures two methods of analysis are permitted: (1) plastic analysis if the members meet stringent conditions of compactness and lateral bracing, and (2) elastic analysis if these conditions are not fulfilled. In the elastic analysis the redundant forces are determined by the conditions of equilibrium and compatibility. In plastic analysis the formation of a kinematic mechanism is the condition that determines the cross-sectional forces. Elastic analysis may be used in lieu of plastic analysis even if the latter is permitted. This is a more conservative approach and it is most often used by designers. The elastic and plastic methods of indeterminate analysis are discussed further in Chapter 8.

3.2 Elastic Bending of Steel Beams

A knowledge of elementary beam theory as presented in texts on mechanics or strength of materials is an essential preliminary to the study of beam design (see Refs. 3.2 and 3.3). Figure 3.6 shows a *unit* length of beam imagined as cut out of the complete beam at any location along the beam. It is acted upon by bending moment M and shear V,

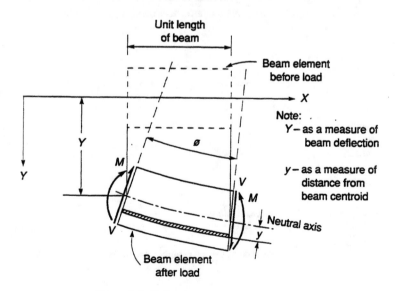

Figure 3.6 Deformation of beam element.

positive as indicated, and is shown in its undeflected straight position before loading and in its deflected and bent position after loading. Note that Y and y, positive as shown, are used to signify two different distances: deflection of the beam axis and distance within the beam cross section from the centroid, respectively. The *curvature* of the beam, or change in slope per unit length of beam, is denoted by ϕ, and the unit longitudinal *strain*, or change in length per unit length, of a horizontal beam fiber is therefore equal to

$$\varepsilon = \phi y \qquad (3.1)$$

Since normal stress (f_b) is equal to the modulus of elasticity (E) multiplied by strain (ε), the stress due to bending is equal, by Eq. (3.1), to

$$f_b = E\phi y \qquad (3.2)$$

Thus the stress due to bending of a beam is known if the curvature is known. This fact could be of interest to a wire manufacturer who wished to know the diameter of a drum or reel on which drawn wire could be wound without inducing any permanent bend. Suppose, for example, that a wire with a diameter of 0.10 in. is wound on a reel having a diameter of 60 in. The curvature of the wire is equal to a unit length (1 in.) divided by the reel radius (30 in.). Thus

$$\phi = \frac{1}{30}$$

The maximum strain in the wire, by Eq. (3.1), is

$$\varepsilon = \frac{1}{30}(0.05) = 0.00167$$

The maximum stress, then, for steel wire, due to bending, for $E = 29{,}000$ ksi, is equal to

$$f_b = (29{,}000)(0.00167) = 48.3 \text{ ksi}$$

The stress of 48.3 ksi is greater than the yield point of carbon structural steel, but less than the elastic limit of most cold-drawn high-strength steel wires, for which the reel diameter would be satisfactory as it would not induce permanent bending deformation in the wire.

Equation (3.2) is convenient in the problem facing the wire manufacturer, but for the design of steel beams the stress due to bending is usually calculated as a function of the bending moment, which is proportional to the curvature. The constant of proportionality between moment and curvature is EI, I being the moment of inertia of the cross section, as tabulated for all rolled sections in AISCM. Thus the bending moment is

$$M = EI\phi \qquad (3.3)$$

Combining Eqs. (3.2) and (3.3), the formula for stress in terms of bending moment is obtained:

$$f_b = \frac{My}{I} \qquad (3.4)$$

Equation (3.4) is sometimes called the *beam equation*, and its use is restricted to the simple bending theory described previously. The maximum stress due to bending $y = c$, where c is the maximum y distance from the centroidal (also neutral) axis of the beam to the extreme top or bottom fiber of the cross section. If the beam section is symmetrical about its x axis, c will be the same for the compression and tension extremities.

$$(f_b)_{max} = \frac{M_{max}c}{I} \tag{3.5}$$

To expedite the design selection of a beam for maximum bending moment, I and c are combined into a single parameter, the *elastic section modulus*, denoted by S and equal to I/c. Equation (3.5) then becomes simply

$$(f_b)_{max} = \frac{M_{max}}{S} \tag{3.6}$$

The maximum stress due to bending (f_b) must be less than the factored yield stress ϕF_y ($\phi = 0.9$, resistance factor); thus the required elastic section modulus is

$$S_{reqd} = \frac{M_{max}}{\phi F_y} \tag{3.7}$$

Unless the beam is extremely short, it should be *chosen* for moment and *checked* for shear. The shear stress as computed by simple bending theory at any location in the beam web is given by

$$f_v = \frac{VQ}{It} \tag{3.8}$$

where V = total resultant shear force on cross section
Q = static moment, taken about the neutral axis, of that portion of the beam area beyond the point at which the shear stress is to be calculated
t = web thickness where the stress is computed

The designer may use Eq. (3.8) for certain shear-dependent details, such as the welds connecting the web and flange of built-up beams, or the webs of unsymmetrical sections. But simpler expressions for the web shear stress are specified for the usual beam design situation. For a W or C beam section, in simple bending, the shear stress is approximated by dividing the resultant shear at any location by the product of the web thickness times the full depth of the beam. In the design of built-up girders the area is based on the depth of the girder web plate between flange plates. Figure 3.7 illustrates the three alternatives just discussed.

The following example illustrates the calculation of the elastic flexural and shear stresses. A W24 × 62 rolled wide-flange steel beam is subject to a maximum moment and maximum shear, respectively, of

$$M_{max} = 315 \text{ kip-ft} \qquad V_{max} = 42 \text{ kips}$$

From Sec. 1 of AISCM we look up the following geometric properties:

$$d = 23.74 \text{ in.} \qquad t_w = 0.43 \text{ in.} \qquad S_x = 131 \text{ in}^3$$

Thus the stresses are

$$f_b = \frac{M_{max}}{S_x} = 28.855 \text{ ksi} \qquad f_v = \frac{V_{max}}{dt_w} = 4.114 \text{ ksi}$$

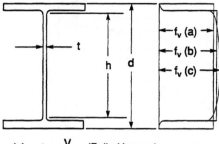

(a) $f_v = \dfrac{V}{dt}$ (Rolled beams)

(b) $f_v = \dfrac{V}{ht}$ (Plate girders – Chapter 7)

(c) $f_v = \dfrac{VQ}{It}$ (Shear dependent details or nonstandard shapes)   **Figure 3.7** Three alternatives for the estimate of web shear stress due to binding.

3.3 Inelastic Behavior of Steel Beams

The maximum moment that can be supported without yielding is $M_y = SF_y$. If the beam is laterally braced and its flange and web are *compact*, the cross section of a steel beam can become fully plastic and thus support more moment. This added capacity is utilized in AISCS, as discussed in this section.

Referring back to Fig. 1.2, let it be assumed that the stress–strain diagram for structural steel consists simply of the two straight-line portions out to the initiation of strain hardening and labeled "elastic range" and "plastic range," respectively. In this case, the maximum stress due to bending in a steel beam would not rise above the yield stress (F_y). Figure 3.8(b) shows the stress distribution for this case. If the bending moment is increased to be equal to F_y, the stress diagram will continue to have the elastic, linear distribution shown in Fig. 3.8(b), and the bending moment will have reached M_y, the yield moment. Above the yield moment, the stress distribution will be as shown in Fig. 3.8(c), finally approaching in the limit the rectangular shape shown

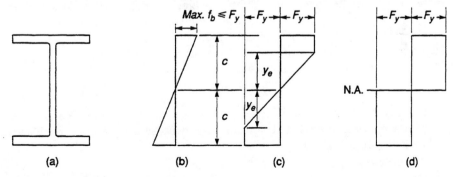

Figure 3.8 Distribution of normal stress due to bending in elastic and inelastic ranges.

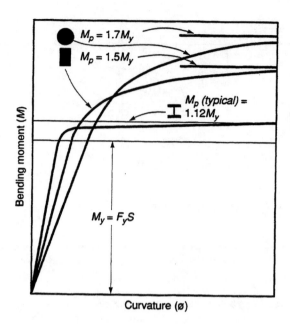

Figure 3.9   Inelastic $M - \phi$ curves for different cross sections.

in Fig. 3.8(d), which corresponds to the plastic moment, M_p, the maximum attainable if no strain hardening were to occur.

The inelastic behavior of a steel beam is best illustrated by a plot of bending moment versus curvature. Above M_y the curvature is no longer linearly related to moment by the elastic relationship of Eq. (3.3). At M_y the curvature $\phi_y = F_y/Ec = M_y/EI$. Above M_y, at moments less than M_p, and using the notation shown in Fig. 3.8(c), the curvature is equal to F_y/Ey_e. M versus ϕ curves for circular, rectangular, and W beam shapes are shown in Fig. 3.9. In these plots each shape has the same section modulus and the same M_y. The ratio M_p/M_y is called the *shape factor*, a measure of the increase in plastic moment strength in comparison with the plastic moment. The circular section is the most inefficient from an elastic point of view, although it absorbs more energy than the other shapes before reaching the yield moment. W shapes are proportioned to provide elastic design economy; consequently, they have low shape factors. To ensure an approach toward M_p without local flange buckling or lateral buckling, the wide-flange shape must be compact and laterally braced.

The plastic moment M_p is calculated from the equation

$$M_p = ZF_y \tag{3.9}$$

where Z is the *plastic section modulus*. The *shape factor* is the ratio Z/S and it characterizes the increased amount of moment capacity because of plastification. Typical values to the shape factor are:

Solid circular cross section, $Z/S = 1.7$

Rectangular section, $Z/S = 1.5$

Compact W shape, major-axis bending, $Z/S = 1.12$ (average)

Compact W shape, minor-axis bending, $Z/S = 1.6$ (average)

Values of the elastic and plastic section moduli for rolled shapes are tabulated in Part 1 of AISCM.

3.4 Flexural Design of Beams

The design of beams is covered in Chapter F of AISCS. This chapter has two sections: F1, "Design for Flexure," and F2, "Design for Shear." The flowchart of Fig. 3.10 illustrates the types of flexural members designed: beams and plate girders. These are distinguished by the value of the web slenderness ratio h/t_w, where the dimension h is the depth of the web between the toes of the flange fillets (e.g., the distance h in Fig. 3.7 for a doubly symmetric I section), and $t_w = t$, the web depth (Figs. 3.2 and 3.7). Plate girders are designed according to the rules of Chapter G of AISCS, and we discuss them in Chapter 7. Such members are constructed with transverse web stiffeners in regions of high shear.

Beams must be designed for limit states of serviceability and ultimate strength. The serviceability limit states relate to excessive flexibility (i.e., the beam deflects too much or it vibrates noticeably under service loads). The designer can anticipate such serviceability problems already at the stage of preliminary design by assuring that the span/depth ratio does not exceed 24. This is an empirical limit that derives from successful practice. Live-load deflection and possibly vibration must be checked after the beam has been proportioned.

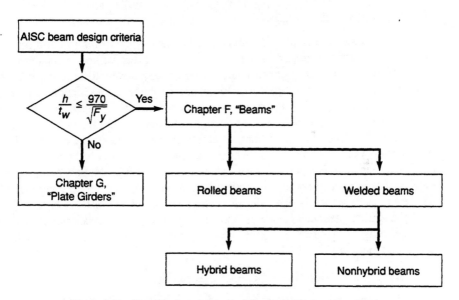

Figure 3.10   Classification of noncomposite symmetric flexural members.

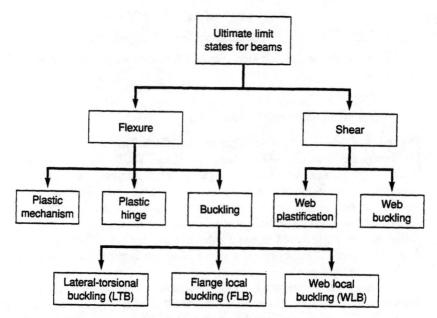

Figure 3.11   Limit states for beams.

The *ultimate limit states* (also called *strength limit states*) are the limits of load-carrying capacity due to excessive yielding or to buckling. The applicable limit states for beams are shown in Fig. 3.11. These are the limit states that Chapter F in AISCS is concerned with.

3.4.1 Flexural Limit States

Continuous closely braced compact beams fail by the formation of a plastic mechanism. Plastic design utilizes this limit state, and this method is discussed in Chapter 8. The present chapter deals with the design of statically determinate beams. When the cross section is compact and closely braced, the beam can support its plastic moment capacity M_p. When lateral bracing is spread out and/or the cross section is noncompact, the member will buckle at a moment that is smaller than M_p.

Whether a cross section is compact or not depends on the flange and web slenderness ratio. For a wide-flange shape the flange slenderness is defined by the ratio $b_f/2t_f$. The dimensions b_f and t_f are the width and thickness of the flange, as shown in Fig. 3.2. The greater this ratio, the smaller the moment under which the compression flange will buckle locally. We call this method of failure *flange local buckling* or (FLB). A similar phenomenon occurs in the web, where the appropriate slenderness is h/t_w. Buckling will take place in the compressed part of the web, and the limit state is called *web local buckling* (WLB).

3.4.2 Classification of Cross Sections

If we denote the slenderness ratio governing either FLB or WLB by the symbol λ, the relationship between the nominal flexural strength M_n and λ can be idealized as the

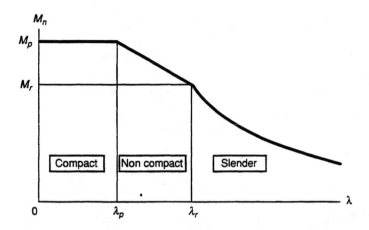

Figure 3.12 Classification of cross sections for local plate buckling.

graph shown in Fig. 3.12. This is a simplification of much more complex behavior, but it captures the essential features of the problem: As long as $\lambda \le \lambda_p$, the capacity is equal to M_p and the shape is compact. Fortunately, most of the rolled cross sections in the AISCM fall into this category. When $\lambda > \lambda_r$, the plate elements will buckle in the elastic range where the strength is inversely proportional to the square of the slenderness ratio. Such shapes are called *slender*. In the range between the compact and the slender domains buckling will occur after some part of the plate has yielded due to the sum of the applied stress and the preexisting residual stress. The buckling curve in this region is assumed to vary linearly with λ, and the shapes that fall into this region are called *noncompact*.

The simplified relationships of Fig. 3.12 are the basis of the local buckling criteria of AISCS. The flowchart of the checking procedure is given in Fig. 3.13, and the applicable parameters for rolled and welded shapes are listed in Table 3.1. With this information it is possible to determine the nominal moment capacity M_n of wide-flange and channel shapes as controlled by FLB and WLB. Tables B5.1 and A.F1.1 of

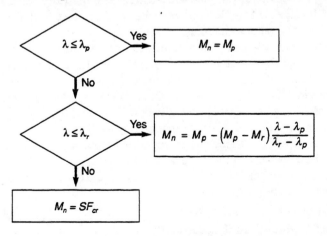

Figure 3.13 Flowchart for determining M_n when the limit states are FLB or WLB.

Table 3.1 Parameters for Determining the Flexural Capacity M_n
of Nonhybrid Compact, Noncompact, and Slender Doubly Symmetric
I-Shaped Beams.

Cross Section	Slenderness Parameter, λ	λ_p	λ_r	M_r	F_{cr}	Limit State
Rolled	$\dfrac{b_f}{2t_f}$	$\dfrac{65}{\sqrt{F_y}}$	$\dfrac{141}{\sqrt{F_y-10}}$	$(F_y - 10)S_x$	$\dfrac{20,000}{\lambda^2}$	FLB
Rolled	$\dfrac{b_f}{2t_f}$	$\dfrac{65}{\sqrt{F_y}}$	$\dfrac{141}{\sqrt{F_y-10}}$	$F_y S_y$	$\dfrac{20,000}{\lambda^2}$	FLB
Welded	$\dfrac{b_f}{2t_f}$	$\dfrac{65}{\sqrt{F_y}}$	$\dfrac{162}{\sqrt{(F_y-16.5)/k_c}}$	$(F_y - 16.6)S_x$	$\dfrac{26,200k_c}{\lambda^2}$	FLB
Welded	$\dfrac{b_f}{2t_f}$	$\dfrac{65}{\sqrt{F_y}}$	$\dfrac{162}{\sqrt{(F_y-16.5)/k_c}}$	$F_y S_y$	$\dfrac{26,200k_c}{\lambda^2}$	FLB
	$\dfrac{h}{t_w}$	$\dfrac{640}{\sqrt{F_y}}$	$\dfrac{970}{\sqrt{F_y}}$	$F_y S_x$	Plate Girder, Sec. G	WLB

[a] $0.35 \leq k_c = 4/\sqrt{h/t_w} \leq 0.763$.

AISCS give similar information for box and circular tubes as well as for hybrid and singly symmetric shapes. The formulas in Table 3.1 have been derived from empirical (test results) and theoretical (elastic and plastic buckling theory) considerations and were simplified for use in AISCS. Their background can be traced by consulting the steel design text of Salmon and Johnson (Ref. 3.1).

At first glance it would seem that the design of beams is a complex and formidable task. It should be realized, however, that almost all rolled beam shapes are compact and that the reductions, where applicable, are already incorporated into the design tables in AISCM. Thus most rolled beams are able to be designed simply for the condition $M_u \leq \phi M_p$ if they are laterally braced. Lateral bracing is often available by the bracing from the slab or deck supported by the beam.

For example, a 36-ft-long simply supported beam is subjected to a uniformly distributed dead load of 1.2 kips/ft and a uniformly distributed live load of 3.7 kips/ft. Lateral support is provided along the whole length of the beam by a floor slab. Design an A36 steel-rolled beam. Given:

$$w_{dead} = 1.2 \text{ kips/ft} \qquad w_{live} = 3.7 \text{ kips/ft} \qquad L = 36 \text{ ft} \qquad F_y = 36 \text{ ksi}$$

The required uniformly distributed load is

$$w_u = 1.2w_{dead} + 1.6w_{live} = 7.36 \text{ kips/ft}$$

The required design moment is

$$M_u = \frac{w_u L^2}{8} = 1192.32 \text{ kip-ft}$$

Select a cross section using the Load Factor Design Selection Table in AISCM: Use a W33 × 130 beam with a listed $\phi M_p = 1260$ kip-ft. Since the beam is listed in this table, all checks with regard to FLB and WLB are taken care of automatically. The reader may wish to verify this.

Box sections are especially recommended for design situations involving incomplete lateral support. Failure by lateral-torsional buckling involves twisting in combination with lateral bending about the weak axis. Box sections are greatly superior to I or W (open) sections in both of these characteristics. Standard structural (box) tubes, as shown in Fig. 3.14(a), are cataloged in the AISCM. Box beams also may be built up as a welded plate assemblage [Figure 3.14(b)].

W shapes are designed to be highly efficient when loaded in the plane of the web and supported laterally. When the lateral support is insufficient, the designer must decide as to which of several alternatives will afford the greatest economy. These include:

1. Rearrange the framing to provide better lateral support.
2. Change to a box section, at higher cost per pound of material.
3. Select a W or C shape with a moment capacity below M_p.

If the first choice is either not feasible or too costly, the third choice will probably prove the economical one if the reduction in the value of M_n is relatively small.

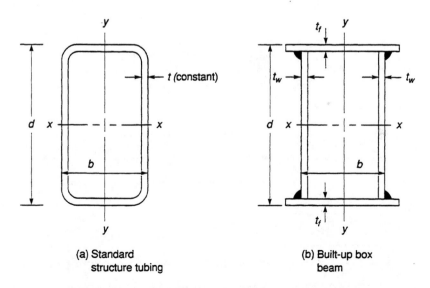

(a) Standard structure tubing

(b) Built-up box beam

Figure 3.14   Box beams.

When a beam fails by lateral-torsional buckling, it bends (buckles) about its *weak* axis, even though loaded normally in the strong plane so as to bend about its *strong* axis—which, indeed, it does, up to the critical load at which it buckles. When the beam buckles laterally about its weak axis, the loads also induce a torsional moment in the beam. The torsional resistance* of a W, S, or C section is made up of two parts: (1) the minimal torsion resistance that would be obtained under uniform torsion alone, plus (2) torsional resistance due to coupled bending of the flanges, inducing shears in the flanges that create a torsion couple. The torsion that accompanies lateral buckling is always nonuniform. Thus the resistance to lateral-torsional buckling of a W beam consists of three parts:

1. Lateral bending about the weak axis
2. Uniform torsion resistance (St. Venant torsion)
3. Nonuniform torsion resistance (warping torsion)

The flowchart governing the lateral-torsional buckling (LTB) limit state in AISCS is presented in Fig. 3.15. The flexural strength is the nominal moment capacity M_n, which is reduced from its maximum value of the full plastic moment M_p as the distance between points of lateral bracing, called the unbraced length L_b, increases. The general relationship between M_n and L_b is also illustrated in Fig. 3.16. The horizontal line $M_n = M_p$ represents the maximum capacity of the beam when the bracing is relatively closely spaced. As the spacing between lateral supports increases, the moment capacity decreases. The curve for $C_b = 1$ in Fig. 3.16 is the lowest of the three curves shown, and it is for an unbraced beam segment under uniform moment. This beam can sustain M_p if $L_b \le L_p$ for a nonhybrid I-shaped section, where

$$L_p = \frac{300r_y}{\sqrt{F_y}} \tag{3.10}$$

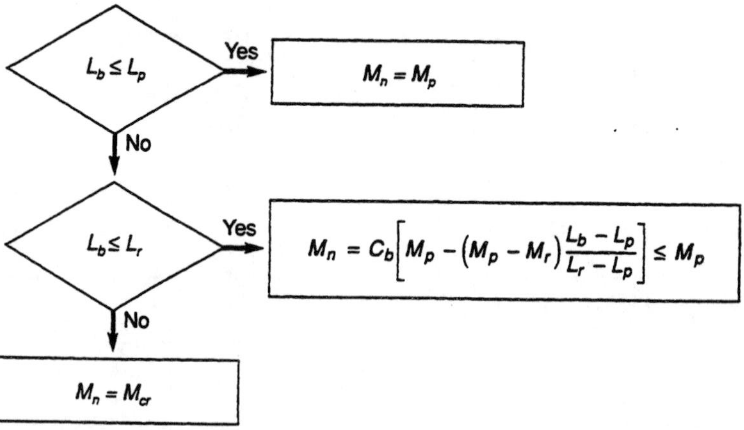

Figure 3.15 Flowchart for determining M_n for the lateral-torsional buckling limit state.

* The torsional resistance of W shapes is explained in greater detail in Chapter 10.

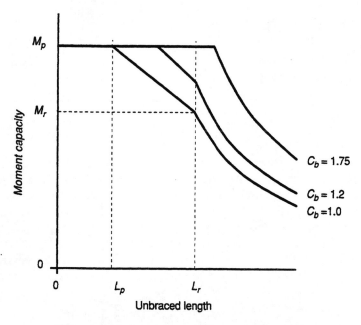

Figure 3.16 Lateral-torsional buckling strength of beams.

A *nonhybrid section* is a welded or a rolled shape for which the yield stress in the flanges and in the web are the same. A *hybrid section* is one for which the flange yield stress is larger than the yield stress of the web. The nominal moment capacity is linearly reduced in the range $L_p < L_b \leq L_r$ from M_p to M_r. This latter moment is where the beam begins to yield in the presence of a compressive residual stress F_r,

$$M_r = S_x(F_y - F_r) \tag{3.11}$$

$F_r = 10$ ksi for rolled I shapes and $F_r = 16.5$ ksi for welded shapes. When the unbraced length exceeds L_r, the beam buckles in the elastic range and $M_n = M_{cr}$ where

$$M_{cr} = \frac{C_b \pi}{L_b} \sqrt{EI_y GJ} \sqrt{1 + \frac{\pi^2 E C_w}{GJL_b^2}} \tag{3.12}$$

The critical moment in lateral buckling [Eq. (3.12)] is that moment when the beam begins to move out of its plane of loading. The equation is derived in Refs. 3.4 and 3.5. The unbraced length L_r is that value of L_b from Eq. (3.12) that corresponds to $M_{cr} = M_r$.

$$L_r = \frac{X_1 r_y}{F_y - F_r} \sqrt{1 + \sqrt{1 + X_2(F_y - F_r)^2}} \tag{3.13}$$

The symbols in the equations above are defined as follows (properties marked with an asterisk are tabulated in Part 1 of AISCM):

$E = 29{,}000$ ksi, elastic modulus

$G = 11{,}200$ ksi, shear modulus

S_x, elastic section modulus*

r_y, y–axis radius of gyration*

J, torsion constant*

A, cross-sectional area*

I_y, y–axis moment of inertia*

C_w, warping constant*

$$X_1 = \frac{\pi}{S_x}\sqrt{\frac{EGJA}{2}}^* \qquad\qquad X_2 = \frac{4C_w}{I_y}\left(\frac{S_x}{GJ}\right)^{2*}$$

The final parameter to be defined is the coefficient C_b. This term accounts for the fact that beams under uniform moment (i.e., where the whole top flange between two lateral supports is in compression due to a moment of uniform magnitude) have a smaller capacity to resist LTB than do beam segments in which the moments cause reversed curvature. The formula for this moment-gradient factor has been changed from previous editions of AISCS to be more general in its applicability. However, the previous formula gives the same results as this new one when there is a linear variation of the moment diagram between the two ends of the unbraced segment. The new equation for C_b is

$$C_b = \frac{1.25M_{max}}{2.5M_{max} + 3M_A + 4M_B + 3M_C} \tag{3.14}$$

where M_{max} = absolute value of the maximum moment in the unbraced length

M_A = absolute value of the moment at the $\frac{1}{4}$ point in the unbraced span

M_B = absolute value of the moment at the $\frac{1}{2}$ point in the unbraced span

M_C = absolute value of the moment at the $\frac{3}{4}$ point in the unbraced span

The previous discussion applies to the limiting lateral-torsional buckling strength of doubly symmetric I-shaped nonhybrid beams. Formulas for singly symmetric I beams (i.e., one flange is larger than the other; see Example 3.6) are given in Fig. 3.17. Chapter F.1 and Appendix A-F.1 in AISCS list further criteria for T beams, box beams, tubular beams, and hybrid beams.

From the great amount of complicated material presented here it would seem that determination of the available beam capacity for laterally unsupported beams is formidable. This is indeed so for unsymmetric welded nonhybrid or hybrid members. However, such situations occur relatively rarely in the usual design day of an engineer. The tables and charts provided in AISCM make the job much simpler when the designer needs to design a rolled steel beam, as illustrated in Example 3.2.

3.4.3 Design for Shear Capacity

Except in the case of very short spans, beams are usually selected on the basis of their bending capacity and then checked for the shear capacity. The design shear capacity is ϕV_n, where $\phi = 0.9$ and

$$V_n = 0.6F_{yw}A_w \tag{3.15}$$

except for cross sections with very slender webs. The terms F_{yw} and A_w are, respectively, the yield stress of the web and the web area. The flowchart for the determination of V_n for unstiffened webs is presented in Fig. 3.18. All rolled shapes with $F_y = 36$

Definition of the terms in the flowchart in Fig 3.15 if the cross section
is a singly symmetric wide-flange section with unequal flanges:

$$L_p = \frac{300r_y}{\sqrt{F_{yf}}}$$

$$M_r = (F_{yw} - F_r) S_{xc} \leq F_{yf} S_{xt}$$

L_r = that value of L_b for which $M_{cr} = M_r$

$$M_{cr} = \frac{57,000C_b}{L_b} \sqrt{I_y J}(B_1 + \sqrt{1 + B_2 + B_1^2}) \leq M_p$$

$$B_1 = 2.25\left(\frac{2I_{yc}}{I_y} - 1\right)\frac{h}{L_b}\sqrt{\frac{I_y}{J}} \qquad B_2 = 25\left(1 - \frac{I_{yc}}{I_y}\right)\frac{I_{yc}}{C}\left(\frac{h}{L_b}\right)^2$$

Where the previously undefined terms are

S_{xt} = elastic section modulus with respect to the tension flange
S_{xc} = elastic section modulus with respect to the compression flange
I_{yc} = moment of inertia of the compression flange about the y axis; if the unbraced beam segment
        is bent in reverse curvature, use the moment of inertia of the smaller flange
h = web depth

The coefficient C_b is to be taken as unity if $I_{yc}/I_y < 0.1$ or $I_{yc}/I_y > 0.9$.

Figure 3.17   Singly symmetric I shapes, limit-state lateral-torsional buckling.

ksi and all but two (M14 × 18 and M12 × 11.8) with $F_y = 50$ ksi are covered by
Eq. (3.15). Members with very slender webs can also be designed as plate girders
with transverse stiffeners to increase their strength. These are dealt with in Chapter 7.

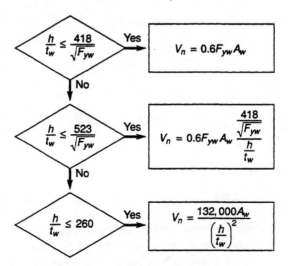

Figure 3.18   Shear capacity of unstiffened webs.

3.5 Lateral Support Requirements

Most beams are designed by simple bending theory with the assumption of full lateral
support and without any required reduction of the full bending capacity M_p. Any beam

with the compression flange attached securely to a floor or roof system that provides continuous or nearly continuous support meets these requirements.

Some conditions for which lateral support may be less than adequate include the following:

1. No positive connection between the beam and the load system that it supports, particularly if loads are vibratory or involve impact.
2. The lateral support system is of a removal type.
3. Lateral support system frames into a parallel system of two or more similarly loaded beams without positive anchorage. This possibility is illustrated in Fig. 3.19, which shows three alternative framing systems in plan view. In Fig. 3.19(a) the lateral support is inadequate for the reason cited. In Fig. 3.19 (b) and (c) it is adequate, because of adjacent beams with wall anchorage or by virtue of K bracing that limits motion, respectively.

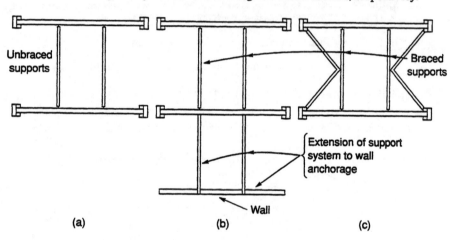

(a)   (b)   (c)

Figure 3.19   Plan view of beams with (a) inadequate and (b,c) adequate lateral support systems.

3.6 Beam Deflection Limitations

Deflections of beams under live loads are often curtailed in order not to impair the serviceability of the floor or roof due to plaster cracking, cracking of the concrete slab, distortion of the partitions, and other unsightly and annoying occurrences. Deflection limits are also imposed when the appearance of deflection could cause the occupants to think that the structure is somehow unsafe. The AISCS states in Sec. L3.1 that "deformations in structural members and systems due to service loads shall not impair the serviceability of the structure." However, no specific limits are given because such limits depend on the function of the structure. It is, therefore, up to the discretion and experience of the design engineer to select a suitable limit. In lieu of experience it is satisfactory to select the live-load deflection limit of span/360. This is approximately the deflection when plaster cracks, and it is also the limit of visible detection by the human eye. It should be realized that this may not be the right limit for all cases, and it is not an assurance that the floor will not vibrate under the common activities occurring on the floor. Thus AISCS advises the designer in Sec. L3 to select a deflection limit carefully and to perform a floor vibration check where in the judgment of the design engineer this is appropriate.

If water on a flat roof accumulates at a rate more rapid than it runs off, additional roof load develops due to the water that is "ponded" as a result of roof deflection, thus further aggravating the imbalance between rainfall and runoff rates. Failure may result. The problem is treated in AISCS Sec. K2 and in AISCS Commentary. In flat floor or roof systems of long span, the supporting girders should be cambered to minimize the ponding hazard if it exists, and to avoid any visually perceptible sag.

3.7 Beams Under Repeated Load

Design of beams for repeated load by the allowable-stress-range procedure is essentially the same as for tension members, which has been covered in Chapter 2. Typical beam category designations are illustrated in Appendix K3 of AISCS.

In beam design it may be possible to effect economy by changing the arrangement of certain welded details so as to place the beam in the most favorable repeated load stress category (see Example 3.4).

3.8 Biaxal Bending of Beams

The W beam, or a modification thereof as shown in Fig. 3.20(b), is frequently used in design situations for which components of bending moment occur simultaneously about both the x–x and y–y axes. If the limiting stress about the two axes were the same, the design could be based simply on the maximum stress calculated by superposing the stresses caused by bending about each of the two principal axes:

$$f_{un} = \pm \frac{M_{ux}}{S_x} \pm \frac{M_{uy}}{S_y} \le \phi F_{\lim} \tag{3.16}$$

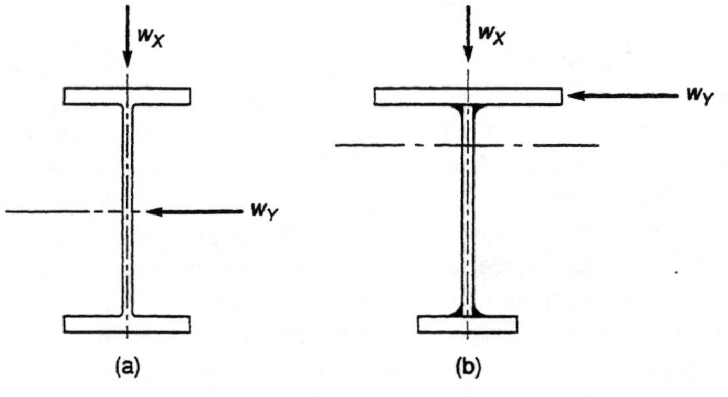

Figure 3.20   Beams under biaxial load.

where Sec. H2 of AISCS defines the limiting stress ϕF_{lim} as either $0.9F_y$ when yielding controls, or $0.85F_{cr}$ when the limit state is buckling.

If the component of load in the weak plane is comparable in magnitude to that in the strong plane, a box section may be a preferred solution. If the lateral loads are relatively small, the W shape may be suitable. In the case that the limiting stresses about the two axes are unequal, an interaction formula should be used. Such an equation is given in Sec. H1.1 of AISCS:

$$\frac{M_{ux}}{\phi_b M_{nx}} + \frac{M_{uy}}{\phi_b M_{ny}} \le 1.00 \tag{3.17}$$

In Eqs. (3.16) and (3.17) the terms are defined as follows:

M_{ux}, M_{uy} = factored required moments about the x and y axes, respectively
M_{nx}, M_{ny} = nominal moment capacities about the x and y axes, respectively
S_x, S_y = elastic section moduli about the x and y axes, respectively
$\phi_b = 0.9$

If the lateral load is applied at or near the top flange, the use of an unsymmetrical section built up by welding three plates, as shown in Fig. 3.20(b), may be the most economical solution. The complex consideration of the torsion that is introduced may be avoided by calculating the stress in the top flange, due to lateral load, under the assumption that the top flange alone carries *all* the lateral load. Alternative approaches to the biaxial beam design problem are presented in Section 3.11.

3.9 Load and Support Details

Individual beams may carry concentrated loads and must be supported at or near their ends. Concentrated loads may be introduced directly into a beam web by means of a riveted, bolted, or welded beam web connection, and the loaded beam may in turn transmit its end reactions to columns or girders by means of similar connections. Beam web connections of the type just described are covered in Chapter 6.

When the end of a beam rests on concrete, it requires a bearing support, such as is shown in Fig. 3.21. Consideration must be given to the local compression in the web, just above the bearing block, and to the required thickness of bearing plate to spread the load to the concrete. The local concentration of compression in the beam web is assumed uniformly distributed over the distance $(N + 2.5k)$, as shown in Fig. 3.21(b). k is tabulated in the W shape detailing information tables in AISCM, and is the distance from the flange face to the termination of the flange to web fillet. The compression stress along this line, equal to $R/(N + 2.5k)t_w$, must be kept below ϕF_y (AISCS, Sec. K1.3, with $\phi = 1.0$). If a support occurs away from the end of a beam, or if a local load is introduced at the top of a beam through a bearing block, the situation is similar, except the spread of load proceeds from each end of the bearing block and the compression stress is assumed to be $R/(N + 5k)t_w$.

In addition to web yielding it is also necessary to check web crippling; that is, the web may crumple into buckled waves near the flange–web juncture. The equations for this limit state are in Sec. K1.4 in AISCS, and they are the following for an end reaction:

$$\phi R_n \ge R \qquad \text{where } \phi = 0.75 \tag{3.18}$$

For $N/d \leq 0.2$:

$$R_n = 68t_w^2 \left[1 + 3\left(\frac{N}{d}\right)\left(\frac{t_w}{t_f}\right)^{1.5} \right] \sqrt{\frac{F_{yw}t_f}{t_w}} \tag{3.19a}$$

For $N/d > 0.2$:

$$R_n = 68t_w^2 \left[1 + \left(4\frac{N}{d} - 0.2\right)\left(\frac{t_w}{t_f}\right)^{1.5} \right] \sqrt{\frac{F_{yw}t_f}{t_w}} \tag{3.19b}$$

In addition to the terms defined in Fig. 3.21 the new definitions are:

$$d = \text{total depth of beam}$$

$$t_f = \text{flange thickness}$$

$$F_{yw} = \text{yield stress of web}$$

In case the reaction is not at the end, AISCS has only one formula, where the number 68 in Eq. (3.19a) is replaced by the number 135.

Bearing plates must be thick enough to spread the reactions or concentrated loads to the concrete at a pressure, F_p, specified in Sec. J9 of AISCS. The required bearing plate thickness is determined by considering the bearing plate as a simple cantilever beam of length n, as shown in Fig. 3.21(a). The beam carries an upward load, assumed uniform, resulting from the concrete pressure.

Following is the derivation of a formula for the bearing plate thickness. Assume a unit width of cantilever beam:

$$M_{max} = F_p(1)n\frac{n}{2} = \frac{F_p n^2}{2}$$

$$M_{max} \leq \phi_b M_p = \phi_b \frac{1 \times t^2}{4} F_y$$

Substituting and solving for t, we obtain

$$t \geq n \sqrt{\frac{2F_p}{\phi_b F_y}}$$

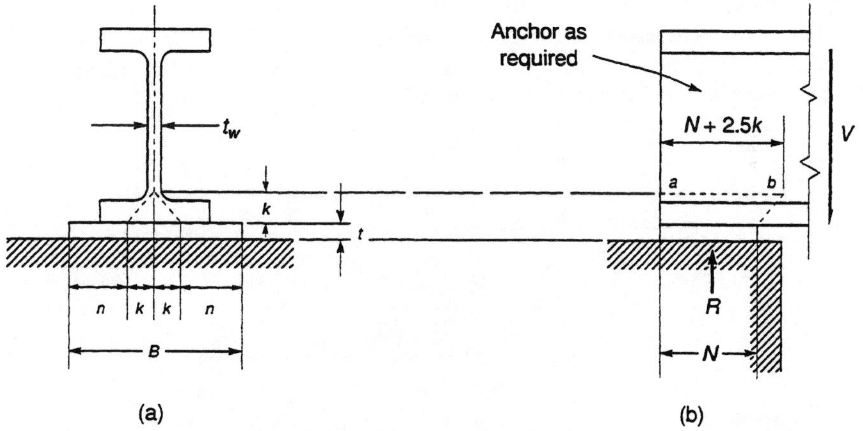

Figure 3.21   Bearing supports at end of a beam.

Section J9 of AISCS specifies that $F_p = \phi_c \times 0.85 f'_c$ where $\phi_c = 0.60$ and f'_c is the crushing strength of concrete in units of ksi. With $\phi_b = 0.9$ the required bearing plate thickness is approximately

$$t_{req} = n \sqrt{\frac{f'_c}{F_y}} \qquad (3.20)$$

3.10 Load Tables for Beams

AISCM provides tables to permit direct selection of beams under uniform load for yield stresses of either 36 or 50 ksi. Information is also supplied regarding maximum unbraced lengths. In addition, design coefficients are provided to expedite calculation of nominal loads for beams having yield points other than 36 or 50 ksi. If the load and/or support conditions are other than a simple beam under uniform load, the allowable load tables may be used in many instances by conversion of the different load situation to an "equivalent (uniform) tabular load." Conversion factors are given in AISCM for a variety of load and support conditions.

   Deflection coefficients for beams in terms of span and maximum stress due to bending permit rapid calculation of center deflection of uniformly loaded beams. Modifying factors permit good approximations for other load conditions.

3.11 Illustrative Examples

The following examples demonstrate the use of AISCS and AISCM in the selection and design of steel beams under some of the conditions that have been covered in this chapter. The examples are intended to do more than merely illustrate procedures; they also show how one may change conditions in some cases to achieve greater economy by altering certain details. The examples, in some cases, demonstrate the logic and judgment that may be applied to "zero in" on a solution when the first try turns out to be unsatisfactory. Readers are urged to make up hypothetical design situations on their own and carry out similar solutions.

Example 3.1

A simply supported beam (below) with a span of 20 ft is to carry a static uniform dead load of 1.0 kip/ft and a live load of 1.5 kips/ft in addition to its own dead weight. The flange is laterally supported by the floor system that it supports. Select the most economical W shape, using A36 steel. The live-load deflection shall not exceed $L/360$.

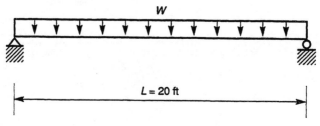

Solution

Select beam type: rolled W shape

Select steel grade: A36, $F_y = 36$ ksi

Loading cycles < 2000; no fatigue check necessary

Static loading; no impact factor needed

Assume a compact laterally braced beam.

Flexural design strength:

$$\phi_b M_n = 0.9 M_p = 0.9 Z_x F_y$$

Assume that the web is compact in shear.

Shear design strength:

$$\phi V_n = 0.9 V_n = 0.9(0.6 F_y A_w)$$

Estimate self-weight: 0.04 kip/ft

Design loads:

Dead load: $\quad w_D = 1.0 + 0.04 = 1.04$ kips/ft

Live load: $\quad w_L = 1.5$ kips/ft

Required design load: $\quad w_u = \max(1.4 w_D, 1.2 w_D + 1.6 w_L)$
$$w_u = \max(1.456, 3.648) = 3.648 \text{ kips/ft}$$

Required design moment:

$$M_u = \frac{w_u L^2}{8} = 3.65 \times 20^2 \times \frac{12}{8}$$

$$M_u = 2189 \text{ kip-in.}$$

Required design shear:

$$V_u = \frac{wL}{2} = \frac{3.65 \times 20}{2} = 36.5 \text{ kips}$$

Required plastic section modulus:

$$Z_x = \frac{M_u}{\phi_b F_y} = \frac{2189}{0.9 \times 36} = 67.6 \text{ in}^3$$

Select beam size: W16 × 40, $Z_x = 72.9$ in³ (from AISCM, Part 3, Load Factor Design Selection Table)

Check weight: 0.04 kip/ft   OK

Check compactness of flange:

$$\frac{b_f}{2t_f} = 6.9 \qquad < \frac{65}{\sqrt{F_y}} = 10.8 \quad \text{OK}$$

Check compactness of web flexure:

$$\frac{h}{t_w} = 46.6 \qquad < \frac{640}{\sqrt{F_y}} = 106.7 \quad \text{OK}$$

Check lateral stability: Compression flange is braced   OK

Check shear compactness:

$$\frac{h}{t_w} = 46.6 \qquad < 260 \quad \text{OK}$$

$$\frac{h}{t_w} = 46.6 \qquad < \frac{418}{\sqrt{F_y}} = 69.7 \quad \text{OK}$$

Design shear strength:

$$0.9 \times 0.6 \times 36 \times d \times t_w = 0.9 \times 0.6 \times 36 \times 16.01 \times 0.305 = 94.9 \text{ kips} \qquad >36.5 \text{ kips} \quad \text{OK}$$

Check live-load deflection:

$$\frac{5w_L L_4}{384 E I_x} = \frac{5 \times (1.5/12) \times (20 \times 12)^4}{384 \times 29{,}000 \times 518} = 0.36 \text{ in.}$$

$$\frac{L}{360} = \frac{20 \times 12}{360} = 0.67 \text{ in.} \qquad > 0.36 \text{ in.} \quad \text{OK}$$

Use W16 × 40.

Note: Preferably select the beam with the boldface type from Part 13 of AISCS, Load Factor Design Selection Table. This is the top entry in each group of beams, and it is the size with the least weight in this group. It is also not necessary to calculate the required section modulus as shown above if $F_y = 36$ or 50 ksi. Instead, calculate the design moment in ft-kip units and pick the size directly from the table. For rolled shapes it is also not necessary to check flexural flange or web local buckling limits because the value of $\phi_b M_p$ is already reduced for these effects, if they are applicable, in the table.

Example 3.2

Rework Example 3.1 with intermittent lateral supports to provide equivalent full lateral support.

Solution   Section F1.2a of AISCS requires that the maximum unbraced length L_b must not exceed L_p if the full plastic capacity of the beam is to be utilized. For the W16 × 40 section, $r_y = 1.57$ in. and thus

$$L_p = \frac{300 r_y}{\sqrt{F_y}} = \frac{300 \times 1.57}{\sqrt{36}} = 78.5 \text{ in.}$$

Alternatively, L_p can be looked up in the tables in Part 3 of AISCM.

If the span 20 ft. = 240 in. is divided into three segments of 80 in. length, the unbraced length is only a small amount larger than the maximum permissible value. This small exceedance of the required value is appropriate in this case because the width of the lateral support is several inches anyway.

Example 3.3

Rework Example 3.1 without any lateral support.

Solution   Estimate weight:

$$w_{sw} = 0.06 \text{ kip/ft}$$

Design loads:

$$w_D = 1.0 + 0.06 = 1.06 \text{ kips/ft}$$
$$w_u = 1.2 \times 1.06 + 1.6 \times 1.5 = 3.67 \text{ kips/ft}$$
$$M_u = \frac{w_u L^2}{8} = 3.67 \times 20^2 \times \frac{12}{8} = 2203 \text{ kip-in.} = 184 \text{ kip-ft}$$

Determine C_b. This coefficient is determined by Eq. (3.14) [Eq. (F1-3) in AISCS]

$$M_{max} = 184 \text{ kip-in.} = M_B \qquad \text{at center of beam}$$
$$M_A = M_C = 138 \text{ kip-in.} \qquad \text{at } \tfrac{1}{4} \text{ and } \tfrac{3}{4} \text{ points}$$
$$C_b = \frac{12.5 \times 184}{2.5 \times 184 + 3 \times 138 + 4 \times 184 + 3 \times 138} = 1.14$$

Preliminary beam selection: From the Beam Design Moments Chart in AISCM for $L_b = 20$ ft and $\phi M_n = 184$ kip-ft, select the closest solid line that exceeds the design values: W14×53.

Check the section. From the Load Factor Design Selection Table in AISCM, for a W14×53 and $F_y = 36$ ksi, look up $L_r = 28.0$ ft, $L_p = 8.0$ ft, $\phi_b M_r = 152$ kip-ft, and $\phi_b M_p = 235$ kip-ft. Since $L_b = 20$ ft lies between 28 and 8 ft (Fig. 3.15),

$$\phi_b M_n = C_b \left[\phi_b M_p - (\phi_b M_p - \phi_b M_r) \frac{L_b - L_r}{L_r - L_p} \right] \le \phi_b M_p$$

$$= 1.14 \left[235 - (235 - 152) \frac{20.0 - 8.0}{28.0 - 8.8} \right] = 209 \text{ kip-ft}$$

Since this is larger than $M_u = 184$ kip-ft, the W14×53 is OK.

Note: The value BF in the AISCM tables is very useful in calculating the design capacity since

$$BF = \frac{\phi_b M_p - \phi_b M_r}{L_r - L_p} = 4.17$$

can be used directly in the equation above.

It is possible that a lighter but deeper W24×44 section would also be adequate. For this beam AISCM gives $L_r = 15.4$ ft, which is less than $L_b = 20$ ft, and therefore from the flowchart in Fig. 3.15 and an alternative form of Eq. (3.12),

$$\phi_b M_n = \phi_b M_{cr} = \frac{\phi_b C_b S_x X_1 \sqrt{2}}{L_b / r_y} \sqrt{1 + \frac{X_1^2 X_2}{2(L_b / r_y)^2}}$$

From the tables in Part 1 of AISCM:

$$X_1 = 1550 \text{ ksi} \qquad X_2 = 0.0366 \text{ ksi}^{-2} \qquad S_x = 81.6 \text{ in}^3 \qquad r_y = 1.26 \text{ in.}$$

Substituting these values into the equation above gives $\phi_b M_n = 119$ kip-ft, which is less than the required moment of 184 kip-ft so the W21×44 is not adequate. Use W14×53.

Example 3.4

Use the same beam selection as in Example 3.3, with uniform load replaced by an equivalent concentrated load at midspan that may be repeated 3,000,000 times.

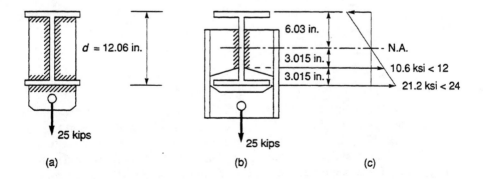

(a)   (b)   (c)

Solution   A midspan concentrated live load of 25 kips is to be suspended with initial design of the loading detail in sketch (a).

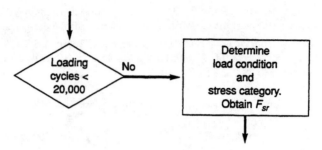

Loading condition is No. 4, as listed in AISCS, Appendix K3, Table A-K3.1, for repetitions of load exceeding 2,000,000 times. Referring to Fig. A-K3.1, the situation is considered to be one of flexural stress adjacent to a welded transverse stiffener (Illustrative Example 7 of Fig. A-K3.1). The *stress category* is now established as C from Table A-K3.2, and from Table A-K3.3, Appendix K3, the *allowable stress range* (F_{sr}) is found to be

$$F_{sr} = 12 \text{ ksi}$$

$$M_x = \frac{PL}{4} = \frac{25 \times 20 \times 12}{4} = 1500 \text{ kip-in.} \qquad \text{(maximum live-load moment)}$$

$$f_{bx} = \frac{1500}{70.6} = 21.2 \text{ ksi} \qquad \text{(W12} \times \text{53)}$$

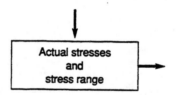

(For design for repeated load, according to Appendix K3 of AISCS, the stress range $f_{sr} = 21.2 - 0 = 21.2$ ksi in this case. The *stress range* is the algebraic difference between the maximum and minimum stresses caused by live load.)

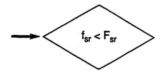

Thus the design as it stands is unsatisfactory. Rather than select a heavier beam section, the problem can be solved by changing the weld detail as shown in sketch (b) to reduce the live-load stress level adjacent to the transverse weld to something less than F_{sr} (12 ksi), as shown in sketch (c). Then the unwelded as-rolled beam surface at the location for maximum tension stress due to flexure has a stress category of A (Table A-K3.2) with an allowable stress range (F_{sr}) of 24 ksi. Category C still holds for the web at the stiffener and $F_{sr} = 12$ ksi. The beam design is OK. The design of the welded connection is deferred to Chapter 6.

Example 3.5
A simply supported beam with a span of 20 ft carries a uniform dead load of 1 kip/ft, including its own weight. A concentrated oblique live load at midspan acts through the centroid, as shown. Determine economical beam selection for A36 steel if there are no intermediate lateral supports.

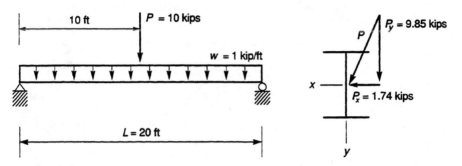

Solution   Design a beam for biaxial bending using the interaction equation (3.17) (modified from Sec. H1 of AISCS):

$$\frac{M_{ux}}{\phi_b M_{nx}} + \frac{M_{uy}}{\phi M_{ny}} \le 1.00$$

For the preliminary design, assume that $M_{nx} = M_{px}$, and $M_{ny} = M_{py}$; that is, the full plastic moment can be attained.

If we let $C_n = M_{px}/M_{py} = Z_x/Z_y$ and note that $\phi_b = 0.9$, the required x-axis plastic-section modulus equals

$$(Z_x)_{reqd} = \frac{M_{ux} + C_n M_{uy}}{0.9 F_y}$$

For the W27 to W36 range, $C_n \approx 5$.
For the W16 to W24 range, $C_n \approx 4$.
For column shapes (square), $C_n \approx 2$.

Factored loads:

$$w_u = 1.2 \times 1 = 1.2 \text{ kips/ft}$$
$$P_{ux} = 1.6 \times 1.74 = 2.784 \text{ kips}$$
$$P_{uy} = 1.6 \times 9.85 = 15.76 \text{ kips}$$

Required moments:

$$M_{ux} = \frac{w_u L^2}{8} + \frac{P_{uy}L}{4} = \frac{1.2 \times 20^2}{8} + \frac{15.76 \times 20}{4} = 138.8 \text{ kip-ft}$$

$$M_{uy} = \frac{P_{ux}L}{4} = \frac{2.784 \times 20}{4} = 13.92 \text{ kip-ft}$$

Preliminary design: Assume that $C_n = 5$.

$$(Z_x)_{reqd} = \frac{(138.8 + 5 \times 13.92) \times 12}{0.9 \times 36} = 77 \text{ in}^3$$

Try W16 × 40, $\phi_b M_p = 197$ ft-kip, $C_n = 5.74$.

$$M_{ux} + C_n M_{uy} = 219 \text{ ft-kip} \qquad > 197 \text{ ft-kip} \quad \text{NG}$$

Try a W12 × 58. From the "Load Factor Design Selection Table" of AISCM, for $F_y = 36$ ksi:

$$L_r = 38.4 \text{ ft} \qquad L_p = 10.5 \text{ ft} \qquad \phi_b M_r = 152 \text{ kip-ft}$$
$$\phi_b M_p = 233 \text{ in.-kip} \qquad Z_x = 86.4 \text{ in}^3$$

From the AISCM Tables of Dimensions and Properties, $Z_y = 32.5$ in^3.

$$\frac{b_f}{t_f} = 7.8 \le \frac{65}{\sqrt{F_y}} = \frac{65}{\sqrt{36}} = 10.83$$

Therefore,

$$\phi_b M_{ny} = \phi_b M_{py} = \phi_b Z_y F_y = 0.9 \times 32.5 \times 36 = 1053 \text{ kip-in.} = 87.75 \text{ kip-ft}$$

Assuming conservatively that $C_b = 1.0$ and noting that $L_p < L_b = 20 \text{ ft} < L_r$,

$$\phi_b M_{nx} = \phi_b M_p - (\phi_b M_p - \phi_b M_r)\frac{L_b - L_p}{L_r - L_p} = 233 - (233 - 152)\frac{20 - 10.5}{38.4 - 10.5} = 205.4 \text{ ft-kip}$$

Substitute into the interaction equation:

$$\frac{M_{ux}}{\phi_b M_{nx}} + \frac{M_{uy}}{\phi M_{ny}} \le 1.00$$

$$\frac{138.8}{205.4} + \frac{13.92}{87.75} = 0.83 \qquad < 1.00 \quad \text{OK}$$

The W12 × 58 is acceptable but conservative. The reader may demonstrate that a W12 × 50 will also be satisfactory if the actual value of $C_b = 1.26$ as computed by Eq. (3.14) is used. In this case the interaction sum equals 0.95. Therefore, use a W12 × 50 beam.

Example 3.6

A simple span crane runway beam supports a moving load, transmitted by two wheels, as shown, of 80 kips, including impact. Using an unsymmetrical section of the type shown in Fig. 3.20(b), determine required plate sizes. Assume that the 16-kip lateral load is applied at the level of the top flange. Use ASTM A36 steel.

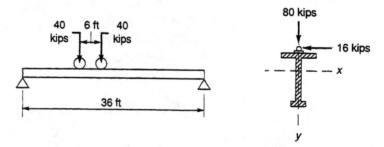

Solution This problem will illustrate successive cut-and-try steps in approaching a final economical section. As discussed in Section 3.8, it will be assumed that all the lateral load is carried by the top flange to compensate for the fact that torsional stresses will not be considered.

As a preliminary guide to determining a suitable size for the top (compression) flange, we make an initial guess that 75% of the top flange capacity about the y axis resists the moment due to the lateral force.

As shown in many texts in structural analysis, the maximum moment due to a moving two-wheel load system is under the wheel nearest the center of the span when it and the center of gravity of the moving load system are equidistant from the center of the span.

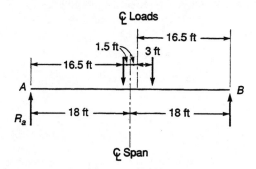

Calculate the maximum moments due to vertical and lateral live loads:

$$R_{ax} = 16\frac{16.5}{36} = 7.33 \text{ kips}$$

$$R_{ay} = 80\frac{16.5}{36} = 36.67 \text{ kips}$$

$$M_x = 16.5 R_{ay} = 605.06 \text{ kip-ft} = 7260 \text{ kip-in.}$$

$$M_y = 16.5 R_{ax} = 121 \text{ kip-ft} = 1452 \text{ kip-in.}$$

Assume that the self-weight of the girder is small compared to the crane loads. The interaction equation to be checked is [Eq. (3.17)]

$$\frac{M_{ux}}{\phi_b M_{nx}} + \frac{M_{uy}}{\phi_b M_{ny}} \le 1.0$$

Assume for the preliminary design that for the top flange alone $\dfrac{M_{uy}}{\phi_b M_{ny}} = 0.75$.

Since the loading is all live load, the load factor is 1.6 and thus

$$M_{uy} = M_y \times 1.6 = 1452 \times 1.6 = 2323 \text{ kip-in.}$$

The moment capacity is the plastic moment for the top flange about its y axis:

$$M_{ny} = M_{py} = \frac{t_f b_f^2 F_y}{4}$$

Setting all the values into the y axis part of the interaction equation gives

$$\frac{2323 \times 4}{0.9 \times t_f b_f^2 \times 50} = 0.75$$

and assuming a flange plate thickness of 1 in., we can solve for $b_f \approx 17$ in.

For the top flange, a trial plate size of 1 in. × 20 in. will be assumed. Since the bottom flange plate will only need to take stresses from x-axis bending, a much smaller plate of 1 in. × 10 in. will be assumed. For the web, a 0.5 in. × 36 in. plate is assumed to give member length/height ratio of approximately $\frac{1}{12}$, which is a reasonable proportion for a crane girder. With these assumed dimensions the x-axis moment capacity M_{xn} is determined according to the requirements of

Sec. F.1 and Appendix F1 of AISCS. This determination is a rather complex and involved procedure and it is best performed on spreadsheet or mathematics programs since it is very likely that
the first assumption of the plate sizes is not going to be correct. Such a program with all of the
requirements of AISCS is given as an appendix to this problem below. The important thing is to
include all the necessary steps so that repeated calculations can be performed.

From the program the following results are obtained.

Given: $b_{f1} = 20$ in., $t_{f1} = 1$ in., $b_{f2} = 10$ in., $t_{f2} = 1$ in., $h = 36$ in., $t_w = 0.5$ in., $L_b = 36$ ft.

$$M_{nx} = 19,973 \text{ kip-in.}$$

$$M_{ny} = \frac{1 \times 20^2 \times 36}{4} = 3600 \text{ kip-in.}$$

Using the interaction equation yields

$$\frac{11,616}{0.9 \times 19,973} + \frac{2323}{0.9 \times 3600} = 1.363 \quad \text{NG}$$

Two new sections were tried until the following section satisfied all the conditions: $b_{f1} = 24$ in.,
$t_{f1} = 1.25$ in., $b_{f2} = 9$ in., $t_{f2} = 1$ in., $h = 36$ in., $t_w = 0.5$ in., $L_b = 36$ ft.
For this section:

$$M_{nx} = 22,683 \text{ kip-in.} \qquad M_{ny} = 6480 \text{ kip-in.} \qquad M_{uy} = 2323 \text{ kip-in.}$$
$$M_{ux} = 11,616 + 452 = 12,068 \text{ kip-in.}$$

The 452 kip-in. value represents the moment from the self-weight of the girder. The interaction
equation sum is 0.99, thus giving a satisfactory design.

Appendix to Example 3.6

Major-Axis Moment Capacity of Welded Singly Symmetric I Shapes

(a) *Cross Section:* compression flange area > tension flange area; nonhybrid beam

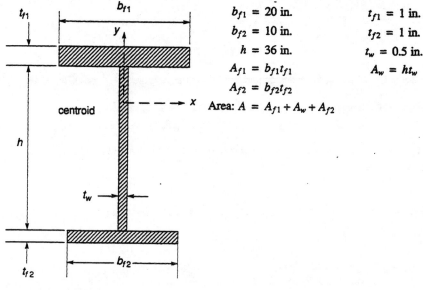

$b_{f1} = 20$ in. $t_{f1} = 1$ in.

$b_{f2} = 10$ in. $t_{f2} = 1$ in.

$h = 36$ in. $t_w = 0.5$ in.

$A_{f1} = b_{f1}t_{f1}$ $A_w = ht_w$

$A_{f2} = b_{f2}t_{f2}$

Area: $A = A_{f1} + A_w + A_{f2}$

(b) *Material properties:*

$$F_y = 36 \text{ ksi}$$
$$E = 29,000 \text{ ksi} \qquad G = 0.385E$$

(c) *Cross-sectional properties:*

Centroidal distance from top of flange 1 to centroid of section:

$$y_e = \frac{1}{A}\left[A_{f1}\frac{t_{f1}}{2} + A_w\left(t_{f1} + \frac{h}{2}\right) + A_{f2}\left(t_{f1} + h + \frac{t_{f2}}{2}\right)\right]$$

x-axis moment of inertia:

$$I_x = A_{f1}\left[\left(y_e - \frac{t_{f1}}{2}\right)^2 + \frac{t_{f1}^2}{12}\right] + A_w\left[\frac{h^2}{12} + \left(t_{f1} + \frac{h}{2} - y_e\right)^2\right] + A_{f2}\left[\left(t_{f1} + h + \frac{t_{f2}}{2} - y_e\right)^2 + \frac{t_{f2}^2}{12}\right]$$

Elastic-section moduli:

$$S_{x1} = \frac{I_x}{y_e} \qquad\qquad S_{x2} = \frac{I_x}{t_{f1} + h + t_{f2} - y_e}$$

$$M_{rx1} = (F_y - 16.5\ \text{ksi})\, S_{x1} \qquad M_{rx2} = F_y S_{x2}$$

$$M_{rx} = \text{if}\,(M_{rx1} \geq M_{rx2}, M_{rx2}, M_{rx1}) \qquad M_{rx} = 14{,}807.313\ \text{kip-in.}\quad \text{Moment at first yield}$$

y-axis moments of inertia:

$$I_{yf1} = A_{f1}\frac{b_{f1}^2}{12} \qquad I_{yf2} = A_{f2}\frac{b_{f2}^2}{12}$$

$$I_y = I_{yf1} + I_{yf2} \qquad r_y = \sqrt{\frac{I_y}{A}}$$

Torsion constant:

$$J = \frac{1}{3}\left(A_{f1}t_{f1}^2 + A_w t_w^2 + A_{f2}t_{f2}^2\right)$$

Summary of cross-sectional properties:

$$y_e = 15.146\ \text{in.} \qquad I_{yf1} = 666.667\ \text{in}^4$$
$$I_x = 11{,}500.979\ \text{in}^4 \qquad I_y = 750\ \text{in}^4$$
$$S_{x1} = 759.349\ \text{in}^3 \qquad r_y = 3.953\ \text{in.}$$
$$S_{x2} = 503.233\ \text{in}^3 \qquad J = 11.5\ \text{in}^4$$

(d) *Plastic moment:* (Note: $A_{f1} > A_{f2}$.) Plastic neutral axis is in flange 1, occurs if $A_{f1} > A_w + A_{f2}$.

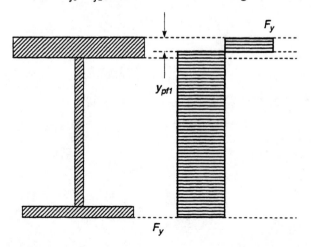

Equilibrium:

$$F_y b_{f1} y_{pf1} = F_y [b_{f1}(t_{f1} - y_{pf1}) + h t_w + b_{f2} t_{f2}]$$

which can be simplified to

$$y_{pf1} = \frac{A}{2b_{f1}}$$

By taking moments of the stress blocks about the plastic neutral axis, the plastic moment is determined:

$$M_{pf1} = F_y \left[b_{f1} \frac{y_{pf1}^2}{2} + b_{f1} \frac{(t_{f1} - y_{pf1})^2}{2} + A_w \left(t_{f1} - y_{pf1} + \frac{h}{2}\right) + A_{f2}\left(t_{f1} - y_{pf1} + h + \frac{t_{f2}}{2}\right) \right]$$

Neutral axis is in the web: $A_{f1} < A_w + A_{f2}$.

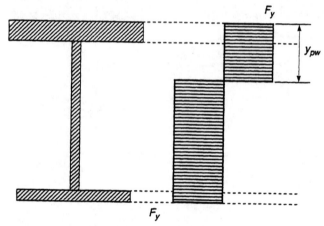

Equilibrium of forces:

$$F_y [A_{f1} + (y_{pw} - t_{f1})t_w] = F_y \{[h - (y_{pw} - t_{f1})]t_w + A_{f2}\}$$

from which we get

$$y_{pw} = \frac{A_w + A_{f2} - A_{f1}}{2t_w} + t_{f1}$$

Plastic moment:

$$M_{pw} = F_y \left[A_{f1}\left(y_{pw} - \frac{t_{f1}}{2}\right) + \frac{(y_{pw} - t_{f1})^2 t_w}{2} + \frac{(h - y_{pw} + t_{f1})^2 t_w}{2} + A_{f2}\left(h - y_{pw} + t_{f1} - \frac{t_{f2}}{2}\right) \right]$$

$$M_{px} = \text{if } (A_{f1} \geq A_w + A_{f2}, M_{pf1}, M_{pw}) = 24{,}012 \text{ kip-in.}$$

(e) *Moment capacity: limit-state flange local buckling (FLB)* subscript c; compression flange; subscript t; tension flange:

$$b_{fc} = b_{f1} \qquad S_{xc} = S_{x1}$$

$$t_{fc} = t_{f1} \qquad S_{xt} = S_{x2}$$

$$\lambda = \frac{b_{fc}}{2t_{fc}} \qquad \lambda_p = 65 \frac{\sqrt{\text{ksi}}}{\sqrt{F_y}}$$

See the flowchart A for k_c.

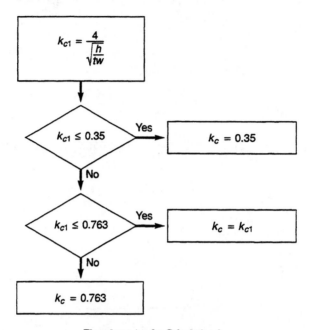

Flowchart A   for Calculating k_c.

$$k_{c1} = \frac{4}{\sqrt{h/t_w}} \qquad k_c = \text{if}\left[k_{c1} \le 0.35, 0.35, \text{if}\,(k_{c1} \ge 0.763, 0.763, k_{c1})\right]$$

$$\lambda_r = \frac{162\sqrt{\text{ksi}}}{\sqrt{\dfrac{F_y - 16.5\ \text{ksi}}{k_c}}} \qquad F_{cr} = 26{,}200\frac{k_c}{\lambda^2}\ \text{ksi}$$

See the flowchart B.

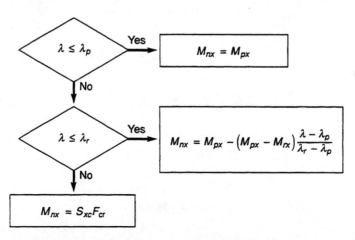

Flowchart B   for M_{nx}, limit-state flange local buckling (FLB)

$$M_{FLB} = \text{if}\left\{\lambda \le \lambda_p, M_{px}, \text{if}\left[\lambda \le \lambda_r, M_{px} - (M_{px} - M_{rx})\frac{\lambda - \lambda_p}{\lambda_r - \lambda_p}, S_{xc}F_{cr}\right]\right\}$$

Results for FLB:

$\lambda = 10 \qquad \lambda_p = 10.833 \qquad k_c = 0.471 \qquad \lambda_r = 25.188 \qquad M_{FLB} = 24{,}012 \text{ kip-in.}$

(f) *Moment capacity, limit-state web local buckling (WLB):*

$$\lambda = \frac{h}{t_w} \qquad \lambda_p = \frac{640\sqrt{\text{ksi}}}{\sqrt{F_y}}$$

$$h_c = 2(y_e - t_{fc}) \qquad \lambda_c = \text{if}\left[\frac{h}{h_c} \le 0.75, 0.75, \text{if}\left(\frac{h}{h_c} \ge 1.5, (1.5)\frac{h}{h_c}\right)\right]$$

$$\lambda_r = \frac{253\sqrt{\text{ksi}}}{\sqrt{F_y}}(1 + 2.83\lambda_c) \qquad \text{(Flowchart C)}$$

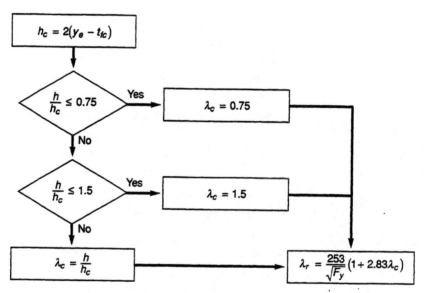

Flowchart C   for λ_r, WLB (Appendix B5.1 AISCS)

$$M_{WLB} = \text{if}\left\{\lambda \le \lambda_p, M_{px}, \left[M_{px} - (M_{px} - M_{rx})\frac{\lambda - \lambda_p}{\lambda_r - \lambda_p}\right]\right\} \qquad \text{(Flowchart D)}$$

Results for WLB:

$$\frac{h}{h_c} = 1.272 \qquad\qquad \lambda = 72$$

$$\lambda_p = 106.667 \qquad\qquad \lambda_r = 194.011$$

$$M_{WLB} = 24{,}012 \text{ kip-in.}$$

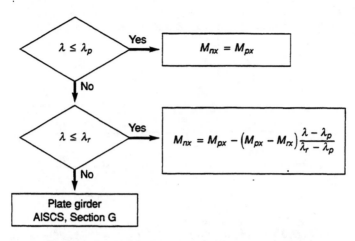

Flowchart D for M_{nx}, limit-state web local buckling (WLB)

(g) *Moment capacity: limit state lateral-torsional buckling (LTB):* See Figs. 3.15 and 3.17 for explanation of operations.

$$L_b = 36 \text{ ft} \qquad I_{yc} = I_{yf1}$$

$$C_b = 1 \qquad L_p = \frac{300 r_y \sqrt{\text{ksi}}}{\sqrt{F_y}}$$

$$T_1 = \pi E \sqrt{I_y J \frac{G}{E}} \qquad T_2 = 2.25 h \left(\frac{2 I_{yc}}{I_y} - 1 \right) \sqrt{\frac{I_y}{J}} \qquad T_3 = 25 h^2 \left[\left(1 - \frac{I_{yc}}{I_y} \right) \left(\frac{I_{yc}}{J} \right) \right]$$

$$L_r = \sqrt{\frac{(2 M_{rx} T_2/T_1) + 1 + \sqrt{[\,(2 M_{rx} T_2/T_1) + 1^2\,] + 4(M_{rx}^2 T_3/T_1^2)}}{2 M_{rx}^2/T_1^2}}$$

$$B_1 = \frac{T_2}{L_b} \qquad B_2 = \frac{T_3}{L_b^2}$$

$$M_{cr1} = \frac{T_1}{L_b} \left(B_1 + \sqrt{1 + B_2 + B_1^2} \right) C_b \qquad M_{cr} = \text{if} \,(M_{cr1} \le M_{px}, M_{cr1}, M_{px})$$

$$M_{cr} = 24{,}012 \text{ kip-in.} \qquad L_p = 16.47 \text{ ft} \qquad L_r = 60.98 \text{ ft}$$

$$M_{\text{LTB}1} = \text{if} \left(L_b \le L_p, M_{px}, \text{if} \left\{ L_b \le L_r, C_b \cdot \left[M_{px} - (M_{px} - M_{rx}) \frac{L_b - L_p}{L_r - L_p} \right], M_{cr} \right\} \right)$$

$$M_{\text{LTB}1} = \text{if} \left(M_{\text{LTB}1} \ge M_{px}, M_{px}, M_{\text{LTB}1} \right)$$

$M_{\text{LTB}} = 19{,}973.219 \text{ kip-in.}$ (lowest; thus this is the 'x-axis moment capacity)

$M_{\text{FLB}} = 24{,}012 \text{ kip-in.}$

$M_{\text{WLB}} = 24{,}012 \text{ kip-in.}$

$M_{nx} = 19{,}973 \text{ kip-in.}$

Example 3.7

A simply supported beam is subjected to the loads shown. Design a welded symmetric I beam. Use SI units.

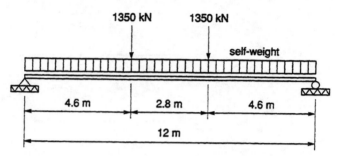

Given: $F_y = 345$ MPa $L = 12$ m
         $E = 200,000$ MPa $P_u = 1350$ kN   [this is a factored (ultimate) load]
         Estimated self-weight of girder: 450 kg/m (mass), or $450 \times 9.81/1000 = 4.4$ kN/m
         (force)

Solution

$$w_d = 4.4 \text{ kN/m}$$

$$M_u = P_u \cdot 4.6m + \frac{1.2w_d L^2}{8} = 6305.04 \text{ kN·m}$$

$$Z_{reqd} = \frac{M_u}{0.9F_y} = 2.031 \times 10^7 \text{ mm}^3$$

Try the following cross section:

Flanges: $b_f = 400$ mm $t_f = 40$ mm
Web: $h = 1200$ mm $t_w = 15$ mm

$$Z_x = 2b_f t_f\left(\frac{h}{2}+\frac{t_f}{2}\right)+\frac{h^2 t w}{4} = 2.524 \times 10^7 \text{ mm}^3 \text{ which is larger than } Z_{req} = 2.031 \times 10^7 \text{ mm}^3$$

Check compactness of flange:

$$\lambda = \frac{b_f}{2t_f} \qquad \lambda_p = \frac{65}{\sqrt{29,000}}\sqrt{\frac{E}{F_y}}$$ Table B5.1 in AISCS,
                                        in nondimensional form
$$\lambda = 5 \qquad < \lambda_p = 9.19 \quad \text{OK}$$ for use with any unit system

Check compactness of web:

$$\lambda = \frac{h}{t_w} = 80 \qquad \lambda_p = \frac{640}{\sqrt{29,000}}\sqrt{\frac{E}{F_y}} = 90.487 \qquad < 80 \quad \text{OK}$$

Cross section is compact. Determine required spacing of the lateral bracing:

$$I_y = \frac{b_f^3 t_f}{12} + \frac{h t_w^3}{12} \qquad A = 2b_f t_f + h t_w$$

$$r_y = \sqrt{\frac{I_y}{A}} = 65.371 \text{ mm}$$

$$L_p = \frac{300}{\sqrt{29,000}}\sqrt{\frac{E}{F_y}}r_y = 2.773 \text{ mm}$$

Lateral braces are needed under the two concentrated loads, at the ends and 2.3 m from each end. The central segment of 2.8 m is somewhat longer than 2.77 m; however, this is only a 1% exceedance, so it is OK.

REFERENCES

3.1. C. G. SALMON AND J. E. JOHNSON, *Steel Structures Design and Behavior*, 3rd ed., Harper & Row, New York, 1990.

3.2. E. P. POPOV, *Introduction to Mechanics of Solids*, Prentice Hall, Englewood Cliffs, NJ, 1968.

3.3. R. C. HIBBELER, *Structural Analysis*, 3rd ed., Prentice Hall, Englewood Cliffs, NJ, 1995.

3.4. A. CHAJES, *Principles of Structural Stability Theory*, Prentice Hall, Englewood Cliffs, NJ, 1974.

3.5. W. F. CHEN AND E. M. LUI, *Structural Stability Theory and Implementation*, Elsevier, New York, 1987.

PROBLEMS

3.1. Assuming that full lateral support is provided, determine the maximum moment and shear in each of the following cases and make the most economical beam selection. Neglect beam dead weight in initial selection, then check the beam with the self-weight included. The span is 27 ft in each case. The beams are simply supported at each end. The steel is ASTM A572 with $F_y = 42$ ksi.

   (a) Uniform load of 3 kips/ft dead load and 6 kips/ft live load.

   (b) Concentrated center load of 100 kips dead load and 175 kips live load.

   (c) Three loads at $\frac{1}{4}$ points of 90 kips dead load and 50 kips live load.

3.2. A simply supported beam with a span of 32 ft carries a static uniform ultimate load of 8 kips/ft. The compression flange is laterally supported by the floor slab. Select the most economical W shape for a steel with $F_y = 50$ ksi (refer to Example 3.1).

3.3. Rework Problem 3.2 with intermittent lateral supports, as may be required, to provide the equivalent of full lateral support (see Example 3.2).

3.4. Rework Problem 3.2 without any lateral support (see Example 3.3).

3.5. Design a simple beam for a span of 25 ft, using A36 steel, under an ultimate uniform load of 4 kips/ft and a concentrated load of 150 kips 2 ft from one end. Intermittent lateral supports are provided at 5-ft intervals.

3.6. Design a simple bearing block support for the heavily loaded end of the beam selected in Problem 3.5 (refer to Section 3.9). Assume support to be on full area of the concrete. $f'_c = 3000$ psi.

3.7. For the beam selected in Problem 3.5, compare (a) the maximum shear stress at the end of the beam nearest the concentrated load, using Eq. (3.8), (b) the average shear stress at the same location by AISCS, and (c) the average shear stress based on the clear depth web area, that is, the web area between the inner faces of the flanges.

3.8. Rework Problem 3.1 with only a single lateral support at midspan.

3.9. Rework Problem 3.1 with no lateral support between the ends of the beams.

3.10. Rework problem 3.1(b) with the 175-kip load repeated 1,200,000 times. Beam is fully supported laterally and it is assumed that the concentration of load at the center will require the use of a vertical stiffener similar to that sketched in Case 7 of AISCS. Use Example 3.4 as a guide. If the stress range permitted by the repeated load requirements controls the design, give special attention to a modification of stiffener details to improve the stress category as may be required.

3.11. A standard rectangular structural tube is used to span a highway with an effective simple span of 56 ft. In addition to its own dead weight, it supports a highway road direction sign having a total weight of 2000 lb, which is attached to the tube at its third points. The sign, measuring 4 ft in height by 18 ft in length, is centrally located and transmits to the tube a horizontal wind force that may be as great as 40 lb/ft². This force is also introduced at the third points. Select a tube, using A36 steel. Do not neglect the wind force on the tube alone at each end of the span. Assume simple end supports and omit design of columns.

3.12. By use of Eq. (3.9) and the values of section moduli (S) and plastic moduli (Z) as provided in the AISCM, determine the shape factor for the following sections:

$W\,33 \times 118$
$W\,18 \times 50$
$S\,24 \times 106$
$W\,14 \times 665$

3.13. Using A36 steel, select the required size for the simply supported beam with overhang loaded as shown in the figure. Note that lateral support is provided only at the location of the 60-kip load and at the reaction supports.

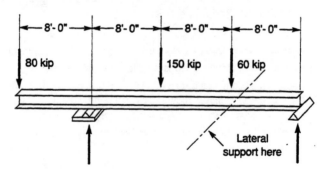

3.14. Select a W shape, using A36 steel, for a span of 34 ft, to support a vertical live load of 600 lb/ft in addition to its own dead weight. The beam, being exposed must also resist a horizontal wind force of 30 psf and is laterally unsupported.

3.15. A simple beam with a span of 26 ft carries a uniform vertical live load of 1.2 kip/ft, the dead weight of beam, and a concentrated live load at the center of 100 kips. Using A572 steel ($F_y = 42$ ksi) select:

(a) A W shape for the vertical load alone.

(b) A W shape for the vertical load plus a concentrated horizontal wind force of 6 kips, at the center of the span, and applied at the neutral axis of the beam.

(c) Same as part (b), but with the horizontal force applied at the level of the top flange. (Assume that the top flange alone acts as a rectangular beam to resist the lateral force and keep the combined compressive stress within the value of $0.9F_y$.)

3.16. Same as Problem 3.15(c), but using a built-up welded member made of three plates similar to the section used in Example 3.6.

3.17. Similar to Example 3.6, except change span to 40 ft, total wheel loads (without impact) 46 kips vertical and 12 kips horizontal for each wheel. Apply the AISCS impact allowance for crane runway support girders to both the vertical and horizontal loads. The two wheel locations are 8 ft apart. Use A36 steel. Study Example 3.6 before attempting this problem, and try to use it as a guide in making a better initial selection of flange plate sizes.

4

Columns Under Axial Load

4.1 Introduction

Originally, the term *column* referred to a vertically upright compression member, such as are found in Egyptian, Greek, or Roman temples, built of hand-hewn segments of rock or marble. In today's usage, a column is not necessarily upright, and any compression member, horizontal, vertical, or inclined, is termed a *column* if the compression that it transmits is the primary factor determining its structural behavior. If bending is also a major factor, the term *beam-column* may be used, and such members will be considered in Chapter 5.

For engineering design purposes, the *axially loaded* column is defined as one that transmits a compressive force whose resultant at each end is approximately coincident with the longitudinal centroidal axis of the member. Although there are no design loads that produce bending moment, there may be moments due to initial imperfections, accidental curvature, or unintentional end eccentricity. Such accidental bending moments reduce the strength of the member, and this fact is reflected in the column formulas of the steel design specifications as well as in the magnitude of the resistance factor ϕ.

The use of the column in structures has at times jumped ahead of design know-how. The tragic failure of the first attempt to build the Quebec Bridge, in 1907, has been attributed to faulty column design in which proportions and sizes were extrapolated beyond the range of previous experience. Unfortunately, the failure of compression members in structures is still too prevalent. An understanding of column behavior is of vital importance to the structural engineer, as it provides an aid to the intelligent use of design specifications.

Column failure involves the phenomenon of *buckling*, during which a member experiences deflections of a totally different character than those associated with the initial loading. Thus, when an axially loaded column is first loaded, it simply shortens or compresses in the direction of the load. If and when a buckling load is reached, the shortening deformation stops, and a sudden lateral and/or twisting deformation occurs in a direction normal to the column axis, thus limiting the axial load capacity.

The strength of a tension member is independent of its length, whereas for the column both the strength and the mode of failure are markedly dependent on length. A very short and compact column built of any one of the common metals will develop about the same strength in compression as it will in tension. But if the column is long, it will fail at the load that is proportional to the bending rigidity of the member, EI, and independent of the strength of the material. Thus a very slender column of steel with a yield stress of 100 ksi has no more column strength than one with a yield stress of 36 ksi. In each case the strength is determined by the Euler column formula that was developed more than 250 years ago.

4.2 Basic Column Strength

The buckling strength of a column decreases with increasing length. Beyond a certain length the buckling stress falls below the proportional limit of the material and buckling for any longer column is elastic. For such a slender column the buckling load is given by the Euler formula,

$$P_e = \frac{\pi^2 EI}{l^2} \qquad (4.1)$$

The yield stress, F_y, does not appear in Eq. (4.1). It plays no part in determining the strength of a very long column. Thus a slender aluminum alloy column, according to Eq. (4.1), will buckle at about one-third of the load of its steel counterpart—not because of any weakness in the material, but simply because the elastic modulus E of aluminum alloy is about one-third that of steel. The aluminum column would also weigh about one-third that of steel, and by reshaping the design of the cross section the strength can be increased, up to a limit, with *no increase in the weight of the member*. The limit of such increase occurs when the material gets so spread out, and correspondingly thin, that *local buckling* occurs prior to general buckling of the entire member.

The Euler load, P_e, is a load that will just hold the column in the deflected shape shown in Fig. 4.1. At any point along the column the external applied moment Py is

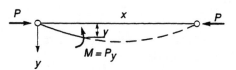

Figure 4.1 Buckled shape of a hinged-end column.

equal to the internal resisting moment $EI\phi$, where ϕ is the column curvature at the corresponding point (refer to Section 3.2).

If both sides of Eq. (4.1) are divided by A and the relationship $I = Ar^2$ is introduced, r being the radius of gyration of the cross section, the buckling load is expressed in terms of the buckling stress, F_e:

$$F_e = \frac{P_e}{A} = \frac{\pi^2 EI}{Al^2} = \frac{\pi^2 E}{(l/r)^2} \qquad (4.2)$$

Equation (4.2) can be modified so as to apply to other end conditions, such as free or fixed, by use of the effective length factor K. For purely flexural buckling, Kl is the length between inflection points and is known as the effective length. Thus Eq. (4.2) becomes

$$F_e = \frac{\pi^2 E}{(Kl/r)^2} \qquad (4.3)$$

For example, if a column is fixed against rotation and translation (lateral movement) at each end, it will buckle with points of inflection at the quarter points, as shown in Fig. 4.2(a), and the effective length factor is 0.5. Consequently, according to Eq. (4.3), a very slender column with fixed ends that buckles elastically will be four times stronger than the same column with hinged ends. But if one end is fixed and the other free both with respect to rotation and translation, as shown in Fig. 4.2(b), there is an imaginary point of inflection at a distance l below the column base, and the effective length factor is 2.0. Such a column has only one-fourth the *elastic* strength of the same member with hinged ends. Table 4.1 shows these and other cases and lists recommended modified values for design usage.

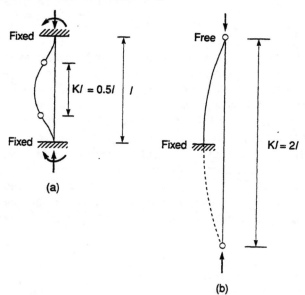

Figure 4.2   Example illustrating the effective length concept.

TABLE 4.1 Effective Length Factors K for Centrally Loaded Columns
with Various Idealized End Conditions

Buckled shape of column is shown by dashed line.	(a)	(b)	(c)	(d)	(e)	(f)
Theoretical K value	0.5	0.7	1.0	1.0	2.0	2.0
Recommended K value when ideal conditions are approximated	0.65	0.80	1.2	1.0	2.10	2.0

End-condition code		
	Rotation fixed	Translation fixed
	Rotation free	Translation fixed
	Rotation fixed	Translation free
	Rotation free	Translation free

l/r is termed the *slenderness ratio* and is almost universally used as a parameter in terms of which the column-strength curve may be drawn graphically or expressed analytically by a column-strength formula. Figure 4.3 shows typical column-strength curves for steel. The strengths of the very short and very long column are expressed by F_y and F_e, respectively. In the intermediate range, the transition from F_y to F_e depends on a complex mix of factors—initial curvature, accidental end eccentricity, and residual stress—and is usually expressed empirically by means of parabolic, straight-line, or more complex expressions.

In the case of structural steel, the presence of residual stress in rolled shapes has been shown to be an important factor influencing the shape of the transition curve between very short and very long columns.[*] Residual stresses are locked in a member as a result of uneven cooling after rolling, welding, oxygen cutting, or by cold-straightening operations.

On the basis of column tests, as well as measurements of residual stresses in rolled shapes, the Column Research Council[†] (CRC) proposed in 1960 an empirical transition for the short and intermediate range, from F_y for $Kl/r = 0$, to the Euler curve (F_e) at $F_y/2$ as given by Eq. (4.4). For $Kl/r < C_c$,

---

[*] For a more complete treatment of inelastic buckling and the residual stress effect, refer to *Guide to Stability Design Criteria for Metal Structures* (Ref. 4.1).

[†] Now renamed *Structural Stability Research Council* (SSRC).

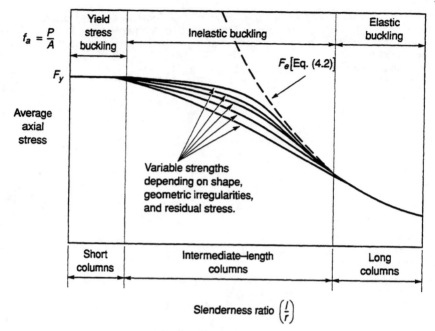

Figure 4.3 Column strength curves.

$$F_c = \left[1 - \frac{1}{2C_c^2}\left(\frac{Kl}{r}\right)\right]F_y \qquad (4.4)$$

where F_c is the column strength, ksi, and $C_c = \sqrt{2\pi^2 E/F_y}$. For $Kl/r \geq C_c$, the Euler formula, Eq. (4.3), should be applied.

Equations (4.3) and (4.4) are predicated on the premise that the basis of column strength is the strength of a perfectly straight column which has only the residual stress as its *initial imperfection* explicitly accounted for in the column design formula. The other "initial imperfections," such as the unavoidable out-of-straightness of the member as it leaves the fabricating shop, and which is less than or equal to the prescribed manufacturing tolerance, are taken care of by a suitable factor of safety by which the critical stress is divided to obtain an *allowable stress*. Some current (1995) steel specifications still use this approach (e.g., the AISC allowable stress specifications and the bridge specification for highway and railway bridges), but many other design standards, including the AISC-LRFD specification, follow the latest recommendation of the Structural Stability Research Council (SSRC) in the 4th edition of its *Guide* (Ref. 4.1), which recommends that the basis of column strength is the column with an initial out-of-straightness *and* with residual stresses. The AISCS column formulas are based on an initial sinusoidal out-of-straightness of 1/1500 of the column length. These formulas are presented in Sec. E2 and they are reproduced as follows:

Effective nondimensional slenderness ratio:

$$\lambda_c = \frac{Kl}{r\pi}\sqrt{\frac{F_y}{E}} \tag{4.5}$$

Column design strength $= \phi_c P_n$

where $\phi_c = 0.85$, the resistance factor to account for uncertainties of column strength prediction, and $P_n = AF_{cr}$, the nominal column strength. A is the gross cross-sectional area of the column.

$$\text{For } \lambda_c < 1.5, F_{cr} = 0.658^{\lambda_c^2} F_y \tag{4.6a}$$

$$\text{For } \lambda_c \geq 1.5, F_{cr} = \frac{0.877}{\lambda_c^2} \tag{4.6b}$$

The comparison between the ratios F_c/F_y from Eqs. (4.3) and (4.4), respectively, and F_{cr}/F_y from Eq. (4.6) and the nonlinear slenderness ratio λ_c is shown in Fig. 4.4. Note that Eqs. (4.3) and (4.4) can be expressed also as

$$\text{For } \lambda_c < \sqrt{2}: \qquad F_c = \left(1 - \frac{\lambda_c^2}{4}\right)F_y$$

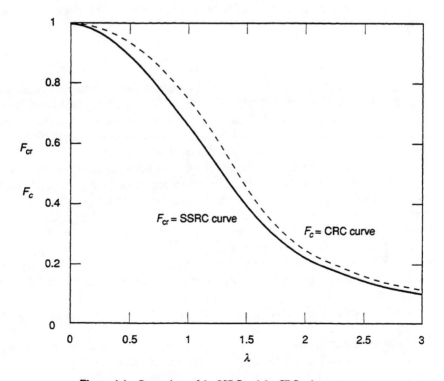

Figure 4.4   Comparison of the SSRC and the CRC column curves.

$$\text{For } \lambda_c \geq \sqrt{2}: \qquad F_c = \frac{F_y}{\lambda_c^2}$$

It is evident from Fig. 4.4 that the initial out-of-straightness results in a reduction of column strength. This does not mean that specifications based on the initially straight column hypothesis are unsafe. These specification account for this effect by adjusting the factors of safety.

4.3 Effective Length of Columns

The basic concept of effective length has been explained in Section 4.2 and certain special cases were shown in Table 4.1. A more general evaluation of effective length factors for columns in continuous frames is available through the use of the alignment charts shown in Fig. 4.5, as presented in the Structural Stability Research Council (SSRC) *Guide*. These charts are functions of the I/L values of adjacent girders (beams), which are assumed to be rigidly attached to the columns. A conservative assumption is made that all columns in the portion of the framework under consideration reach their individual buckling loads simultaneously. The charts are based upon a slope-deflection analysis that includes the effect of column load. In Fig. 4.4(a) and (b) the subscripts A and B refer to the joints at the two ends of the column that is under consideration. G is defined as

$$G = \frac{\Sigma I_c/L_c}{\Sigma I_g/L_g} \tag{4.7}$$

In Eq. (4.7) the Σ indicates a summation of all members rigidly connected to that joint (A or B) and lying in the plane in which buckling of the column is being considered. I_c is the moment of inertia and L_c is the length of a column section. I_g is the moment of inertia and L_g is the length of a girder (beam) or other restraining member. I_c and I_g are taken about the axis perpendicular to the plane of buckling.

In connection with the use of these charts, the following recommendations are made by the SSRC *Guide*. For a column base connected to a footing or foundation by a frictionless hinge, G is theoretically infinite, but should be taken as 10 in design practice. If the column base is rigidly attached to a properly designed footing, G approaches a theoretical value of zero, but should be taken as 1.0.

For greater accuracy, the girder stiffness I_g/L_g in Eq. (4.7) should be multiplied by a factor when certain conditions at the far end are known to exist. For the cases with sidesway prevented [Fig. 4.5(a)], the appropriate multiplying factors are 1.5 for far end of girder (beam) hinged, and 2.0 for far end of girder fixed. For the case with sidesway not prevented [Fig. 4.5(b)], the multiplying factor is 0.5 for far end of girder hinged.

Having determined G_A and G_B for a column, K is obtained by constructing a straight line between the appropriate points on the scales for G_A and G_B. For example, in Fig. 4.5(a) if G_A is 0.5 and G_B is 1.0, then K is found to be 0.73.

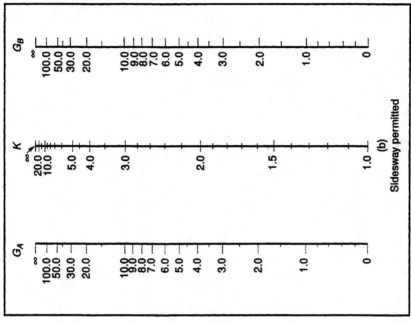

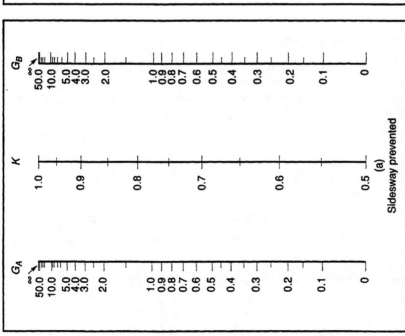

Figure 4.5   Charts for effective length of columns in continuous frames. (From the Structural Stability Research Council *Guide*, therein by courtesy of Jackson and Moreland Division of United Engineers and Constructors, Inc.)

The following simple example illustrates the calculation of the effective length for the columns in a single-story single-bay rigid frame. At the top of the column:

$$G_t = \frac{\Sigma I_c/L_c}{\Sigma I_g/L_g} = \frac{I/18}{2I/40} = 1.11$$

At the bottom of the column:

$$G_b = 1.0 \text{ (fixed end)}$$

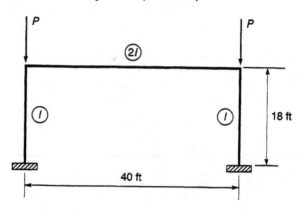

From the sidesway permitted nomograph [Fig. 4.5(b)]: $K = 1.33$.

4.4 Types of Steel Columns

Cross sections of various shapes are shown in Fig. 4.6. The column cross section to be used will be conditioned by the magnitude of the load and by the type of end framing or connections that are most convenient for the particular structural application. In general, within the limits of available clearance and with an eye to thickness limitations, the designer uses a section with the largest possible radius of gyration, thus

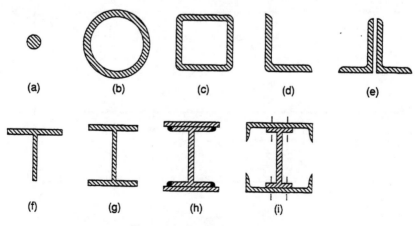

(a)   (b)   (c)   (d)   (e)

(f)   (g)   (h)   (i)

Figure 4.6   Types of steel columns.

reducing the slenderness ratio and increasing the critical stress. In building design, of course, with rather small lengths and the need to maximize the available occupancy space, heavy compact sections are used for heavy loads. Similarly, in exposed areas it may be desirable to use a small section to minimize wind loads.

4.4.1 Round, Solid Bars

The radius of gyration (r) for a solid round cross section is equal to

$$r = \frac{d}{4} = \frac{A}{\pi d} \qquad (4.8)$$

where d is the diameter of the solid round bar and A is the cross-sectional area of the bar. Solid round bars of high strength steel have found particular use as main column elements of tall TV or radio towers, which have been built to heights of more than 2000 ft (682 m). Aside from its increased strength, high-strength steel reduces dead weight, which is particularly beneficial when seismic forces are being considered. Similarly, the use of a round section of relatively small diameter minimizes wind force and reduces added weight due to ice formation. Obviously, lateral force is a problem in the design of the individual column element acting as a beam, as in an exposed tower structure. The weight and area of round and square solid bars are tabulated in AISCM.

4.4.2 Steel Pipes

The steel pipe as shown in Fig. 4.6(b) is more efficient than the solid round bar, since the radius of gyration may be increased almost independently of the cross-sectional area, thus reducing l/r and increasing the critical stress. The possibility of local buckling must be considered if the wall thickness in comparison to the pipe diameter becomes overly small. The full allowable column stress may be used as long as the D/t ratio is less than $3300/F_y$ (AISCS, (Table B5.1). (D is the outside diameter of the pipe and t is the wall thickness.)

Material cost per pound for tubular shapes almost always exceeds that of standard rolled shapes, and the end connections in a framed structure will require special consideration. Advantages include those listed for solid round bars. If the ends are hermetically sealed to prevent access of air, the pipe interiors need not be treated to prevent corrosion. Reference should be made to AISCM for dimensions and properties of available standard sizes of steel pipe, ranging from the smallest up to 12 in. (305 mm) in diameter. Such pipe are available for a number of steel grades and methods of manufacture. Circular tubes are very popular in architecturally exposed trusses in airport halls and other public buildings. Welded tubes are used almost exclusively in fixed offshore drilling platforms.

4.4.3 Box Sections and Structural Tubes

The box section shown in Fig. 4.6(c) is one of the standard types available, square up to 16 in. × 16 in. (406 mm × 406 mm) or rectangular up to 20 in. × 12 in., (508 mm × 305 mm) with properties of cross section as listed in AISCM. Larger sizes may be built up by welding various combinations of plates, angles, or channels. As a compression member, the square tube combines the effectiveness of the hollow steel pipe with the advantage of simpler end connection details in usual building-frame applications.

4.4.4 Angle Struts

Single-angle struts, as shown in Fig. 4.6(d), are satisfactory as secondary members for light loads. If they can be used in a situation where the load can be brought uniformly into both angle legs at each end or where the ends can be prevented from rotation or twist by a rigid end connection attachment to a heavy member or footing, they may be designed by usual procedures for centrally loaded columns, provided width/thickness ratio limitations are met. A special part of AISCM gives detailed design criteria for single-angle compression members.

Double-angle struts, as shown in Fig. 4.6(e), are often used in single plane trusses. Frequent "stitching" must be provided by means of bolts or rivets to ensure that the two angles act as a single unit (see AISCS, Sec. E4). End connections should be designed so as to result in uniform distribution of load. The cross-sectional properties of selected double-angle sections are tabulated in AISCM for members made up either of two unequal leg angles or two equal leg angles.

Table B5.1 of AISCS provides that single-angle struts as well as double-angle struts with separators subject to axial compression shall be considered as fully effective (i.e., the leg of the angle will not buckle locally before it yields) when the ratio of width to thickness is not greater than $76/\sqrt{F_y}$ (for struts comprising double angles in contact the ratio shall be not greater than $95/\sqrt{F_y}$); otherwise, the critical compression stress shall be modified by the appropriate reduction factor as provided in Appendix B5.3a of AISCS.

4.4.5 Structural Ts

Structural Ts (WT shapes), as shown in Fig. 4.6(f), are often used as chord sections in light welded trusses, with double-angle struts welded to the T web. Structural Ts are made by splitting W shapes, or standard beams, longitudinally; hence they have stems (webs) considerably thinner than the flange. The restraint of the flange with respect to buckling of the stem permits the use of greater width/thickness ratios for the stem of the T in comparison with the angle struts, as also provided by Table B5.1 of AISCS, permitting a width/thickness ratio in the stem up to $127/\sqrt{F_y}$. For ratios greater than this limit, the allowable compression stress is modified in accordance with Appendix B5.3a of AISCS.

4.4.6 Wide-Flange Shapes

Wide-flange W, M, or HP shapes are doubly symmetric, as shown in Fig. 4.6(g), and are rolled in a wide range of weights and sizes. These are therefore suited to a correspondingly great range of column loads and lengths and find frequent use in building construction. The W14 series provides a wide range of coverage of shapes (some with exceptionally wide flanges to balance r about the two axes), which are especially suited to column requirements for tall multistory building frames.

In areas remote from mills that produce heavy W shapes, equivalent sections are produced by means of continuous longitudinal welds joining three plate segments and designated WW. Combination shapes may also be made up by means of a cover-plated W member, as shown in Fig. 4.6(h), or an S (standard beam) shape together with two channel (C) shapes. The latter shape might be desirable for a very long member with not so great loads, for which the cross section should be spread as much as possible to reduce the l/r ratio. Although Figs. 4.6(h) and (i) are shown as welded and bolted (or riveted), respectively, either fabrication method could be used for either shape.

Figure 4.7 shows a heavy W14 column shape shortly after erection in one of the lower floors of an office building. The heavy bolted column splice has been completed at the lower end, and the splice plates at the upper end are ready to receive the next two-story column segment. Note the beam-connecting plates that are welded to the columns and field bolted to the beams.

Figure 4.7   W14 steel column in substructure of a multistory building. (Courtesy of American Institute of Steel Construction.)

4.4.7 Columns with Lacing, Battens, or Perforated Cover Plates

In a situation where a very long column is required, it may be necessary to spread the cross section to a degree that makes a laced column economical, as shown for example in Fig. 4.8(a). Before the advent of rolled W shapes, such laced members were also used for more usual lengths found in bridges and buildings, where they may still be seen in older structures. Laced columns are used today in derrick booms and TV or radio towers. In such applications the four-angle member shown in Fig. 4.8(a) together with the lacing bars may be replaced by solid round bars with welded end connections, which help reduce wind and icing loads on exposed members.

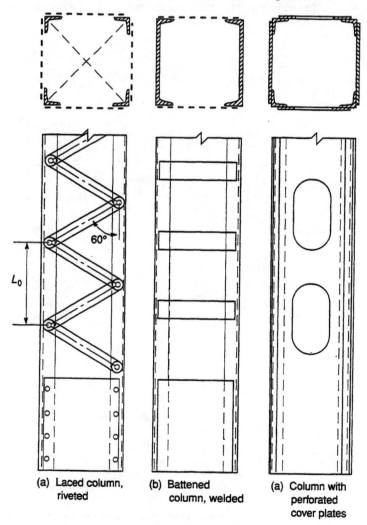

(a) Laced column,   (b) Battened   (a) Column with
    riveted   column, welded   perforated
                                                    cover plates

Figure 4.8 Types of built-up columns.

Lacing bars carry no column load, but they do perform the following functions:

1. They hold the component parts of the laced column in position so as to maintain the shape of the overall column cross section. For the same purpose, cross bracing in a plane normal to the column axis must be provided intermittently, as shown in the sectional view at the top of Fig. 4.8(a).

2. The lacing provides lateral support for the component column segments at each connecting point. For example, in Fig. 4.8(a), the l_0/r_0 of each individual angle, between support points, should be less than the overall l/r of the whole member. The r_0 of the individual angle should be the minimum value, as tabulated in the AISCM for the inclined $z–z$ axis.

3. The lacing acts as a web replacement to resist shear and provides for the corresponding transfer of longitudinal variations of stress in the component longitudinal elements. In a centrally loaded column, shear force arises from accidental end eccentricity of the load and from curvature of the member under load. The shear force provides the basis for lacing bar design, and under Sec. E4 of AISCS, the resultant shear for design is taken as 2% of the axial load in the member. Lacing bars must act either in tension or compression and be designed for both loading conditions. Complete design rules for built-up members, including laced members and their end tie plates, are provided in Sec. E4 of AISCS.

Battened columns [Fig. 4.8(b)] are not covered by AISCS, but are used occasionally. They present the same design problems as do laced columns, but in comparison with the truss action of the lacing members, the battens resist shear by the less effective and more complex continuous-frame action.

Tie plates at the ends of both laced and battened members are particularly important to distribute the applied end loads. In weakly battened columns they may add materially to the overall column strength. For a more complete coverage of design of both laced and battened columns, reference should be made to Chapter 11 of the SSRC *Guide* (4th ed.).

Columns with perforated cover plates [Fig. 4.8(c)] are used chiefly in bridge construction (see also Fig. 2.6). The net section of such plates may be included in the column area, and they resist shear more effectively than either the laced or battened column. The perforations are provided primarily for drainage in exposed locations and to provide access for cleaning and painting the interior surfaces. Simple design rules for such members are provided in all modern bridge specifications.

4.5 Width/Thickness Ratios

Width/thickness limitations are established to ensure that overall column buckling rather than local buckling governs the critical design stress. When the limitations are not exceeded, the full cross section of the column may be considered to be effective. Limits on width/thickness ratios have been discussed in Sec. 4.4 for the case of the

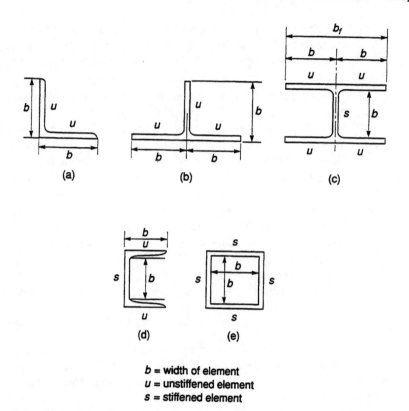

b = width of element
u = unstiffened element
s = stiffened element

Figure 4.9 Stiffened and unstiffened elements of structural shapes as defined for AISCS (Sec. B5) width/thickness limitations.

angle and tee section. The theoretical background for local buckling is given in Chapter 4 of the SSRC *Guide* (Ref. 4.1).

Limitations are as specified in Sec. B5 of AISCS. Width/thickness limits are established under two broad categories, *unstiffened elements* and *stiffened elements*, as defined in Sec. B5 of AISCS, and illustrated herein in Fig. 4.9 and Table 4.2. For equal width/thickness ratios, a stiffened element is much more effective than an unstiffened element, and much greater ratio limits are allowed for the stiffened element. As the yield stress increases, a more stocky element (smaller width/thickness ratio) is required to prevent premature local buckling under the increased allowable stress.

When a thin-walled member does double duty both as a column and partition, it may be desirable to exceed the width/thickness limits. Such members may be used provided a "reduced effective width" and/or reduced allowable stresses are employed, as covered in Appendix B5 of AISCS or in the AISI specifications. Reference also may be made to Chapter 14 of the SSRC *Guide* (4th ed.).

TABLE 4.2 Plate Slenderness Limits (AISCS, Table B5)

Type of element	λ_r	
Outstanding elements in welded built-up compression members	$\dfrac{109}{\sqrt{F_y/k_c}}$	$k_c = \dfrac{4}{\sqrt{h/t_w}}, \ 0.35 \le k_c \le 0.763$
Outstanding elements in rolled beams	$\dfrac{95}{\sqrt{F_y}}$	
Legs of angles	$\dfrac{76}{\sqrt{F_y}}$	
Stems of Ts	$\dfrac{127}{\sqrt{F_y}}$	
Flanges of box columns of uniform thickness	$\dfrac{238}{\sqrt{F_y}}$	
Plates supported along both edges	$\dfrac{253}{\sqrt{F_y}}$	
Circular hollow section	$\dfrac{3300}{\sqrt{F_y}}$	

4.6 Column Base Plates and Splices

Columns on footings must be provided with base plates to distribute load to the concrete within the nominal bearing capacity of the concrete. Nominal bearing pressures are specified in Sec. J9 of AISCS. For very heavy columns in tall buildings, individual base plates may not be sufficient, and a grillage system may be required. For information on base-plate design, including recommended design procedures, reference should be made to AISCM.

4.7 Design Compressive Strength

Insofar as general buckling under axial load governs the design solution of a steel column, the design strength is determined by the formulas in Sec. E2 of AISCS. These were presented in Sec. 4.2 and their use is illustrated in the worked examples in Sec. 4.8.

The solution of a column design problem, using either calculator or computer, illustrates very well the basic difference between a problem of *analysis* of a given structure and the problem of *design*, for which the rules (specifications) are given and the structure is the end product. In column design the design strength is a function of the slenderness ratio and is therefore generally unknown in advance of a trial design selection. Exceptions to this statement are represented by the tube, of a given diameter, and the built-up column of four angles. In these cases the suitable general dimensions of the cross section can usually be established in advance, thus pinpointing the slenderness ratio, and the final design is arrived at by simply changing the wall or angle thickness to provide the required area that is equal to the load divided by the allowable stress. Similarly, in heavy tier building construction, using W14 shapes, the radius of gyration of the cross section will be known approximately in advance. But in many cases it may be necessary simply to guess at some preliminary trial selection and arrive at the final design by means of one or two additional trial selections. One may either guess the radius of gyration, or, more directly, guess the design strength. The analysis of such a trial selection will quickly lead to a much better trial. Most of the examples presented herein involve the check of a final trial selection—in each case readers should ask themselves: How best could this trial have been arrived at?

Flowchart 4.1 systematizes the use of AISCS for the selection of a column. The general sequence of column design consists of the preliminary selection and the detailed checking procedures. The preliminary design steps are:

1. Select the column type: for example,
   (a) Doubly symmetric section (wide-flange, tube, box, etc.).
   (b) Singly symmetric section (channel, equal-leg angle, etc.).
   (c) Unsymmetrical section (unequal-leg angle, etc.).
   (d) T or double-angle section.
   (e) Thin-walled section (C-shaped, hat-shaped, etc.)
2. Select the grade of steel and manufacture (rolled, welded, built-up, etc.).
3. Select the column size.
4. Calculate the effective lengths about both principal axes.
5. Perform a detailed column check.

4.8 Illustrative Examples

Example 4.1

A portion of a trussed TV antenna tower, as shown, has main longitudinal elements that carry an axial load P and are laterally braced at 6-ft intervals. No advantage will be taken of rotation restraint provided by lateral bracing at ends. The dashed line indicates a natural buckling mode of the main compression elements at failure. Select a solid round bar for the main compression elements to satisfy AISCS. Use A36 steel.

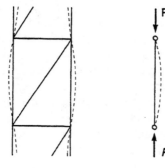

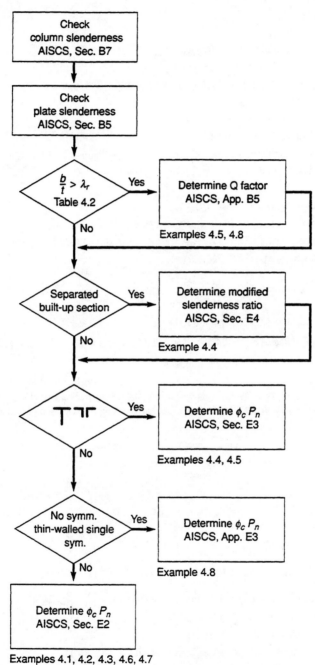

Flowchart 4.1   Column Checking Procedure

Solution

Axial forces:

Due to dead load: $P_d = 107$ kips

Due to wind load: $P_w = 200$ kips

Required strength: $P_u = 1.2 \times 107 + 1.3 \times 200 = 388$ kips

Column type: solid circular bar

Member size: diameter $d = 4.5$ in.

Steel grade: A36, $F_y = 36$ ksi

Effective length: pinned-end column, K = 1.0 (see Table 4.1), $KL = 72$ in.

Area: $A = 15.9$ in^2

Weight/ft: 54.1 lb/ft

Column weight: $6 \times 0.0541 = 0.3$ kip

Radius of gyration: $r = d/4 = 4.5/4 = 1.125$ in.

Slenderness ratio: $KL/r = 72/1.125 = 64.0$ < 200 OK (AISCS, Sec. B7)

Design strength (AISCS, Sec. E2):

$$\lambda_c = \frac{KL}{r\pi}\sqrt{\frac{F_y}{E}} = \frac{64}{\pi}\sqrt{\frac{36}{29,000}} = 0.718 \qquad\qquad < 1.5$$

$$F_{cr} = F_y\left(0.658^{\lambda_c^2}\right) = 36 \times 0.658^{0.718^2} = 29.02 \text{ ksi}$$

$$\phi_c P_n = \phi_c A F_{cr} = 0.85 \times 15.9 \times 29.02 = 392 \text{ kips} \qquad < 388 \text{ kips}$$

Capacity check is OK. Use a 4.5-in.-diameter bar.

Example 4.2

For an unbraced and effectively hinged-end length of 8-ft, design a steel pipe column for an axial load of 78.4 kips, using A36 steel, AISCS, and limiting the selection to standard pipe sizes.

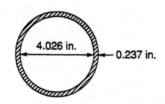

4.026 in.   0.237 in.

Solution

Axial forces:

Due to dead load: $P_d = 12$ kips   includes 0.1-kip self-weight

Due to snow load: $P_s = 35$ kips

Due to wind load: $P_w = 10$ kips

Required force: controlling combination (AISCS Sec. A4.1)

$$P_u = 1.2 P_d + 1.6 P_s + 0.8 P_w = 78.4 \text{ kips}$$

Column type: standard weight circular pipe (AISCM)

Column size: 4 in. nominal diameter

$$A = 3.17 \text{ in}^2 \qquad \text{area}$$
$$r = 1.51 \text{ in.} \qquad \text{radius of gyration}$$
$$D_o = 4.5 \text{ in.} \qquad \text{outside diameter}$$
$$t = 0.237 \text{ in.} \qquad \text{wall thickness}$$
$$w = 10.79 \text{ lb/ft} \qquad \text{weight/ft}$$
$$L = 8 \text{ ft} \qquad \text{column length}$$

Column weight: $W = wL = 0.086$ kip   OK

Steel grade A36: $F_y = 36$ ksi

Effective length factor: $K = 1$

Slenderness ratio of column: $K\dfrac{L}{r} = 63.576$ \qquad < 200 OK

Check width/thickness ratio of circular tube (AISCS, Table B5.1):

$$\lambda = \frac{D_o}{t} \qquad \lambda = 18.987 \qquad \lambda_r = \frac{3300}{F_y} = 91.667$$

Since $\lambda < \lambda_r$, the ratio is OK.

Design strength (AISCS, Sec. E2):

$$\lambda_c = \frac{KL}{r\pi}\sqrt{\frac{F_y}{E}} = 0.713 \qquad < 1.5$$
$$F_{cr} = F_y \cdot 0.658^{\lambda_c^2} = 29.1 \text{ ksi}$$
$$P_n = AF_{cr} \qquad \phi_c = 0.85 \qquad \phi_c P_n = 78.41 \text{ kips}$$

Since $\phi_c P_n = P_u$, the design is OK.

Example 4.3

Design a square structural tube for the same conditions as in Example 4.2.

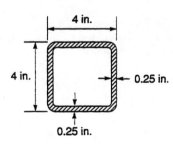

Solution

Axial forces:

Due to dead load: $P_d = 12$ kips   includes 0.1-kip self-weight

Due to snow load: $P_s = 35$ kips

Due to wind load: $P_w = 10$ kips

Required force: controlling combination (AISCS Sec. A4.1)

$$P_u = 1.2P_d + 1.6P_s + 0.8P_w = 78.4 \text{ kips}$$

Column type: hot-rolled square hollow tube (AISCM)
Column size: $4 \times 4 \times 0.25$

$$A = 3.59 \text{ in}^2 \qquad \text{area}$$
$$r = 1.51 \text{ in.} \qquad \text{radius of gyration}$$
$$D = 4 \text{ in.} \qquad \text{width of outside face}$$
$$t = 0.237 \text{ in.} \qquad \text{wall thickness}$$
$$w = 12.21 \text{ lb/ft} \qquad \text{weight/ft}$$
$$L = 8 \text{ ft} \qquad \text{column length}$$

Column weight: $W = wL = 0.098$ kip   OK
Steel grade A36: $F_y = 36$ ksi $E = 29{,}000$ ksi
Effective length factor: $K = 1$

Slenderness ratio of column: $K\dfrac{L}{r} = 63.576 \qquad < 200$ OK

Check width/thickness ratio of plate (AISCS, Table B5.1):

$$\lambda = \frac{D-2t}{t} = 14.878 \qquad \lambda_r = \frac{238}{\sqrt{F_y}} = 39.667$$

Since $\lambda < \lambda_r$, the ratio is OK

Check design strength (AISCS, Sec. E2):

$$\lambda_c = \frac{KL}{r\pi}\sqrt{\frac{F_y}{E}} = 0.713 \qquad < 1.5$$
$$F_{cr} = F_y \cdot 0.658^{\lambda_c^2} = 29.1 \text{ ksi}$$
$$P_n = AF_{cr} \qquad \phi_c = 0.85 \qquad \phi_c P_n = 88.798 \text{ kips}$$

Since $\phi_c P_n > P_u$, the design is OK.

Example 4.4

Design a double-angle strut for the same conditions as in Example 4.2 with the added stipulation that special attention must be given to end connections to bring the axial load uniformly into both legs at each end.

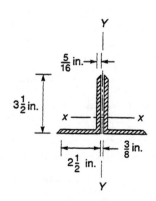

Solution

Axial forces:

Due to dead load: $P_d = 12$ kips   includes 0.1-kip self-weight
Due to snow load: $P_s = 35$ kips
Due to wind load: $P_w = 10$ kips

Required force: controlling combination (AISCS Sec. A4.1)

$$P_u = 1.2P_d + 1.6P_s + 0.8P_w = 78.4 \text{ kips}$$

Column type: double-angles separated by $\frac{3}{8}$-in. gusset plates at the ends and by filler plates at equal intervals along the column

Column size: $3.5 \times 2.5 \times \frac{5}{16}$ angles, long legs back to back (AISCM)

$$A = 3.55 \text{ in}^2 \qquad \text{area}$$
$$r_x = 1.11 \text{ in.} \qquad \text{radius of gyration about } x\text{-axis}$$
$$r_y = 1.10 \text{ in.} \qquad \text{radius of gyration about } y\text{-axis}$$
$$r_z = 0.54 \text{ in.} \qquad \text{radius of gyration about } z\text{-axis}$$
$$w = 12.2 \text{ lb/ft} \qquad \text{weight/ft}$$
$$L = 8 \text{ ft} \qquad \text{column length}$$

Column weight: $W = wL = 0.098$ kip   OK

Steel grade A36: $F_y = 36$ ksi $E = 29{,}000$ ksi $G = 0.385 E$

Effective length factor: $K = 1$

Slenderness ratios of column: $K\dfrac{L}{r_x} = 86.486$ $K\dfrac{L}{r_y} = 87.273$ < 200 OK

y-axis buckling controls since $KL/r_y > KL/r_x$.

Check width/thickness ratio of plate (AISCS, Table B5.1):

$$b = 3.5 \text{ in.} \qquad t = \frac{5}{16} \text{ in.}$$
$$\lambda = \frac{b}{t} = 11.2 \qquad \lambda_r = \frac{76}{\sqrt{F_y}} = 12.667$$

Since $\lambda < \lambda_r$, the ratio is OK.

The two angles are separated, so we must consider Sec. E4 of AISCS. Spacing of fillers $= a$. Governing slenderness ratio is $KL/r_y = 87$. $0.75\, KL/r_y > a/r_z$.

$$a_{max} = 0.75\frac{L}{r_y}r_z = 35.345 \text{ in.}$$

Use two intermediate fillers spaced at $a = L/3$ $a = 32$ in. Specify that gusset and filler plates are welded to the angles.

Modified slenderness ratio for KL/r_y (no modification is needed for KL/r_x because there is no gap between the two angles along the y axis).

From AISCM:

$$x = 0.637 \text{ in.}$$
$$y = 1.14 \text{ in.}$$
$$r_{ib} = 0.727 \text{ in.} \quad (r_{ib} \text{ is the same as } r_y)$$
$$g = \tfrac{3}{8} \text{ in.}$$

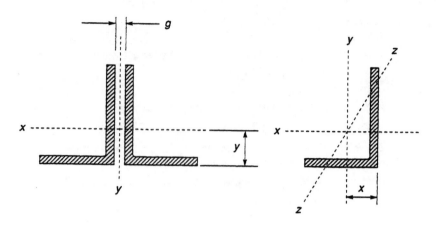

$$\alpha = \frac{2x+g}{2r_{ib}} = 1.134$$

Modified slenderness ratio:

$$\lambda_c\text{mod} = \frac{1}{\pi}\sqrt{\frac{F_y}{E}}\sqrt{\left(K\frac{L}{r_y}\right)^2 + 0.82\frac{\alpha^2}{1+\alpha^2}\left(\frac{a}{r_{ib}}\right)^2} = 1.035$$

Critical stress about the y axis:

$$F_{cry} = F_y \cdot 0.658^{(\lambda_c)^2_{\text{mod}}} = 23 \text{ ksi}$$

Critical torsional buckling stress:

$$J = 2 \cdot 0.0611 \text{ in}^4 \qquad \text{(AISCM, Part 1)}$$

$$y_0 = y - \frac{t}{2} \quad x_o = 0 \text{ in.} \quad \text{(shear center location)}$$

$$r_o = \sqrt{x_o^2 + y_o^2 + r_x^2 + r_y^2} = 1.847 \text{ in.}$$

$$H = 1 - \frac{x_o^2 + y_o^2}{r_o^2} = 0.716$$

$$F_{crz} = \frac{GJ}{Ar_o^2} = 111.455 \text{ ksi}$$

Flexural-torsional buckling stress (AISCS, Sec. E3)

$$F_{crft} = \frac{F_{cry} + F_{crz}}{2H}\left[1 - \sqrt{1 - \frac{4F_{cry}F_{crz}H}{(F_{cry} + F_{crz})^2}}\right]$$

$$\phi_c = 0.85 \qquad P_n = AF_{crft} (P_n) \text{factored} = \phi_c P_n = 65.0 \text{ kips} \qquad < 78.4 \text{ kips} \qquad \text{NG}$$

Use of the tables in Part 3 of AISCM can simplify this complicated design calculation considerably in selecting an acceptable design.

Example 4.5

Design a structural T column for the same conditions as in Example 4.2.

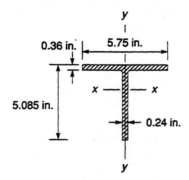

Solution

Axial forces:

Due to dead load: $P_d = 12$ kips   includes 0.1-kips self-weight

Due to snow load: $P_s = 35$ kips

Due to wind load: $P_w = 10$ kips

Required force: controlling combination (AISCS Sec. A4.1)

$$P_u = 1.2P_d + 1.6P_s + 0.8P_w = 78.4 \text{ kips}$$

Column type: T section

Column size: WT5 × 11 (AISCM)

$A = 3.24$ in^2	area
$r_x = 1.46$ in.	radius of gyration about x-axis
$r_y = 1.33$ in.	radius of gyration about y-axis
$d = 5.085$ in.	
$t_w = 0.24$ in.	
$b_f = 5.75$ in.	
$t_f = 0.36$ in.	
$Q_s = 0.836$ for $F_y = 50$ ksi	
$J = 0.119$ in^4	
$r_o = 2.17$ in.	
$H = 0.831$	
$w = 11$ lb/ft	weight/ft
$L = 8$ ft	column length

Column weight: $W = wL = 0.088$ kip   OK

$E = 29,000$ ksi $G = 0.385E$

Effective length factor: $K = 1$

Slenderness ratio of column: $K\dfrac{L}{r_x} = 65.753$ $K\dfrac{L}{r_y} = 72.18$ < 200 OK

y-axis buckling controls

Check width/thickness ratio of plates: (AISCS, Table B5.1):

Flange:

$$\lambda = \frac{b_f}{2t_f} = 7.986 \qquad \lambda_r = \frac{95}{\sqrt{F_y}} = 13.435 \quad \text{OK}$$

Web:

$$\lambda = \frac{d}{t_w} = 21.188 \qquad \lambda_r = \frac{127}{\sqrt{F_y}} = 17.961$$

Since λ_r is less than λ, we must go to Appendix B5.1 of AISCS to calculate Q. However, this value of Q is also tabulated in Part 1 of AISCM and it is listed above.

Determine the column capacity according to Sec. E3 of AISCS:

$$\lambda_{cy} = \frac{KL}{r_y \pi}\sqrt{\frac{F_y}{E}} = 0.954 \qquad \lambda_{cyq} = \lambda_{cy}\sqrt{Q_s} \doteq 0.872$$

$$F_{cry} = Q_s F_y \cdot 0.658^{\lambda_{cyq}^2} = 30.4 \text{ ksi} \qquad F_{crz} = \frac{GJ}{Ar_o^2}$$

$$F_{crft} = \frac{F_{cry} + F_{crz}}{2H}\left[1 - \sqrt{1 - \frac{4F_{cry}F_{crz}H}{(F_{cry}+F_{crz})^2}}\right] = 28.131 \text{ ksi}$$

$$\phi_c = 0.85 \qquad P_n = AF_{crft} \qquad (P_n)_{factored} = \phi_c P_n = 77.473 \text{ kips} \qquad < 78.4 \text{ kips} \quad \text{NG}$$

Use tables in Part 3 of AISCM to find a satisfactory column size.

Example 4.6

Design a wide-flange (W) column 20 ft in length to support an axial dead load of 180 and a live load of 220 kips in the interior of a building. The column base is rigidly fixed to the footing and the top of the column is rigidly framed to very stiff girders. Assume that bracing is provided to prevent sidesway in the weak deflection plane of the column, but the sidesway in the strong plane is not prevented. Select an economical W shape to satisfy AISCS. Use A36 steel.

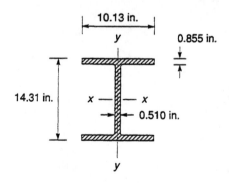

Solution

Axial forces:

Due to dead load: $P_d = 180$ kips   includes 2 kips self-weight

Due to live load: $P_l = 220$ kips

Required force [controlling combination (AISCS Sec. A4.1)]

$$P_u = 1.2P_d + 1.6P_l = 568 \text{ kips}$$

Column type: rolled wide-flange shape

Column size: W14 × 82

Grade of steel: A36

Effective length factor (see Table 4.1): $K_x = 1.2$, $K_y = 0.65$

Length: $L = 20$ ft

Material properties: $F_y = 36$ ksi $E = 29,000$ ksi

Cross-sectional properties: $A = 24.1$ in^2 $r_x = 6.05$ in. $r_y = 2.48$ in.

Flange slenderness: $\lambda_f = 5.9$ $\lambda_r = \dfrac{95}{\sqrt{F_y}} = 15.833$ OK

Web slenderness: $\lambda_w = 22.4$ $\lambda_r = \dfrac{253}{\sqrt{F_y}} = 42.167$ OK

Column slenderness ratios: $\lambda_{cx} = \dfrac{K_x L}{r_x \pi}\sqrt{\dfrac{F_y}{E}} = 0.534$ $\lambda_{cy} = \dfrac{K_y L}{r_y \pi}\sqrt{\dfrac{F_y}{E}} = 0.705$

controls

$F_{cr} = F_y \cdot 0.658^{\lambda_{cy}^2} = 29.231$ ksi

$\phi_c = 0.85$ $P_n = A F_{cr}$ $P_n(\text{factored}) = \phi_c P_n = 598.788$ kips

Use W14 × 82 (A36).

Example 4.7*

Rework Example 4.6, but now the top of the column is rigidly framed to two-way floor beams, as shown in the figure. Determine the axial load capacity of the selected W14 × 82 column in Example 4.6. Assume that all stories are spaced at 20 ft and adjacent stories have the same column section. Use AISCS and A36 steel.

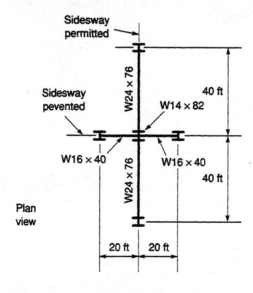

Sidesway permitted

Sidesway pevented

W24 × 76

W14 × 82

40 ft

W16 × 40

W16 × 40

W24 × 76

40 ft

Plan view

20 ft   20 ft

---

* See AISCM for an explicit illustration of the use of the alignment chart prior to study of this example.

Member properties:

W14 × 82 column: $A = 24.1$ in^2 $I_x = 882$ in^4 $I_y = 148$ in^4
                    $r_x = 6.05$ in. $r_y = 2.48$ in.
W24 × 76 beam: $I_x = 2100$ in^4
W16 × 40 beam: $I_x = 518$ in^4

Solution Determine the governing slenderness ratio.

Case 1: Major x axis (sidesway permitted). At the column top,

$$G_A = \frac{\Sigma I_c/L_c}{\Sigma I_g/L_g} = \frac{(882/20) + (882/20)}{(2100/40) + (2100/40)} = 0.840$$

At the column bottom $G_B = 1.0$ (AISCS, Commentary) for column base rigidly attached to a footing. Apply Fig. 4.5(b) for $G_A = 0.840$, $G_B = 1.0$; then

$$K_x = 1.29$$

$$\frac{K_x l}{r_x} = \frac{1.29(20 \times 12)}{6.05} = 51.2$$

Case 2: Minor y axis (sidesway prevented). At the column top,

$$G_A = \frac{(148/20) + (148/20)}{(518/20) + (518/20)} = 0.286$$

At the column bottom, $G_B = 1.0$. Apply Fig. 4.5(a) for $G_A = 0.286$, $G_B = 1.0$; then

$$K_y = 0.695$$

$$\frac{K_y l}{r_y} = \frac{0.695(20 \times 12)}{2.48} = 67.3$$

Therefore, the slenderness ratio is governed by Case 2, for which the buckling will occur in the column about the minor y axis:

$$\frac{Kl}{r} = 67.3$$

Therefore,

$$F_{cr} = 28.36 \text{ ksi} \qquad \text{(from AISCS, Sec. E2)}$$

The axial load capacity of the W14 × 82 column is

$$\phi_c P_n = 0.85 \times 24.1 \times 28.36 = 581 \text{ kips}$$

Example 4.8

Determine the axial design capacity of a doubly symmetric wide-flange column made by welding. The member is a pinned-end column of 10 ft length. The material has a yield stress of 50 ksi. This example illustrates the checking of a column that has slender plate elements, using the provisions of Appendix B5 of AISCS.

Given:

Flange dimensions: $b_f = 10$ in. $t_f = 0.375$ in.
Web dimensions: $h = 12$ in. $t_w = 0.25$ in.
Material properties: $F_y = 50$ ksi $E = 29,000$ ksi
Length: $L = 10$ ft

Solution   Cross-sectional properties:
Area:

$$A = 2b_f t_f + h t_w$$

Moments of inertia and radii of gyration:

$$I_x = 2\left(\frac{b_f t_f^3}{12}\right) + 2b_f t_f \left(\frac{h + t_f}{2}\right)^2 + \frac{h^3 t_w}{12}$$

$$r_x = \sqrt{\frac{I_x}{A}} = 5.548 \text{ in.} \qquad I_y = \frac{2 b_f^3 t_f}{12} + \frac{h t_w^3}{12}$$

$$r_y = \sqrt{\frac{I_y}{A}} = 2.44 \text{ in.}$$

Since $r_x > r_y$, y-axis buckling controls, so the slenderness parameter

$$\lambda_c = \frac{L}{r_y \pi} \sqrt{\frac{F_y}{E}} = 0.65$$

Plate slenderness limits (Table 4.2 or AISCS, Table B5.1): Flange slenderness:

$$\lambda_f = \frac{b_f}{2 t_f} = 13.333 \qquad k_c = \frac{4}{\sqrt{\dfrac{h}{t_w}}} = 0.577$$

Since k_c is between 0.35 and 0.763, the value above is the correct value to use. The limiting slenderness parameter defining slender elements is

$$\lambda_{fr} = \frac{109}{\sqrt{\dfrac{F_y}{k_c}}} = 11.713$$

Since this limit is less than the actual plate slenderness, we must go to Appendix B5 of AISCS [Eq. (AB-5-7)] to calculate a plate-buckling reduction factor Q_s:

$$\lambda_{fu} = \frac{200}{\sqrt{\dfrac{F_y}{k_c}}} = 21.491$$

The actual flange plate slenderness lies between λ_{fr} and λ_{fu}, and therefore

$$Q_s = 1.415 - 0.0038 \lambda_f \sqrt{\frac{F_y}{k_c}} = 0.943$$

Web slenderness:

$$\lambda_w = \frac{h}{t_w} = 48$$

Limiting web slenderness (AISCS, Table B5.1, or Table 4.2 in this chapter):

$$\lambda_{wr} = \frac{253}{\sqrt{\dfrac{ksi}{F_y}}} = 35.78$$

The actual slenderness exceeds the limiting value, and therefore we must go to Appendix B5 of AISCS to determine an effective width and an effective area [Eq. (A-B5-12) applies]. To use this equation we must know the stress on the column cross section. As a first approximation we will use the stress computed with the assumption that the web is not slender (i.e., using $Q = Q_s = 0.942$ calculated for the flanges previously). This is unconservative and the correct value of the critical stress must be determined by iteration.

$$Q = Q_s \qquad \lambda_{cQ} = \sqrt{Q}\lambda_c = 0.631 \qquad < 1.5$$
$$F_{cr} = QF_y \cdot 0.658^{\lambda_{cQ}^2} = 39.925 \text{ ksi}$$

Effective width:

$$b_e = \frac{326t_w}{\sqrt{F_{cr}}}\left(1 - \frac{57.2}{\lambda_w\sqrt{F_{cr}}}\right) = 10.466 \text{ in.} \qquad < h = 12 \text{ in.}$$

Effective area:

$$A_e = A - (h - b_e)t_w = 10.116 \text{ in}^2$$

Reduction factor:

$$Q_a = \frac{A_e}{A} = 0.963$$

The total Q factor is

$$Q = Q_sQ_a = 0.909 \qquad \lambda_{cQ} = \sqrt{Q}\lambda_c = 0.62 \qquad < 1.5$$
$$F_{cr} = QF_y \cdot 0.658^{\lambda_{cQ}^2} = 38.702 \text{ ksi}$$

The new critical stress is less than the previous value, and thus the next iteration will give a Q value that is somewhat larger than 0.909, so a somewhat larger critical stress. The difference between the two critical stresses computed so far is about 3%. One more iteration will yield only a very small change, so the iteration will be stopped here. Thus

$$F_{cr} = 38.702 \text{ ksi}$$

The design capacity can now be computed as follows:

$$\phi_c = 0.85$$
$$P_{design} = \phi_c A F_{cr} = 345.415 \text{ kips}$$

The factored design capacity = 345.4 kips.

Example 4.9
Calculate the design capacity of an axially loaded pinned end column of 2.3 m length consisting of an unequal leg angle. Use SI units and properties from AISC *Metric Properties of Structural Shapes* (Ref. 4.2).

Given:
      Angle size: L203 × 152 × 19
      Material properties:

$$F_y = 345 \text{ MPa} \qquad E = 200,000 \text{ MPa} \qquad G = 0.385E$$

Cross-sectional properties, from AISCM:

$$A = 6410 \text{ mm}^2 \qquad t = 19 \text{ mm} \qquad J = 7.91 \times 10^5 \text{ mm}^4$$

$$I_x = 26.2 \times 10^6 \text{ mm}^4 \qquad x = 39.6 \text{ mm} \qquad \alpha = \tan^{-1} 0.551$$

$$I_y = 12.7 \times 10^6 \text{ mm}^4 \qquad y = 65.1 \text{ mm} \qquad Q = 1.0$$

$$r_z = 32.8 \text{ mm} \qquad r_o = 101 \text{ mm} \qquad L = 2.3 \text{ m}$$

Solution   Cross-sectional properties about the principal axes z–w:

$$I_z = r_z^2 A$$

$$I_w = I_x + I_y - I_z$$

$$r_w = \sqrt{\frac{I_w}{A}}$$

$$I_z = 6.896 \times 10^6 \text{ mm}^4$$

$$I_w = 3.2 \times 10^7 \text{ mm}^4$$

$$r_w = 70.66 \text{ mm}$$

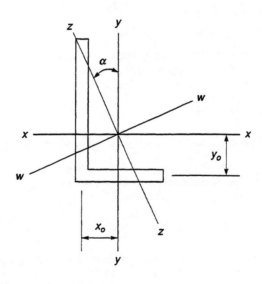

Location of the shear center in the x–y coordinate system (the shear center is at the intersection of the centerlines of the legs of the angle):

$$x_0 = x - \frac{t}{2} \qquad y_0 = y - \frac{t}{2}$$

Location of the shear center in the w–z coordinate system:

$$z_0 = y_0 \cos\alpha - x_0 \sin\alpha = 34.171 \text{ mm}$$

$$w_0 = y_0 \sin\alpha - x_0 \cos\alpha = 53.195 \text{ mm}$$

Calculation of the design capacity according to Appendix E3 of AISCM:

$$F_{ez} = \frac{\pi^2 E}{(L/r_z)^2} = 401.441 \text{ MPa} \qquad\qquad \text{buckling about } z \text{ axis}$$

$$F_{ew} = \frac{\pi^2 E}{(L/r_w)^2} = 1.863 \times 10^3 \text{ MPa} \qquad \text{buckling about } w \text{ axis}$$

$$F_{et} = \frac{GJ}{Ar_0^2} = 931.465 \text{ MPa} \qquad\qquad \text{torsional buckling,}$$
$$\text{assuming that } C_w = 0$$

The elastic flexural-torsional buckling stress is the lowest root of the cubic equation $F_{un} = 0$, and it can be determined by trial and error to be equal to

$$F_e = 371.8 \text{ MPa}$$

$$F_{un} = (F_e - F_{ez})(F_e - F_{ew})(F_e - F_{et}) - F_e^2(F_e - F_{ew})\frac{z_0^2}{r_0^2} - \left[F_e^2(F_e - F_{ez})\frac{w_0^2}{r_0^2} \right]$$

Equivalent slenderness parameter:

$$\lambda_e = \sqrt{\frac{F_y}{F_e}} = 0.963 \qquad < 1.5 \qquad F_{cr} = F_y \cdot 0.658^{\lambda_e^2} = 233.963 \text{ MPa}$$

$$\phi_c = 0.85 \ P_{\text{design}} = \phi_c A F_{cr} = 1.275 \times 10^3 \text{ kN}$$

The solution given above is a fairly lengthy and complex one, and it is best accomplished by using a spreadsheet. A much less complex but also somewhat unconservative method is to calculate the critical stress for the least radius of gyration, which is about the z axis.

$$\lambda_c = \frac{L}{r_z \pi} \sqrt{\frac{F_y}{E}} = 0.927 \qquad < 1.5$$

$$F_{cr} = F_y \cdot 0.658^{\lambda_c^2} = 240.77 \text{ MPa} \qquad P_{\text{design}} = \phi_c A F_{cr} = 1.312 \times 10^3 \text{ kN}$$

The two methods give answers which are within 2% of each other.

Further study of problems relating to the design of columns in frames and the interaction between framed columns is provided in Section 8.4.

REFERENCES

4.1. T.V. GALAMBOS, ed. *Guide to Stability Design Criteria for Metal Structures*, 4th ed., John Wiley, New York, (1988).

4.2. AISC, *Metric Properties of Structural Shapes with Dimensions according to ASTM A6M*, American Institute of Steel Construction, Chicago, IL, 1996.

PROBLEMS

4.1. Determine the diameter of a solid round A36 steel bar, 9 ft in length, that will support a factored axial force of 250 kips. Choose an available size. Lateral support is assumed at each end. The ends are assumed to be hinged; hence the effective column length is the same as the actual length. Redesign, using A441 steel. See AISCM for F_y corresponding to available size. Compare the weights.

4.2. Select standard A36 steel shapes having an area as near as possible to that of the requirement determined for Problem 4.1 and compare the column load capacities for the following:

(a) A W8 section.

(b) A single angle with equal legs.

(c) A double angle with long legs spaced $\frac{3}{8}$ in. apart.

(d) A structural tee.

(e) A round pipe.*

(f) A square structural tube.*

4.3. Assume for the purposes of this problem that the weight of a 20-story building, excluding exterior walls, is 80 lb/ft², and its contents (vertical live load) equal 80 lb/ft² of floor area per floor. The exterior walls are assumed to weigh 15 lb/ft². For a floor plan 72 ft by 144 ft, encompassing 40 columns, make a preliminary first-floor trial-column-size selection for an effective column length of 14 ft. There are four rows spaced 24 ft apart, of 10 columns each. Columns are assumed to be axially loaded and the choice can be made with the aid of any available column load tables—but the actual capacity of the column for axial load should be checked by the AISCS formula. Assume each column to be loaded by floor areas tributary to it.

4.4. Rework Problem 4.3 on the assumption that the columns will be braced laterally in their weak direction.

4.5. A WT7 × 41 is welded to a PL $\frac{3}{4}$ × 14 as shown. The overall length is 22 ft and the column is fixed at the base, hinged at the top. Use recommended effective length factor from Table 4.1. Determine the design column load $\phi_c P_n$ for (a) A36 steel, and (b) a steel with $F_y = 50$ ksi.

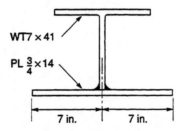

WT7 × 41

PL $\frac{3}{4}$ × 14

|← 7 in. →|← 7 in. →|

4.6. A hinged end column 28 ft in length is supported in its weak direction only, at the midpoint, in which case it may be assumed to behave as if the effective length for buckling about the y–y axis is 14 ft. Select a section to carry a factored load of 240 kips, using A36 steel. Provide as nearly equal slenderness ratios as possible about the two axes.

4.7. (See the illustration at top of next page.) Considering the column cross section, length, local details, and bracing, as shown, what is the permissible load P utilizing A36 steel? See Sec. E4 of AISCS for design assumptions and select an adequate size for the lacing bars, which may be assumed to have adequate welded end connections and which need to be selected on the basis of adequate compressive strength.

4.8. (See the illustration at bottom of next page.) Design a column 46 ft in length to carry an ultimate load of 1000 kips without any intermediate lateral support. Use a cross section consisting of four corner angles with double lacing similar to that used in Problem 4.7 on all four sides. (See the sketch of the cross section.) Design all details, including size of lacing, except for welded connections between lacing and angles. Follow requirements of Sec. E4 of AISCS and use a steel with $F_y = 50$ ksi. Take $a = 32.0$ in.

---

* Use the largest available outside dimension for the given area.

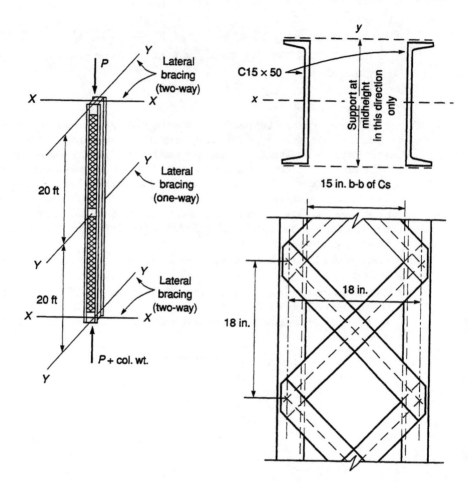

4.9. Rework Example 4.6, changing steel from A36 to $F_y = 50$ ksi.
4.10. Rework Example 4.7, changing steel from A36 to $F_y = 50$ ksi.

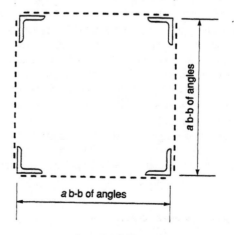

5

Members under Combined Forces

5.1 Introduction

In Chapter 3 we treated steel beam design and in Chapter 4 the design of axially loaded columns. In actual structures, most columns, in addition to axial load, must support lateral loads and/or transmit moments between their ends, and are thus subjected to combined stress due both to axial load and moment. Such members are termed *beam-columns*. The end moments may be caused by continuous frame action and/or by the effective eccentricity of the longitudinal loads. For example, columns in a tall building frame, in addition to the live and dead loads of the structure above any given level, often must transmit bending moments that result from wind load or lateral inertia forces due to earthquake. They must also resist the end moments introduced through the continuous frame action of the loaded adjacent connecting beams. In building frames the moments induced by the sway deflection of the structure must also be included.

When a column is part of a frame, the ideal solution would be based on the interaction of the complete structure. There is a trend toward such a design procedure, but at this time the traditional method of isolating the individual member as a basis for design prevails.

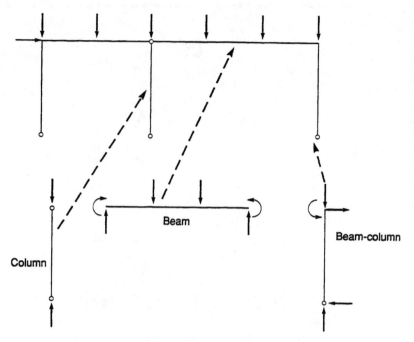

Figure 5.1 Column, beam, and beam-column members in a frame.

This chapter deals with the design of such isolated beam-columns. The design of beam-columns as part of a frame is discussed in detail in Chapter 8. The way in which various components of a frame are related to the individual member types of beam, column, and beam-column·is illustrated schematically in Fig. 5.1. In this illustration it is assumed that the forces and deformations all act in the plane of the frame. Very often, bending is about both principal axes of the cross section. This is shown in Fig. 5.2, where moments M_x and M_y cause biaxial bending. The design may then proceed along one of the following three paths:

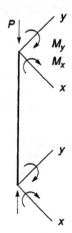

Figure 5.2 Biaxially loaded beam-column.

1. The load at which the maximum stress reaches the yield stress is determined and that load is divided by a factor of safety to give an allowable load. Moment due to deflection is included. The elastic limit method is rarely used anymore since it is generally very conservative. However, it is useful when a complex cross section, such as an unequal-leg angle, is subjected to biaxial end moments.

2. The AISCS interaction formula may be used. This provides for an empirical transition in the member selection, in accordance with AISCS requirements, from the beam (as the column load approaches zero) to the axially loaded column (as the bending moment approaches zero).

3. A computer-based numerical method may be used to obtain a more accurate evaluation of the ultimate strength of the beam-column. An eccentrically loaded member of the type shown in Fig. 5.3(a) will deflect typically as illustrated by the schematic load–deflection curve shown at the right of this figure. If the beam-column were to respond elastically, and if the added moment due to the force P times the deflection Δ were ignored, the response would be linear, as indicated by the thin straight line. However, the added moment $P\Delta$ causes more deflection, and this in turn causes more moment, so that the load–deflection curve becomes nonlinear (i.e., for each increment of P a larger Δ results). Eventually, the outer fibers of the member begin to yield, adding yet another component to the nonlinear behavior. Finally, the internal resistance can no longer keep up with the externally applied force P and the load–deflection curve reaches a peak. This is the maximum force that can be determined by numerical means. It is also conservatively approximated by the AISC interaction equation.

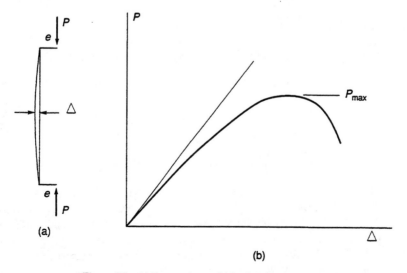

Figure 5.3   (a) Beam-column; (b) load–deflection curve.

Following we concentrate on the use of the interaction equation approach for the design of beam-columns.

5.2 Design by Use of Interaction Formulas

An interaction equation is the simplest way to design a beam-column. Such a member is subject to an axial force P_u and moments M_{ux} and M_{uy}. When only one of these effects is present, the ultimate capacity is the corresponding individual nominal capacity P_n, M_{nx}, or M_{ny}, respectively. The following linear interaction relationship has been used for many years by AISCS and many other specifications for the design of beam-columns still use it:

$$\frac{P_u}{P_n} + \frac{M_{ux}}{M_{nx}} + \frac{M_{uy}}{M_{ny}} \leq 1.0 \tag{5.1}$$

When the predictions of this equation are compared to test results and to numerically determined ultimate strengths, it is evident that they are quite conservative, especially for wide-flange shapes bent about the minor axis. The interaction curves shown in Fig. 5.4 illustrate this fact for compact short columns. The equations that are plotted are given below:

H column, bending about y axis:

$$\frac{M_{pcy}}{M_{py}} = 1.19 \left[1 - \left(\frac{P}{P_y} \right)^2 \right] \leq 1.0 \tag{5.2}$$

H column, bending about x axis:

$$\frac{M_{pcx}}{M_{px}} = 1.18 \left[1 - \left(\frac{P}{P_y} \right) \right] \leq 1.0 \tag{5.3}$$

Rectangular column:

$$\frac{M_{pc}}{M_p} = 1 - \left(\frac{P}{P_y} \right)^2 \tag{5.4}$$

AISCS lower-bound curve:

$$\frac{P}{P_y} + \frac{8}{9} \left(\frac{M_{pcx}}{M_{px}} \right) \leq 1.0 \qquad \text{for } \frac{P}{P_y} \geq 0.2 \tag{5.5a}$$

$$\frac{P}{2P_y} + \left(\frac{M_{pcx}}{M_{px}} \right) \leq 1.0 \qquad \text{for } \frac{P}{P_y} < 0.2 \tag{5.5b}$$

The AISCS curve is seen to be a lower bound and thus it is conservative for most cross sections. The curves in Fig. 5.4 and Eqs. (5.2) to (5.4) are applicable only for short beam-columns of compact shape, which can support the axial yield load P_y when only P acts, and the plastic moment M_p if only the moment M_{pc} is present. Equations (5.2) and (5.3) for H-shaped cross sections are themselves approximations of more complex equations.

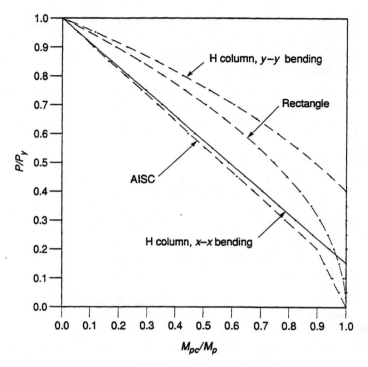

Figure 5.4   Interaction curves for different shapes.

The AISCS interaction equations can be generalized to be very versatile tools for long beam-columns and for biaxial bending. These equations are presented in flowchart form in Fig. 5.5. They are contained in Sec. H of AISCS, Members Under Combined Forces and Torsion, as Eqs. (H1-1). The various terms in these equations will now be explained.

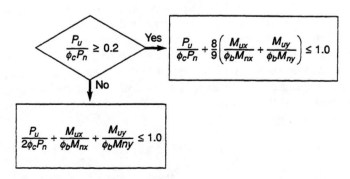

Figure 5.5   AISCM interaction equations.

P_u: the required force determined by structural analysis for the factored loads.

$\phi_c = 0.85$, the resistance factor for columns if the axial force is in compression. If the axial force is in tension, this factor is replaced by the appropriate value for tension (i.e., $\phi = 0.9$).

P_n: the nominal axial capacity, including the influence of the slenderness ratio and local buckling. This is the load the member can sustain if only axial load were to act. It is determined according to Chapter 4 of this book or according to Sec. E of AISCS.

$\phi_b = 0.9$, the resistance factor in bending.

M_{nx}, M_{ny}: the nominal bending capacity, including the effects of local buckling and lateral-torsional buckling if appropriate. These are the moments the member can sustain if only bending moment were present. The values are determined by the methods given in Chapter 3 of this book and Sec. F of AISCS.

M_{ux}, M_{uy}: the required moments about the x and y axes, as determined by structural analysis, including the second-order effects due to the additional moments caused by the axial force times the deflection.

There are two kinds of second-order moments: (1) the amplification of the moment between the ends of the beam-column due to the axial force times the deflection of the member (as illustrated in Fig. 5.3.), and (2) the amplification of the end moments due to the story deflection times the story gravity loads. The former is a property of the member, while the latter is a characteristic of the frame. Each can be calculated rather precisely by making a second-order analysis of the member or of the frame, respectively. Tools in the form of computer programs are beginning to be available to permit a practical second-order analysis of this kind. Such analyses are too advanced for a beginning course in steel design, and many designers will also prefer the simpler but more approximate procedure permitted by Sec. C of AISCS. This will now be presented. Only one plane is discussed, but the procedure for the other plane is analogous.

$$M_{ux} = B_1 M_{nt} + B_2 M_{lt} \qquad (5.6)$$

The first term is the second-order moment for the member, while the second term relates to frame moment magnification. This is discussed in Chapter 8. The moment M_{nt} is the moment that results from an analysis which assumes that the top of the column cannot deflect with respect to its bottom (i.e., there is no story sway deformation). The parameter B_1 is the magnification factor that accounts for the increase in the moment due to the axial force times the maximum member deflection between the column ends. When the moment is maximum at either end (i.e., there is no moment inside the span which is larger than the given end moments), $B_1 = 1$. Otherwise,

$$B_1 = \frac{C_m}{1 - \dfrac{P_u}{P_{e1}}} \geq 1.0 \qquad (5.7)$$

where

$$C_m = 0.6 - 0.4\left(\frac{M_1}{M_2}\right) \tag{5.8}$$

$$P_{e1} = \frac{AF_y}{\lambda_c^2} \tag{5.9}$$

$$\lambda_c = \frac{L}{r_x \pi}\sqrt{\frac{F_y}{E}} \tag{5.10}$$

In Eq. (5.8), M_1 is the numerically smaller end moment and M_2 is the numerically larger one. The ratio is positive if the moments cause reverse curvature, and it is negative when they cause single curvature. Thus C_m can take on values between 1.0 (equal end moments causing single-curvature bending) and 0.2 (equal end moments causing double-curvature bending, i.e., an S-shape). In Eq. (5.9), A is the cross-sectional area of the member. In Eq. (5.10), L is the actual length of the beam-column. AISCS permits an effective length determined for the nonsway case (i.e., $K < 1.0$) to be used. For most practical cases it is sufficient to use the conservative value $K = 1.0$. The examples at the end of this chapter illustrate the various applications of the interaction equation approach to the design of beam-columns.

5.3 Equivalent Axial Compression Load

The preliminary selection of a beam-column, proportioned by the AISCS interaction formulas, may be expedited by conversion to an equivalent axial load, thereby making use of the tabulated loads provided in AISCM. These tables are derived from the interaction equations in the following manner, as illustrated for the first equation:

$$\frac{P_u}{\phi_c P_n} + \frac{8}{9}\left(\frac{M_{ux}}{\phi_b M_{nx}} + \frac{M_{uy}}{\phi_b M_{ny}}\right) = 1 \tag{5.11}$$

Rearranging this equation gives

$$P_u + \frac{8}{9}\frac{\phi_c P_n}{\phi_b}\left(\frac{M_{ux}}{M_{nx}} + \frac{M_{uy}}{M_{ny}}\right) = \phi_c P_n = P_{\text{eff}} \tag{5.12}$$

This equation can be also written as

$$P_{\text{eff}} = P_u + mM_{ux} + mUM_{uy} \tag{5.13}$$

By making a number of simplifying assumptions AISCS derived and tabulated the coefficients m and U so that for preliminary design, one can calculate P_{eff} and then look up the corresponding column size, which has the same axial capacity. The member must, of course, be checked by the interaction formulas. This procedure is also illustrated in the examples.

5.4 Illustrative Examples

Example 5.1

A 30-ft-long column is subjected to an axial compression load, $P = 400$ kips, and lateral uniform load, $w = 0.40$ kip/ft, that causes bending about the column weak axis as shown. Select an economical W shape to satisfy the AISCS. Use A36 steel.

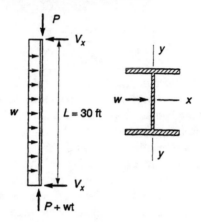

Given initial data:

Required axial capacity: $P_u = 400$ kips
Effective length: $L = 30$ ft
Required distributed load capacity: $w_u = 0.4$ kip/ft
Yield stress and modulus of elasticity: $F_y = 50$ ksi $E = 29{,}000$ ksi

$$M_{max} = \frac{w_u L^2}{8} = 45 \text{ kip-ft} = 540 \text{ kip-in.}$$

Solution

(a) *Preliminary design:* Bending is about y axis, hence effective axial force: $P_{ueff} = P_u + M_{uy} mU$, where $M_{uy} = M_{max}$ from Part 3, Table B of AISCM:

$$m = 1.8 \qquad U = 2 \text{ ft}^{-1}$$
$$P_{ueff} = P_u + M_{max} mU = 562 \text{ kips}$$

Try W14 × 90, factored capacity = 564 kips. Second iteration:

$$U = 1.51\frac{1}{\text{ft}} \qquad m = \frac{1.8}{0.85} \qquad P_{ueff} = P_u + M_{max} mU = 543.894 \text{ kips}$$

Not much change, so investigate the trial section W14 × 90. Cross-sectional properties:

$$A = 26.5 \text{ in}^2 \qquad b_f = 14.52 \text{ in.} \qquad Z_y = 75.6 \text{ in}^3$$
$$d = 14.02 \text{ in.} \qquad t_f = 0.71 \text{ in.} \qquad S_y = 49.9 \text{ in}^3$$
$$t_w = 0.44 \text{ in.} \qquad r_y = 3.7 \text{ in.}$$

(b) *Detailed analysis*—Axial capacity: check flange slenderness (AISCS, Table 5.1):

$$\lambda = \frac{b_f}{2t_f} = 10.225 \qquad \lambda_r = \frac{95}{\sqrt{F_y}} = 13.435$$

Since the actual value is smaller than the limiting value, there is no reduction for flange local buckling (i.e., $Q = 1$).

Axial capacity (AISCS, Sec. E2): Buckling will be about the y axis (minor axis).

$$\lambda_{cy} = \frac{L}{r_y \pi} \sqrt{\frac{F_y}{E}} = 1.286 \qquad < 1.5 \qquad F_{cr} = F_y \cdot 0.658^{\lambda_{cy}^2} = 25.024 \text{ ksi}$$

$$\phi_c = 0.85 \qquad P_n = F_{cr}A = 663.138 \text{ kips}$$

$$\frac{P_u}{\phi_c P_n} = 0.71 \qquad > 0.2 \quad \text{use Eq. (H1-1a) in AISCS}$$

Bending capacity (AISCS, Sec. F2):

$$M_{n1} = F_y Z_y = 3780 \text{ kip-in.} \qquad M_{n2} = 1.5 F_y S_y = 3742.5 \text{ kip-in.}$$

Since $M_{n2} < M_{n1}$,

$$M_{ny} < M_{n2} = 3742.5 \text{ kip-in.}$$

Amplified required moment (AISCS, Sec. C1):

$$M_{nt} = M_{max} \qquad M_{1t} = 0 \qquad C_m = 1 \qquad \text{(Table C-C1.1 in Commentary)}$$

$$P_{e1} = \frac{AF_y}{\lambda_{cy}^2} \qquad B_1 = \frac{C_m}{1 - P_u/P_{e1}} = 1.997 \qquad M_{uy} = B_1 M_{nt} = 1078.382 \text{ kip-in.}$$

Interaction equation:

$$\phi_b = 0.9 \qquad \frac{P_u}{\phi_c P_n} + \frac{8}{9} \frac{M_{uy}}{\phi_b M_{ny}} = 0.994 \qquad < 1.000 \quad \text{OK}$$

Use W14 × 90.

Example 5.2

Rework Example 5.1, but now the lateral uniform load causes bending about the column strong axis.

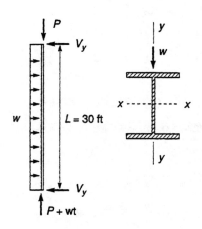

Given initial data:

Required axial capacity:	$P_u = 400$ kips
Effective length:	$L = 30$ ft
Required distributed load capacity:	$w_u = 0.4$ kip/ft
Yield stress and modulus of elasticity:	$F_y = 50$ ksi $E = 29{,}000$ ksi

$$M_{max} = \frac{w_u L^2}{8} = 45 \text{ kip-ft.} = 540 \text{ kip-in.}$$

Solution   Try W14 × 90. Cross-sectional properties:

$A = 26.5$ in^2	$b_f = 14.52$ in.	$X_1 = 2900$ ksi	$S_x = 143$ in^3
$d = 14.02$ in.	$t_f = 0.71$ in.	$X_2 = 1750 \times 10^{-6}$ ksi^{-2}	$Z_x = 157$ in^3
$t_w = 0.44$ in.	$r_y = 3.7$ in.		
	$r_x = 6.14$ in.		

Detailed analysis: Check flange slenderness (AISCS, Table 5.1):

$$\lambda = \frac{b_f}{2t_f} = 10.225 \qquad \lambda_r = \frac{95}{\sqrt{F_y}} = 13.435$$

Since the actual value is smaller than the limiting value, there is no reduction for flange local buckling (i.e., $Q = 1$).

Axial capacity (AISCS, Sec. E2): Buckling will be about the y axis (minor axis).

$$\lambda_{cy} = \frac{L}{r_y \pi} \sqrt{\frac{F_y}{E}} = 1.286 \qquad < 1.5$$

$$F_{cr} = F_y \cdot 0.658^{\lambda_{cy}^2} = 25.024 \text{ ksi}$$

$$\phi_c = 0.85 \qquad P_n = F_{cr}A = 663.138 \text{ kips}$$

$$\frac{P_u}{\phi_c P_n} = 0.71 \qquad > 0.2 \quad \text{use Eq. (H1-1a) in AISCS}$$

Bending capacity (AISCS, Sec. F2):

$$M_p = Z_x F_y = 7850 \text{ kip-in.} \qquad F_L = F_y - 10 \text{ ksi}$$

$$M_r = S_x F_L = 5720 \text{ kip-in.} \qquad L_p = \frac{300 r_y}{\sqrt{F_y}} = 156.978 \text{ in.} \qquad L_b = 360 \text{ in.} \qquad > L_p$$

$$L_r = \frac{r_y X_1}{F_L} \cdot \sqrt{1 + \sqrt{1 + X_2 F_L^2}} = 460.684 \text{ in.} \qquad > L_b$$

Moments at $\frac{1}{4}$ and $\frac{3}{4}$ points on beam: $M_A = 405$ kip-in. $M_C = M_A$

Moment at center of beam: $M_B = M_{max}$

$$C_b = \frac{12.5 M_{max}}{2.5 M_{max} + 3M_A + 4M_B + 3M_C} = 1.136$$

$$M_{nx} = C_b \left[M_p - (M_p - M_r) \frac{L_b - L_p}{L_r - L_p} \right] = 7302.426 \text{ kip-in.} \qquad < M_p$$

Check flange local buckling (AISCS, Table B5.1):

$$\lambda = \frac{b_f}{2t_f} = 10.225 \qquad \lambda_p = \frac{65}{\sqrt{F_y}} = 9.192 \ < 10.225 \qquad \lambda_r = \frac{141}{\sqrt{F_L}} = 22.294 \ > 10.225$$

$$M_n = M_p - (M_p - M_r)\frac{\lambda - \lambda_p}{\lambda_r - \lambda_p} = 7682.066 \text{ kip-in.} \qquad > M_{nx} = 7303 \text{ kip-in.}$$

Check web local buckling (AISCS, Table B5.1):

$$\lambda = 25.9 = \frac{h}{t_w} \qquad P_y = AF_y = 1325 \text{ kips} \qquad \phi_b = 0.9 \qquad \frac{P_u}{\phi_b P_y} = 0.335 \qquad > 0.125$$

$$\lambda_p = \frac{191}{\sqrt{F_y}}\left(2.33 - \frac{P_u}{\phi_b P_y}\right) = 53.876 \qquad > 25.9 \quad \text{compact web}$$

Amplified required moment (AISCS, Sec. C1):

$$M_{nt} = M_{max} \qquad M_{lt} = 0$$

$$C_m = 1 \quad \text{(Table C-C1.1 in Commentary)} \qquad \lambda_{cx} = \frac{L}{\pi r_x}\sqrt{\frac{F_y}{E}} = 0.775$$

$$P_{e1} = \frac{AF_y}{\lambda_{cx}^2} \qquad B_1 = \frac{C_m}{1 - P_u/P_{e1}} = 1.221 \qquad M_{ux} = B_1 M_{nt} = 659.578 \text{ kip-in.}$$

Interaction equation:

$$\frac{P_u}{\phi_c P_n} + \frac{8}{9}\frac{M_{ux}}{\phi_b M_{nx}} = 0.799 \qquad < 1.000 \quad \text{OK}$$

Use W14 × 90. The reader is asked to check if a W14 × 82 will work.

Example 5.3

A 30-ft-long wide-flange section is used as the tension chord in a truss. The beam-column is subjected to a tensile ultimate force of $P_u = 300$ kips. A bending ultimate force of $Q_u = 20$ kips is also applied at the center of the member, causing bending about the x axis of the cross section. Lateral bracing exists at the ends and at the center. The material is A36 steel. Design the member.

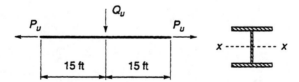

Given: Unbraced length in x direction: $L_x = 30$ ft
Unbraced length in the y direction: $L_y = 15$ ft
Yield stress: $F_y = 36$ ksi
Tensile ultimate force: $P_u = 300$ kips
Bending ultimate force: $Q_u = 20$ kips

Solution   Maximum bending moment:

$$M_{ux} = \frac{Q_u L_x}{4} = 150 \text{ kip-ft} \qquad C_b = 1.75$$

Starting point:

(a) Section needed if $M = 0$:
$$\phi = 0.9$$
$$A_{min} = \frac{P_u}{\phi F_y} = 9.259 \text{ in}^2$$

(b) Section needed if $P = 0$, assuming a compact shape:
$$Z_{min} = \frac{M_{ux}}{\phi F_y} = 55.556 \text{ in}^3$$

(c) Maximum slenderness ratio, 300:
$$r_{min} = \frac{L_y}{300} = 0.6 \text{ in.}$$

(d) Maximum length/depth ratio, 24:
$$d_{min} = \frac{L_x}{24} = 15 \text{ in.}$$

Try W18 × 55. Bending capacity: From AISCM,

$$\phi M_p = 302 \text{ kip-ft.} \qquad \phi M_r = 192 \text{ kip-ft.}$$
$$L_r = 21.4 \text{ ft} \qquad L_p = 7.0 \text{ ft}$$

Since $L_b = L_y = 15$ ft is between L_r and L_p,

$$\phi M_n = C_b \left[\phi M_p - (\phi M_p - \phi M_r) \frac{L_y - L_p}{L_r - L_p} \right] = 421.556 \text{ kip-ft}$$

This exceeds the plastic moment; thus

$$\phi M_n = \phi M_p = 302 \text{ kip-ft}$$

Axial capacity:

$$A = 16.2 \text{ in}^2 \qquad \phi P_n = \phi A F_y = 524.88 \text{ kips}$$
$$\frac{P_u}{\phi P_n} = 0.572 \qquad > 0.2$$

Interaction equation (AISCS, Eq. H1-1a):

$$\frac{P_u}{\phi P_n} + \frac{8}{9} \frac{M_{ux}}{\phi M_n} = 1.013 \qquad > 1.0 \quad \text{NG}$$

Another trial with a W24 × 55 section, which weighs the same as the W18 × 55, shows that the interaction sum equals 0.94. Use W24 × 55.

Example 5.4

Select a beam-column in a building frame for a 15-ft story height to support the ultimate forces shown below. The member is braced against sidesway buckling at its ends in both the x and y planes. There is no lateral brace between the ends of the member. Bending is about the x axis.

Given: $L = 15$ ft
$P_u = 586$ kips
$M_{u2} = 301$ kip-ft
$M_{u1} = 236$ kip-ft
$F_y = 50$ ksi
$E = 29,000$ ksi

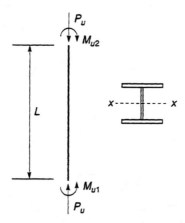

Solution   Preliminary design: Use the equivalent axial force method. From AISCM, Part 3, "Columns," Table B, $m = 2.6$ ft^{-1} for $L = 15$ ft.

$$M_x = M_{u2} \qquad P_{eff} = P_u + mM_x = 1368.6 \text{ kips}$$

From the AISCM Column Selection Table, try a W14 column. As a second iteration, find from Table B,

$$m = 2.0 \text{ ft}^{-1} \qquad P_{eff} = P_u + mM_x = 1188 \text{ kips}$$

From the Column Selection Table, try W14 × 109.

$$\phi_c P_n = 1150 \text{ kips} \qquad \text{based on } y\text{-axis buckling}$$

$$\frac{P_u}{\phi_c P_n} = 0.51 \qquad \text{use Eq. (H1-1a)}$$

Cross-sectional and member properties for W14 × 109 from AISCM:

$$A = 32 \text{ in}^2 \qquad r_x = 6.22 \text{ in.} \qquad L_p = 13.2 \text{ ft} \qquad L_r = 43.2 \text{ ft}$$
$$\phi_b M_p = 720 \text{ kip-ft} \qquad \phi_b M_r = 519 \text{ kip-ft}$$

Determine M_{ux}:

$$\lambda_x = \frac{L}{r_x \pi} \sqrt{\frac{F_y}{E}} = 0.382$$

$$P_{ex} = \frac{AF_y}{\lambda_x^2} = 10{,}936.629 \text{ kips}$$

$$C_m = 0.6 - 0.4 \cdot \left(\frac{M_{u1}}{M_{u2}}\right) = 0.914$$

$$B_1 = \frac{C_m}{1 - P_u/P_{ex}} = 0.965 \qquad < 1.0, \text{ hence } B_1 = 1.0$$

$$M_{ux} = B_1 M_{u2} = 301 \text{ kip-ft}$$

Flexural capacity: If B_1 is less than or equal to 1.0, C_b may be larger than 1.0. If $B_1 > 1.0$, then $C_b = 1$.

$$C_b = 1.75 + 1.05\left(\frac{M_{u1}}{M_{u2}}\right) + 0.3\left(\frac{M_{u1}}{M_{u2}}\right)^2 = 1.111$$

$$\phi_b M_n = C_b\left[\phi_b M_p - (\phi_b M_p - \phi_b M_r)\frac{L-L_p}{L_r-L_p}\right] = 786.639 \text{ kip-ft} \quad > \phi_b M_p = 720 \text{ kip-ft}$$

Hence

$$\phi_b M_n = \phi_b M_p = 720 \text{ kip-ft.}$$

Interaction equation check:

$$\frac{P_u}{\phi_c P_n} + \frac{8}{9}\frac{M_{ux}}{\phi_b M_n} = 0.881 \qquad < 1.00 \quad \text{OK}$$

A W14 × 99 is also acceptable, giving an interaction equation sum of 0.98. The reader may verify this calculation. Use W14 × 99.

Example 5.5

Check the beam-column which is subjected to an axial load and bending moments about both the x and y axes. The given forces are required ultimate values. The member is braced against sidesway buckling in both the x and y directions. There is no lateral brace between the ends of the beam-column. Use SI units.

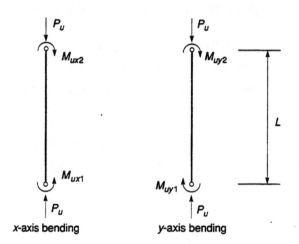

x-axis bending   y-axis bending

Given: Shape: W360 × 196 (W14 × 132)

F_y = 345 MPa (50 ksi)

E = 200,000 MPa

P_u = 2000 kN L = 5 m

M_{ux1} = −360 kN · m M_{uy1} = 50 kN · m

M_{ux2} = 500 kN · m M_{uy2} = 144 kN · m

A = 25,000 mm²   area

Z_y = 1860 × 10³ mm³   plastic section modulus

r_y = 95.7 mm r_x = 159 mm   radii of gyration

Solution Flexural resistance parameters used to calculate the nominal moment capacity (from AISCM, Load Factor Selection Table, converted from U.S. units):

$$\phi M_{px} = 1190 \text{ kN·'m} \qquad \phi M_{rx} = 850 \text{ kN·'m}$$
$$L_p = 4.05 \text{ m} \qquad L_r = 5.88 \text{ m}$$

Calculation of the moment gradient factor C_b, using the more convenient form of the expression from the 1986 AISCS, which gives the same answer as the formula in the 1993 AISCS but is more convenient if only end moments are applied to the member:

$$C_b = 1.75 + 1.05\frac{M_{ux1}}{M_{ux2}} + 0.3\left(\frac{M_{ux1}}{M_{ux2}}\right)^2 = 1.15$$

Because $L_p < L < L_r$, the factored moment capacity about the x axis is equal to

$$\phi M_{nx} = C_b\left[\phi M_{px} - (\phi M_{px} - \phi M_{rx})\frac{L-L_p}{L_r-L_p}\right] = 1165.035 \text{ kN·'m} \qquad < \phi M_p = 1190 \text{ kN·'m}$$

Flexural resistance about the y axis:

$$\phi M_{ny} = 0.9Z_y F_y = 577.53 \text{ kN·'m}$$

Axial capacity: L/r_y will govern.

$$\lambda_y = \frac{L}{\pi r_y}\sqrt{\frac{F_y}{E}} = 0.691$$

$$F_{cr} = F_y \cdot 0.658^{\lambda_y^2} = 282.55 \text{ MPa}$$
$$P_n = AF_{cr} = 7063.752 \text{ kN}$$

Calculation of B_1:

$$\lambda_x = \frac{L}{\pi r_x}\sqrt{\frac{F_y}{E}} = 0.416$$

$$P_{ex1} = \frac{AF_y}{\lambda_x^2} = 49,902.694 \text{ kN}$$

$$C_{mx} = 0.6 - 0.4\frac{M_{ux1}}{M_{ux2}} = 0.888$$

$$B_{1x} = \frac{C_{mx}}{1 - P_u/P_{ex1}} = 0.925 \qquad < 1.0, \quad \text{hence } B_{1x} = 1.0$$

$$P_{ey1} = \frac{AF_y}{\lambda_y^2} = 18,078.135 \text{ kN}$$

$$C_{my} = 0.6 - 0.4 \cdot \frac{M_{uy1}}{M_{uy2}} = 0.461$$

$$B_{1y} = \frac{C_{my}}{1 - P_u/P_{ey1}} = 0.518 \qquad < 1.0, \quad \text{hence } B_{1y} = 1.0$$

$$B_{1x} = 1.0 \qquad B_{1y} = 1.0$$
$$M_{ux} = B_{1x}M_{ux2} = 500 \text{ kN·'m} \qquad M_{uy} = B_{1y}M_{uy2} = 144 \text{ kN·'m}$$

$$\phi_c = 0.85 \qquad \frac{P_u}{\phi_c P_n} = 0.333 \qquad > 0.2$$

Applicable interaction equation:

$$\frac{P_u}{\phi_c P_n} + \frac{8}{9}\left(\frac{M_{ux}}{\phi M_{nx}} + \frac{M_{uy}}{\phi M_{ny}}\right) = 0.936 \qquad < 1.0$$

W360 × 196 is OK.

PROBLEMS

5.1. A square box column, 38 ft in length, with sidewalls 24 in. apart (dimensioned as shown to the middle planes of the plates), carries an ultimate axial load of 1400 kips and a lateral load of 900 lb/ft, acting normal to one of the flat sides. Determine the plate thickness to the nearest $\frac{1}{16}$ in. if A36 steel is used.

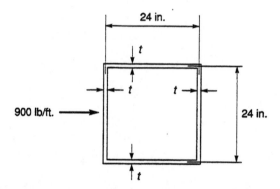

5.2. Redesign for the conditions of Problem 3.14 if the beam, in addition to the specified vertical and horizontal beam loads, carries a compressive force of 500 kips applied through the centroid at each end.

5.3. The beam in the figure is restrained laterally in the weak direction by ties as shown. For bending in the weak (horizontal) directions, it may be assumed as hinged at the center. Select an adequate member of A36 steel for this biaxial bending situation.

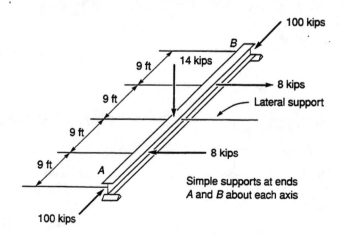

5.4. Select a column for a 24-ft story height to support the following forces:

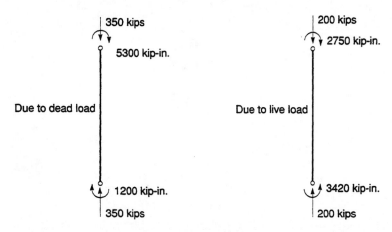

Bending is about the x axis only. The column is prevented from sidesway in both the x and the y directions. Lateral support against movement in the y direction, and against twist, is provided at the ends and at the center. Use A36 steel:

$$F_y = 36 \text{ ksi} \qquad E = 29,000 \text{ ksi}$$

6

Connections

6.1 Introduction

The component rolled plates and shapes of a steel structure are held together by means of fasteners (rivets or bolts) or by welds, which may either fuse the parts together into an integral unit or stitch them together intermittently as do fasteners. The fasteners and welds are used in the shop fabrication processes to build up members and again in the field erection to connect the separate members together to form the completed structural frame. If a member is too large to be shipped as a single unit, field connections must also be used to splice the member segments. Because of the relatively greater cost of making field connections, such splices are kept to a minimum.

In truss construction, the tension and compression members meeting at a joint may be attached separately by fasteners to a *gusset plate* (see Fig. 2.3), or, if welding is used, it may be possible to join them directly without the use of an auxiliary plate.

In building frame construction, Sec. A2.2 of AISCS recognizes two basic types of beam-to-column framing:

1. *Type FR (fully restrained):* also called *rigid frame* or *continuous connection*; assumes that the beam-to-column or beam-to-girder connection transmits the calculated moment and shear and has sufficient rigidity to

provide the full continuity that has been assumed in the structural force analysis. This means that the original angle between the members joined will be retained after the loads are applied.

2. *Type PR (partially restrained):* assumes that the connections have insufficient rigidity to maintain the original angles between the intersecting members. When the rotational restraint is ignored, the connection is called a *simple connection.* When the actual moment–rotation relationship is utilized by the designer, the connection is a *semirigid joint.* The simple connections are designed purposely to permit beam-end rotations relative to column or girder to a degree that will permit one to ignore the incidental bending moments and small inelastic yielding that may be developed. The semirigid connections possess a known moment–rotation curve intermediate in degree between the rigid and the simple framing. This relationship must be documented in the design calculations, and the resulting method of structural design makes use of more advanced methods of analysis than will be considered in this book. This added complication is justified, however, because such connections have many technical, economical, and constructional advantages over the FR joints.

The relationship between the rigid, semirigid, and simple framing is illustrated in Fig. 6.1, which shows schematically the variation of the angle discontinuity and the applied moment on the joint. The fact is that it is almost impossible to manufacture a FR or a simple joint. The fully rigid joint will always have a small change in the original angle, while the simple joint will always be able to support some moment. Practical joints have evolved which for all practical purposes do act as these idealized joints, and example designs are presented in later portions of this chapter. Semirigid connections are joints where the actual moment–rotation must be accounted for in design. The regions between the three types of connections are indicated in Fig. 6.1.

In addition to the material on connections in AISCS and the Commentary, Vol. II of AISCM provides extensive design information and tables that facilitate the proportioning of commonly used connections.

A connection is said to be loaded concentrically if the resultant (axial) force passes through the centroid of the fastener group or weld pattern. At low loads the distribution of stress transfer in such a connection is quite nonuniform, but prior to failure the local yielding of material tends to distribute the load uniformly to all parts of the connection. This fact, together with the results of experience and many laboratory tests to failure, permits the assumption in design that all parts of a concentrically loaded connection share equally in resisting the applied force. If the resultant force does not pass through the centroid of the connection, the force system may be reduced to a force and a couple at the connection centroid. Each element of the connection is assumed to resist the axial component of force uniformly and to resist the moment in proportion to the distance of that element from the centroid of the connection. Design rules based on the foregoing assumptions have been found to be safe and workable.

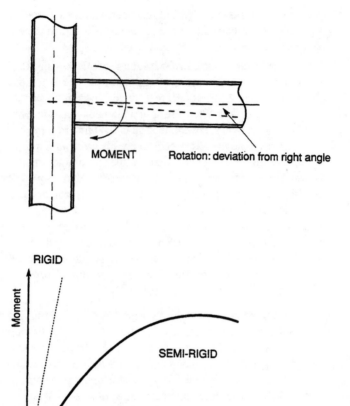

Figure 6.1   Moment–rotaion behavior of joint in a framed structure.

Design recommendations and nominal stresses in connections as supplied in the AISCS are very largely backed up and based on recommendations of the Research Council on Riveted and Bolted Joints and the American Welding Society, for bolted and welded connections, respectively.

Four methods of making structural connections (rivets, bolts, pins, and welds) will be discussed and simple design examples presented. The information available on this subject in AISCM also should be studied in detail.

The overall strength and safety of a structure may be directly dependent on the connections that join main members. Such connections should be shown explicitly and in detail on the design drawings, both in the interest of safe design and for economy in view of bid-price allowances that might otherwise be made for contingencies.

6.2 Riveted and Bolted Connections

For many years riveting was the accepted method for making connections. However, the use of rivets has declined rapidly due to the development and economic advantages of welding and high-strength bolts. The advent of both welding and high-strength bolting has made possible the advantageous use of a combination of these fastening methods, such as shop fabrication by welding followed by the use of high-strength bolts for the field connections. In this way the advantages of each procedure are realized, as the welding is under shop-controlled conditions with the members positioned to produce good welds and economy in fabrication. The advantages of rapid assembly while the members are being held in position in the field are obtained through the use of high-strength bolts.

Rivets and bolts, as shown in Fig. 6.2, transmit force from one plate element to another by either single shear or double shear. At low loads the transfer of force from

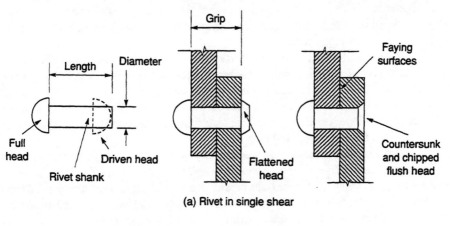

(a) Rivet in single shear

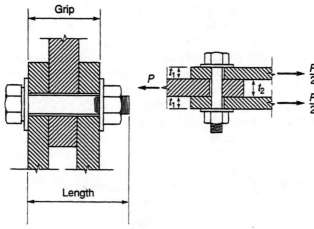

(b) Bolt in double shear

Figure 6.2   Rivets and bolts.

plate to plate is largely by friction. At higher loads, after slip takes place, the fasteners come into direct bearing. In more complex joints with multiple plates interacting, more than two shear planes may occur.

Rivets are manufactured with a special full (acorn) head and are installed in holes that are punched or drilled $\frac{1}{16}$ in. larger in diameter than the rivets. Hot driven rivets are brought to approximately 1800°F before being placed in the hole, and a second head is formed by means of a riveting hammer or a machine pressure type of riveter. During the process of forming this head, the shank is upset so that it should now completely fill the hole (see Fig. 6.2). The formed head can be either a full button head, a flattened head, or a countersunk and chipped head. The chipping is usually done with a chisel and air hammer after the rivet has cooled. Rivets with countersunk and chipped heads have less strength and cost considerably more than those with full or flattened heads. During cooling, the rivet shrinks, setting up tensile forces in the shank which may approach the yield point of the material. This residual tension force is unpredictable and may vary from practically nothing to a stress equal to the yield point. Because of this uncertainty, the clamping force exerted on the joined material is neglected in design calculations.

Rivets shall conform to the provisions of the *Specifications for Structural Rivets*, ASTM A502, grades 1 or 2. The size of rivets used in ordinary steel construction ranges from $\frac{5}{8}$ to $1\frac{1}{2}$ in. in diameter by $\frac{1}{8}$-in. increments. The strength of rivets is tabulated in AISCS Table J3.2. AISCM should now be studied with regard to detailing practice, erection clearances, recommended spacing, and conventional signs for use on drawings for both rivets and bolts.

Unfinished bolts, also known as ordinary or common bolts, shall conform to the *Specifications for Low Carbon Steel Externally and Internally Threaded Standard Fasteners*, ASTM A307. These bolts range from $\frac{5}{8}$ to $1\frac{1}{2}$ in. in diameter by $\frac{1}{8}$-in. increments. The strength is also tabulated in AISCS Table J3.2. Because of the uncertainty as to whether or not the threaded portion of unfinished bolts extends into the shear plane, their strength is considerably less than that for rivets or high-strength bolts. Their use is usually restricted to structures subjected to static loads, and for secondary members such as purlins, girts, and bracing.

Since the initial tension developed by rivets and unfinished bolts is uncertain and possibly very small, no frictional resistance on the faying surfaces is assumed, and slip may occur at low shearing loads. This brings the rivets or bolts into bearing, and the mode of stress transfer is as shown in Fig. 6.3.

The bearing stress f_p on the contact area between the fasteners and the connected plates is defined as the transmitted shear load P divided by the total effective bearing area, dt, where d is the nominal diameter of the rivet or unfinished bolt and t the thickness of the connected material.

The shearing stress f_v in fasteners is defined as the transmitted shear load P divided by the total effective shearing area A_v:

$$f_v = \frac{P}{A_v}$$

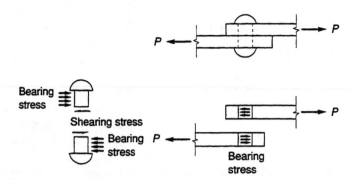

Figure 6.3 Stress transfer by shear and bearing in a riveted or bolted bearing-type connection.

where

$$A_v = \begin{cases} \dfrac{\pi d^2}{4} & \text{for single shear} \\[2ex] \dfrac{\pi d^2}{2} & \text{for double shear} \end{cases}$$

Rivets and unfinished bolts are acceptable in tension-type connections (such as hangers) for static loads and the tensile strength is tabulated in Table J3.2 of AISCS. Stresses for rivets are based on the gross cross-sectional area using the nominal diameter, and for bolts on the area determined by the maximum diameter of the threaded portion.

High-strength bolts are torqued to a high tensile stress in the shank, thus developing a dependable clamping pressure. Shearing stress is transferred by friction at working loads, as illustrated in Fig. 6.4. High-strength bolts are a preferred fastener for field connections, resistance to stress reversal, impact loads, and other applications where joint slip is not desirable. Ease of installation is another desirable attribute.

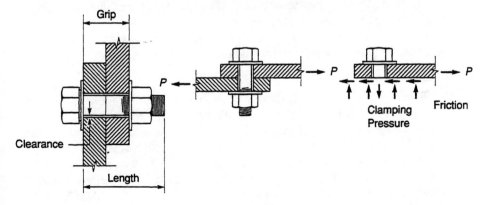

Figure 6.4 Friction-type connection.

High-strength bolts are available in two different strength levels, and are used in accordance with the provisions of the *Specification for Structural Joints Using ASTM A325 or A490 Bolts* as approved by the Research Council on Riveted and Bolted Structural Joints.

In the friction-type, or slip-critical, connection, the bolts are not actually stressed in shear and are not in bearing, since no slip occurs at service loads. However, a shear stress is specified as a matter of convenience, and the number of fasteners determined in the same manner as for other riveted or bolted connections. High-strength bolts are tightened so as to produce a minimum initial tension in the bolt shank equal to the proof load, or approximately 70% of the tensile strength of the bolt. In order to obtain the specified initial tension in the bolts, they are usually tightened with calibrated wrenches or by the turn-of-nut method. A third method of installation was added to the bolt specification in 1972, whereby a direct tension indicator is utilized. This specification also provides for high-strength bolts, which incorporate a design feature intended to indicate when a predetermined tension or torque has been achieved. Specifications require that hardened washers be used under the turned element when using the calibrated-wrench method. Hardened washers are required under both head and nut when using A490 bolts to connect material with a yield point of less than 40 ksi (276 MPa). Beveled washers are required when an outer face of the connection has a slope greater than 1:20. In the turn-of-nut method, nuts are first brought to a "snug" fit, defined as the condition when all surfaces are in good contact and no free rotation of the nut is possible. Bolts under eight diameters or 8 in. in length are then given an additional one-half turn of the nut; over eight diameters or 8 in., two-thirds of a turn; and if both faces have a 1:20 slope, without use of washers, the nuts are to be given a three-quarter turn, regardless of length.

Resistance to slip is determined by the amount of bolt tension and the condition of the contact surfaces in a given connection. Connections having painted contact surfaces or contact surfaces of unrusted mill scale offer the least resistance to slip; rusted surfaces that have been well cleaned may provide up to two times as much resistance.

Rivets and bolts in combined shear and tension are proportioned according to Sec. J3.7 of AISCS. In the case of rivets and bearing-type bolted connections, the limiting tensile stresses are reduced if the simultaneous shear stress exceeds a certain value. For example, the tensile stress limit for A502 grade 1 rivets is

$$F_t = 59 - 1.8 f_v \leq 45 \text{ ksi}$$

This says, in effect, that the tensile stress limit is in no case greater than 45 ksi, and if the shear stress exceeds 7.78 ksi, tensile stress limit will be less than 45 ksi, as given by the formula. Similar formulas for grade 2 rivets, and for unfinished and high-tensile bolts in bearing-type joints, are provided in Table J3.5 of AISCS.

In the case of slip-critical joints with high-tensile bolts, the nominal shear resistance is dependent on adequate initial tension. Hence if additional tension due to external loads is applied, the clamping force is reduced and Sec. J3.9a of AISCS requires that the nominal shear stress prescribed by Table J3.6 be multiplied by the reduction factor $(1 - T/T_b)$, where T is the tension force on the joint due to a direct load applied to all the bolts and T_b is the specified pretension load on the bolt as given in Table J3.1.

It should be noted that AISCS considers loss of friction in a slip-critical joint to be a serviceability limit state and so the loads acting on such a joint are the unfactored service loads. However, in some cases the designer may feel that loss of friction is an ultimate limit state and therefore AISCS provides for such a requirement in Secs. J3.8 and J3.9.

At this point the reader should review in detail the AISCS and AISCM information on rivets and bolts, as follows: Secs. J3.6 to J3.11 of AISCS together with Table J3.2. Also read Secs. J3.1 to J3.5 of AISCS, which provide detailed information on minimum pitch, minimum edge distance, and so forth, and Sec. M5, which defines good fabrication practice for riveted and bolted joints. Finally, note the detailed dimension and weight information regarding rivets and bolts in AISCM, and the tabulation of rivet and bolt strength values. A reason for the textual omission of much of the information just referred to is the desirability of fostering the direct use of specification and manual information in the handling of design problems. Examples 6.1 to 6.3 illustrate the design of simple concentrically loaded riveted and bolted joints.

Example 6.1 *Lap Joints*

A lap joint is simply a joint in which two members overlap and are connected to each other with some type of fastener. The lap joint is not a desirable structural connection and should be used only for minor connections. The eccentricity of the loads causes secondary bending stresses in the members and at least two fasteners should be used on each line. (Edge distance, net section, and spacing checks omitted for these; see Example 6.2B.)

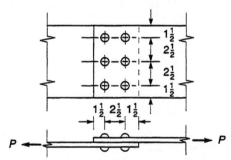

A. Two 0.25 in. × 8 in. A36 plates are to be connected using 0.75-in. A502 grade 1 rivets, as shown in the figure. How many rivets are required to transfer a dead load of 20 kips and a live load of 16 kips?

Solution   Rivets are critical in single shear or bearing (refer to AISCS, Secs. J3.6 and J3.10). Shear:

$$d_{rivet} = 0.75 \text{ in.}$$

$$A_{rivet} = \frac{\pi d_{rivet}^2}{4} = 0.442 \text{ in}^2 \qquad \text{area}$$

$$\phi = 0.75 \qquad \text{from Table J3.2}$$

$$F_v = 25 \text{ ksi} \qquad \text{from Table J3.2}$$

$$P_D = 20 \text{ kips} \qquad P_L = 16 \text{ kips} \qquad \text{service load}$$

$$P_u = 1.2P_D + 1.6P_L = 49.6 \text{ kips} \qquad \text{ultimate load}$$

$$n_{req} = \frac{P_u}{\phi F_v A_{rivet}} = 5.988 \qquad \text{number of rivets}$$

Bearing at bolt holes (AISCS, Sec. J3.10):

$$L_e = 1.5 \text{ in.} \quad \text{edge distance in line of force}$$
$$s = 2.5 \text{ in.} \quad \text{spacing in line of force}$$
$$t = 0.25 \text{ in.} \quad \text{plate thickness}$$
$$F_u = 58 \text{ ksi} \quad \text{tensile strength of plate material}$$

There are two rows of bolts in the line of force

$$\text{Edge distance} = 1.5 \text{ in.} > 1.5 \times \text{rivet diameter} = 1.125 \text{ in.}$$
$$\text{Rivet spacing} = 2.5 \text{ in.} > 3 \times \text{rivet diameter} = 2.25 \text{ in.}$$
$$\phi = 0.75 \quad R_n = 2.4 d_{\text{rivet}} t F_u$$
$$n_{\text{req}} = \frac{P_u}{\phi R_n} = 2.534 \quad < 5.99 \quad \text{shear governs}$$

Use six rivets.

B. Same as part A, except use 0.75-in. A325 bolts in a slip-critical connection (AISCS, Sec. J3.8). Bearing is OK (identical to rivets). Use standard holes and design for limit state of slip at service loads.

Solution

$$P_s = P_D + P_L \quad \phi = 1.0$$
$$F_v = 17 \text{ ksi} \quad (\text{Table J3.6}) \quad A_{\text{bolt}} = A_{\text{rivet}}$$
$$n_{\text{req}} = \frac{P_s}{\phi F_v A_{\text{bolt}}} = 4.793$$

Use six bolts for symmetry. (Symmetry is desired to avoid secondary stresses that add to the already complex stress distribution. Note that the number of required bolts could have been obtained directly from AISCM.)

C. Same as part B, except use bearing-type connection with threads excluded from the shear plane (AISCS, Table J3.2 and Sec. J3.6). Bearing is OK.

Solution

$$\phi = 0.75 \quad F_v = 60 \text{ ksi}$$
$$n_{\text{req}} = \frac{P_u}{\phi F_v A_{\text{bolt}}} = 2.495$$

Use four bolts, in two rows (4 in. apart) and two lines.

Example 6.2 *Butt Splice*
A. A butt splice is to be designed as shown in the figure. Use 0.75-in.-diameter A325 bolts with standard-sized holes in slip-critical connection and A36 steel plates. The unfactored force $P_L = 90$ kips is due to live load only. Determine the number of bolts required on either side of the splice. Bolts are in double shear.

‍

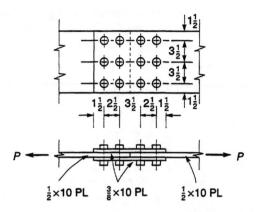

Solution Check gross section; $\frac{1}{2}$ in. plate is critical.

$$\phi = 0.9 \qquad F_y = 36 \text{ ksi} \qquad A_g = 0.5 \text{ in.} \times 10 \text{ in.} \qquad A_g = 5 \text{ in}^2$$

$$P_L = 90 \text{ kips} \qquad P_u = 1.6P_L = 144 \text{ kips}$$

$$\phi F_y A_g = 162 \text{ kips} \qquad > 144 \text{ kips} \quad \text{OK}$$

Check net section:

$$A_n = A_g - 3\left[\left(\frac{3}{4}\right) + \frac{1}{8}\right] \text{ in.} \times 0.5 \text{ in.} = 3.688 \text{ in}^2$$

$$\phi = 0.75 \qquad F_u = 58 \text{ ksi}$$

$$\phi F_u A_n = 160.406 \text{ kips} \qquad > 144 \text{ kips} \quad \text{OK}$$

Number of bolts required in double shear (AISCS, Sec. J3.8): Design slip-critical joint for service load P_L.

$$A_b = 0.442 \text{ in}^2 \qquad F_v = 17 \text{ ksi} \qquad \phi = 1.0$$

$$n = \text{number of bolts required} = \frac{P_L}{2\phi F_v A_b} = 5.989$$

Number of bolts required for bearing capacity:

Minimum bolt spacing (Sec. J3.3): $s_{\min} = 3d = 3 \times 0.75 = 2.25 \text{ in.}$ < 2.5 in. OK

Minimum edge distance (Sec. J3.4): $L_{e\min} = 1.25$ in. (Table J3.4, for sheared edges)

< 1.5 in. OK

$$\phi = 0.75 \qquad d = 0.75 \text{ in.} \qquad t = 0.5 \text{ in.}$$

$$R_n = 2.4dtF_u = 52.2 \text{ kips}$$

$$n = \frac{P_u}{\phi R_n} = 3.678$$

Use six bolts (three bolts per line).

B. Same as part A, except use the bolts in a bearing-type connection with the threads excluded from the shear plane (AISCS, Sec. J3.6).

Solution Four bolts are needed for bearing. Bolts are in double shear.

$$\phi = 0.75 \qquad F_v = 48 \text{ ksi} \quad \text{(Table J3.2)}$$

$$n = \frac{P_u}{2\phi F_v A_b} = 4.525$$

Use six bolts as in part A.

Example 6.3 *Bracket Connection*

A tension member, two L4 × 3 × $\frac{1}{2}$, carries an unfactored tension load P of 110 kips due to wind force with a direction of 30° from the horizontal axis. A bracket utilizing a structural T section will be used to connect the tension member by rivets and will be joined to a column flange by high-strength bolts with a friction-type connection, as shown in the figure. Determine the required size and number of rivets and bolts. Use A502 grade 2 rivets, A490 high-strength bolts, and A36 steel in accordance with AISCS.

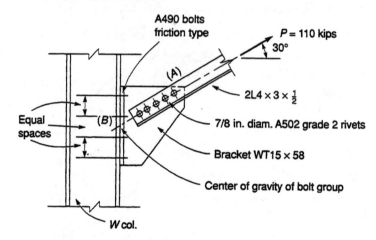

Solution *Connection A:* tension angles to bracket (try WT15 × 58, $t_w = 0.565$ in.). Shear strength of the rivets in double shear (AISCS, Table J3.2).

$$P_w = 110 \text{ kips} \qquad P_u = 1.3P_w = 143 \text{ kips}$$

$$\phi = 0.75 \qquad F_v = 33 \text{ ksi}$$

Use $\frac{7}{8}$ in.-diameter rivets. $A_b = 0.601 \text{ in}^2$

$$n = \text{required number of rivets} = \frac{P_u}{2\phi F_v A_b} = 4.807$$

Bearing strength: Rivet spacing and edge distance are such that Sec. J3.10a applies. Deformation around holes is a design consideration.

$$\phi = 0.75 \qquad d = \frac{7}{8} \text{ in.} \qquad t = 0.565 \text{ in.} \quad \text{web thickness of W15 × 58 bracket}$$

$$F_u = 58 \text{ ksi} \qquad R_n = 2.4dtF_u = 68.817 \text{ kips}$$

Number of rivets required $= n = \dfrac{P_u}{\phi R_n} = 2.771 \qquad < 4.81$

Check gross area of double angles:

$$A_g = 6.5 \text{ in}^2 \qquad \phi = 0.9 \qquad F_y = 36 \text{ ksi}$$
$$\phi F_y A_g = 210.6 \text{ kips} \qquad > 143 \text{ kips } \text{ OK}$$

Check effective net area:

$$A_n = A_g - 2 \cdot 0.5 \text{ in.}\left(\frac{7}{8} + \frac{1}{8}\right)\text{in.} = 5.5 \text{ in}^2$$

$$U = 0.85 \qquad \text{(AISCS Commentary B3)}$$

$$A_e = A_n U = 4.675 \text{ in}^2 \qquad \phi = 0.75$$

$$\phi F_u A_e = 203.362 \text{ kips} \qquad > 143 \text{ kips } \text{ OK}$$

Check block shear rupture strength (AISCM, Sec. J4.3):

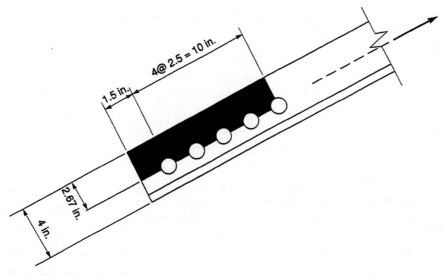

Shaded area is subject to tear-out.

$F_y = 36 \text{ ksi} \qquad F_u = 58 \text{ ksi}$

$\phi = 0.75 \qquad$ resistance factor

$A_{gv} = 0.5 \text{ in.} \times 11.5 \text{ in.} \qquad$ gross area subject to shear

$A_{gt} = 0.5 \text{ in.} \times 2.67 \text{ in.} \qquad$ gross area subject to tension

$$A_{nv} = 0.5 \text{ in.}\left[11.5 \text{ in.} - 4 \cdot \left(\frac{7}{8} \text{ in.} + \frac{1}{16} \text{ in.} + \frac{1}{16} \text{ in.}\right)\right] \qquad \text{net area subject to shear}$$

$$A_{nt} = 0.5 \text{ in.}\left[2.67 \text{ in.} - 0.5 \cdot \left(\frac{7}{8} \text{ in.} + \frac{1}{16} \text{ in.} + \frac{1}{16} \text{ in.}\right)\right] \qquad \text{net area subject to tension}$$

$F_u A_{nt} = 62.93 \text{ kips}$

$0.6 F_u A_{nv} = 130.5 \text{ kips} \qquad > 62.93 \text{ kips}$

$\phi(0.6 F_u A_{nv} + F_y A_{gt}) = 133.92 \text{ kips} \qquad > 143/2 \text{ kips } \text{ OK} \quad \text{[Eq. (J4-3b)]}$

Block shear rupture strength is OK.

Use five rivets. Arrange as shown, at $2\frac{1}{2}$ in. center to center. Detailer to check layout for clearances and edge distances.

Connection B: bracket to column flange. Assume that the tension load passes through the center of gravity of the bolts. Then the shear tension components of the tension are:

Tension component:

$$T = P\cos30° = 110 \times 0.866 = 95.26 \text{ kips}$$

Shear component:

$$V = P\sin30° = 110 \times 0.5 = 55 \text{ kips}$$

This is a slip-critical connection, using standard holes for $\frac{7}{8}$-in.-diameter A490 bolts. For service loads Sec. J3.9a applies.

$$A_b = 0.601 \text{ in}^2$$
$$F_v = 21 \text{ ksi} \qquad \text{nominal resistance in shear (Table J3.6)}$$
$$\phi = 1.0$$
$$T_b = 49 \text{ kips} \qquad \text{minimum bolt pretension (Table J3.1)}$$
$$T = 95.26 \text{ kips}$$

Assume eight bolts in single shear.

$$R_n = 8\phi F_v A_b\left(1 - \frac{T}{T_b \cdot 8}\right) = 76.432 \text{ kips}$$
$$\phi R_n = 76.432 \text{ kips} \qquad > V = 55 \text{ kips} \quad \text{OK}$$

Use eight bolts of $\frac{7}{8}$ diameter.

Note that the prying force on the A490 bolts should also be investigated. The calculations involved are demonstrated in Example 6.13.

6.3 Pinned Connections

Pinned connections are sometimes used in bridge-bearing supports with the purpose of permitting end rotation. They are also used to connect pin-connected members of the types discussed in Chapter 2. Ranging from 2 to 10 in. or more in diameter, they are designed in a manner analogous to bearing connections of bolts, but with the added requirement to check stress due to bending in the pin itself. Details of standard pins and caps or nuts to hold them in position are provided in the connection section of the AISCM. Although the actual distribution of stress in a short circular beam is complex, designs have been found to be satisfactory when based on simple beam theory and on average stress due to shear and bearing. Bending moments may be conservatively calculated on the assumption that forces are concentrated at the centers of bearing areas.

On the basis of assumed locations of acting forces, bending moments and shears may be determined, and a preliminary selection of required pin diameters may then be determined on the basis of whichever condition (bending or shear) is critical. The bearing stress may then be checked, and revision of either the pin diameter or length of bearing may then be made if required.

The following criteria apply for the design of pins:

Bending (AISCS, Table A-F.1, solid symmetric shapes):

$$M_n = Z_x F_y \qquad \text{nominal moment}$$

$$\phi = 0.9$$

$$Z_x = \frac{d^3}{6} \qquad \text{plastic-section modulus}$$

Shear (AISCS, Sec. H2):

$$F_{uv} = 0.6F_y \qquad \text{nominal stress in shear}$$

$$\phi = 0.9$$

$$f_{uv} = \frac{V}{A} \qquad \text{shear stress due to shear } V$$

$$A = \frac{\pi d^2}{4} \qquad \text{cross-sectional area}$$

Bearing (AISCS, Sec. J8):

$$R_n = 1.8F_y A_{pb} \qquad \text{nominal capacity}$$

$$\phi = 0.75$$

$$A_{pb} = \text{projected bearing area}$$

Example 6.4 *Pin Connection*

Although the connection shown is of a type not commonly used in recent years, it illustrates the essential problems in pin design. The shown loads are factored. Use A36 steel.

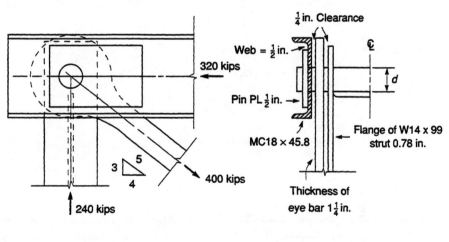

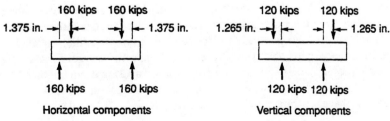

Horizontal components   Vertical components

Bending moment:

$$M_h = 160 \times 1.375 = 220 \text{ kip-in.}$$
$$M_v = 120 \times 1.265 = 152 \text{ kip-in.}$$
$$M = \sqrt{220^2 + 152^2} = 267 \text{ kip-in.}$$

The following criteria apply for the design of the pin. Pin diameter required by bending moment:

$$d_{\text{req}} = \left(\frac{6M_u}{0.9F_y} \right)^{1/3} = 3.672 \text{ in.}$$

Pin diameter required by shear force:

$$V = \frac{400 \text{ kips}}{2} \qquad d_{\text{req}} = \sqrt{\frac{4V}{0.9 \cdot 0.6\pi F_y}} = 3.619 \text{ in.}$$

Pin diameter required by bearing:

Flange of vertical strut:

$$P_b = 120 \text{ kips} \qquad t_b = 0.78 \text{ in.}$$
$$d_{\text{req}} = \frac{P_b}{0.75 \cdot 1.8F_y t_b} = 3.166 \text{ in.}$$

On chord:

$$P_b = 160 \text{ kips} \qquad t_b = 1 \text{ in.}$$
$$d_{\text{req}} = \frac{P_b}{0.75 \cdot 1.8F_y t_b} = 3.292 \text{ in.}$$

On eyebar:

$$P_b = 200 \text{ kips} \qquad t_b = 1.25 \text{ in.}$$
$$d_{\text{req}} = \frac{P_b}{0.75 \cdot 1.8F_y t_b} = 3.292 \text{ in.}$$

Use pin diameter $d = 4.0$ in.

Note: Attachment of the pin plate to the channel web is deferred to Problem 6.5.

6.4 Welded Connections

Structural welds are usually made either by the manual shielded-metal-arc process or by the submerged-arc process, the latter being especially suited to automatic shop welding of built-up members with controlled positioning. In either process the heat of an electric arc simultaneously melts the welding electrode and the adjacent steel in the parts being joined. The electrode is deposited in the weld as filler metal. The wide adoption of welding in recent years has required improved control of steel chemistry in order to provide steels that are *weldable*, that is, steels that can be joined together with sound metal, of adequate strength and ductility, and with minimal metallurgical damage to adjacent parent metal.

In the shielded-metal-arc process, pictured in Fig. 6.5, the electrode coating creates a gaseous shield that protects the molten weld metal from the atmosphere. In the submerged-arc process, the arc occurs underneath a previously deposited fusible powdered flux that blankets the welding zone, and the bare electrode usually is fed automatically from a reel of wire.

The adoption of rules governing the qualification of welders and welding processes together with quality control of all materials have brought welding to the point where it is today permitted for practically all steel fabrication for both shop and field connections. Welding offers many advantages, which may be briefly outlined as follows:

1. Simplicity of design details, efficiency, and minimum weight are achieved because welding provides the most direct transfer of stress from one member to another.

2. Fabrication costs are reduced because fewer parts are handled and operations such as punching, reaming, and drilling are eliminated.

3. There is a saving in weight in main tension members since there is no reduction in area due to rivet and bolt holes. Additional saving is also achieved because of the fewer connecting parts required.

4. Welding provides the only plate-joining procedure that is inherently air- and watertight and hence is ideal for water and oil storage tanks, ships, and so forth.

5. Welding permits the use of fluidly changing lines that enhance the structural and architectural appearance, as well as reduce stress concentrations due to local discontinuities.

6. Simple fabrication becomes practicable for those joints in which a member is joined to a curved or sloping surface, such as structural pipe connections.

7. Welding simplifies the strengthening and repair of existing riveted or welded structures.

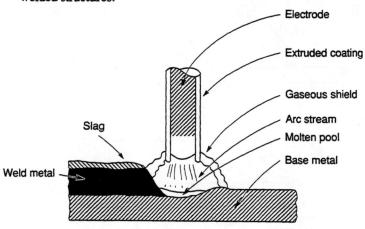

Figure 6.5   Shielded-metal-arc welding process.

The two most common types of welds are fillet welds and groove welds. *Fillet welds* are used to attach a plate to another plate or member in either a parallel (lapped) or protruding (tee) position, as shown in Fig. 6.6(a). *Groove welds*, as shown in Fig. 6.6(b), retain the continuity of plate elements that are butt joined along their edges. Groove welds require special edge preparation and careful fit up, and when welded from both sides, or from one side with a backup strip on the far side, they may be said to achieve *complete penetration* and may be stressed as much as the weakest piece that has been joined. Incomplete-penetration groove welds are used only when the plates are not required to be fully stressed and full continuity is not required. Complete-penetration groove welds are also used for corner or tee joints when full plate development is required.

AISCS, Table J2.5, indicates that nominal strengths in compression, tension, and shear for full-penetration groove welds are the same as for the base metal. Partial-penetration groove welds are permitted full stress effectiveness only for compression normal to the effective throat. They may be stressed in tension on the effective throat at a reduced nominal stress equal to that allowed in shear.

Standard weld symbols to be used in steel detailing are shown in the AISCM with their correct use illustrated in many sketch details.

Fillet welds are more easily made than groove welds because of the greater fit-up tolerances allowed. As shown in Fig. 6.6, fillet welds are commonly used in connecting

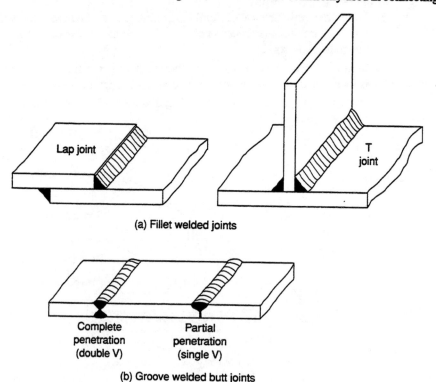

(a) Fillet welded joints

(b) Groove welded butt joints

Figure 6.6 Types of welded joints.

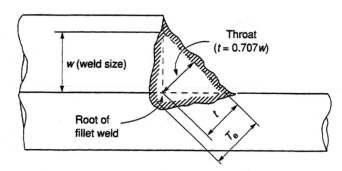

Figure 6.7   Fillet-weld nomenclature.

lapped plates or projecting plate elements to another plate or member. The allowable force transmitted by a unit length of fillet weld is equal to the product of the *effective throat dimension* multiplied by the resistance factor $\phi = 0.75$ and by the nominal strength $0.6F_{Exx}$, where F_{Exx} is the tensile strength of the welding electrode metal, as listed in Table J2.5. The effective throat dimension T_e is illustrated in Fig. 6.7, where w is the nominal weld size and t the throat dimension. When the welded faces of the joined parts are at 90°, as shown,

$$t = 0.707_w$$

When the manual shielded-metal-arc process is used, the effective throat is equal to the dimension t. When the submerged-metal-arc process is used, greater heat input produces a deeper penetration and an effectively greater throat dimension is allowed. Thus the submerged-arc process, Section J2.2a of AISCS, permits the effective throat, T_e, to be taken as equal to the weld size, w, when w is $\frac{3}{8}$ in. or less, and equal to $t + 0.11$ when w is greater than $\frac{3}{8}$ in.

The *resultant stress* on the effective throat is taken to be *equivalent* to shear stress in determining the allowable force on a unit length of weld. This concept is illustrated in Fig. 6.8, where the z axis is arbitrarily shown along the fillet weld throat and the x and y axes are in the surface planes of the tee joint, as shown in Fig. 6.8(b). The region above lines A–A–A has been removed in the pictorial representation of Fig. 6.8(a), in which the section through the fillet throat is exposed. The resultant stress f_r, having components f_x, f_y, and f_z, must be kept below $\phi \times 0.6F_{Exx}$. It may be noted that f_z is a pure shear component, directed along the weld, and that f_x and f_y include both shear and normal components. The stress resultant, f_r, is

$$f_r = \sqrt{f_x^2 + f_y^2 + f_z^2} \tag{6.1}$$

The x, y, and z axes may be oriented in any arbitrary direction. The use of the stress resultant as a strength criterion has been shown to be adequate by many tests of welded connections, including those that are eccentrically loaded.

Fillet welds are specified on drawings and in design calculations by their size, w, which varies from $\frac{3}{16}$ to $\frac{1}{2}$ in. in $\frac{1}{16}$-in. increments, and which varies in $\frac{1}{8}$-in. increments for sizes greater than $\frac{1}{2}$ in. Weld sizes of $\frac{3}{16}$, $\frac{1}{4}$, and $\frac{5}{16}$ in. are favored because they can be made by a single pass of the electrode. The amount of filler metal increases

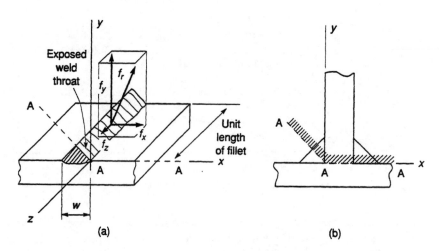

Figure 6.8 Stress components and resultant stress on fillet-weld throat.

as the square of the weld size. Thus time and cost of welding increase disproportion-ately as the weld size increases.

There are other limitations on fillet-weld size. A small weld at the edge of a thick plate is chilled rapidly, which has an embrittling effect, and it may crack as it tries to shrink when it cools while being restrained by the heavy plate. Minimum per-mitted sizes of fillet welds, in reference to the thicker plate joined, are listed in Table J2.4 of AISCS. Fillet welds along the edges of plates thicker than $\frac{1}{4}$ in. are also limited as to *maximum* size, which can be no greater than the plate thickness minus $\frac{1}{16}$ in. (AISCS, Sec. J2.2b). When a fillet weld is terminated, small sections near the ends are not fully effective. Rules of design covering this and other related topics are covered in Sec. J2.2b of AISCS.

Design weld strength values per lineal inch are most conveniently quoted in terms of the weld size, w, even though determined by the resultant stress on the throat. Weld sizes are called for in multiple values of $\frac{1}{16}$ in. It is therefore convenient in design to use the strength value of a $\frac{1}{16}$-in. weld as a basic unit. Thus the design shear strength (kips per inch of weld length and per $\frac{1}{16}$ in. of fillet weld size) will be termed $\bar{q}$, where

$$\bar{q} = \frac{1}{16}(0.707)\phi \times 0.6F_{Exx} \qquad (6.2)$$

Letting N = number of $\frac{1}{16}$-in. in a weld (for example, $N = 4$ for a $\frac{1}{4}$-in. weld), the design shear in kips per inch of fillet weld is termed q_a, where

$$q_a = N\bar{q} \qquad \text{where} \quad N = 16w$$

For submerged-arc welds, the greater effective throat was previously defined (AISCS, Sec. J2.1b). Allowable shears per $\frac{1}{16}$ in. of weld size are equal to $\frac{1}{16} \times \phi \times 0.6F_{Exx}$ for welds of $\frac{3}{8}$ in. or less. For weld sizes over $\frac{3}{8}$ in. the design shear stresses are the same as for the metal-arc process, plus a fixed bonus of $0.11\phi \times 0.6F_{Exx}$. These values are listed and explained in Table 6.1 for six different weld metal strengths.

Table 6.1 Design Shear (kips/in.) of Fillet Welds, $\bar{q}$, per $\frac{1}{16}$ in. of Weld Size.

	Metal-Arc Electrode					
	E60	E70	E80	E90	E100	E110
F_{Exx} (ksi)	60	70	80	90	100	110
Allowable weld shear ($\bar{q}$) (kips/in.) per $\frac{1}{16}$ in. of weld size						
Metal-arc process	1.19	1.39	1.59	1.79	2.00	2.19
Submerged-arc size $\frac{3}{8}$ in. or less	1.69	1.97	2.25	2.53	2.81	3.09
Submerged-arc bonus[a] for size over $\frac{3}{8}$ in., but use metal-arc's allowable shear	2.97	3.46	3.96	4.45	4.95	5.44

[a] For example, for a $\frac{1}{2}$-in. submerged-arc weld and for electrode E80, $q_a = 8 \times 1.59 + 3.96 = 16.68$ kips/in.

The designer has two choices in laying out a suitable concentrically loaded fillet weld to transmit a given connection load P:

1. Select a weld size, determine the weld value q_a in kips per lineal inch of weld, and then determine the total length l of weld required:

$$l = \frac{P}{q_a} \qquad (6.3)$$

In using Table 6.1, $q_a = N\bar{q}$ for the metal-arc process welds and for submerged-arc welds of size $\frac{3}{8}$ in. or less. For submerged-arc weld sizes greater than $\frac{3}{8}$ in., use the footnote to Table 6.1.

2. Alternatively, if a particular weld length is suggested by the geometry of the connected parts, the required weld shear force is determined:

$$q = \frac{P}{l}$$

The required weld size (N) in sixteenths may then be determined by dividing the calculated shear force per inch (q) by $\bar{q}$ as listed in Table 6.1. This procedure must be modified for submerged arc welds of size greater than $\frac{3}{8}$ in. by first subtracting the tabulated bonus and then dividing the calculated remaining force by the *metal-arc electrode* listing for $\bar{q}$.

In designing a double fillet-welded tee joint [Fig. 6.6(a)] the design shear force per inch of two fillet welds may exceed the design shear force per inch of stem of tee, in which case the latter would control the design and impose an effective upper limit on the useful size of the fillet welds. Fillet-welded tee connections to a single plate should always be welded on both sides because of the weakness in bending and susceptibility to shipping damage of a single fillet weld.

For single or double angles under static axial load, Sec. J1.8 of AISCS does not require that fillet welds in end connections be disposed so as to balance the forces about the neutral axis of the member. However, for members subject to repeated stress variation, it is recommended that the fillet welds be so placed as to balance the forces about the neutral axis and eliminate eccentricity. Two common cases occur, as illustrated in Fig. 6.9: case 1, comprised of two longitudinal welds, and case 2, with a transverse weld added.

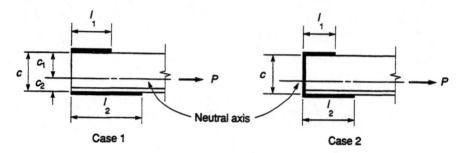

Figure 6.9   Balancing fillet welds for angle connections.

Case 1.   Assume all fillet welds to be the same size. Then the equilibrium conditions of forces and moments about the center-of-gravity axis of the angle member are given, respectively, as follows:

$$l = l_1 + l_2 = \frac{P}{q_a} \tag{6.4}$$

$$c_1(l_1 q_a) = c_2(l_2 q_a) \tag{6.5}$$

from which is obtained

$$l_1 = \frac{c_2}{c}l \tag{6.6}$$

$$l_2 = \frac{c_1}{c}l \tag{6.7}$$

where l_1, l_2 = required lengths of fillet welds
   l = required total length of fillet welds
   c_1, c_2 = distances from neutral axis to extreme fibers of the angle
   P = axial load
   q_a = permissible shear strength of fillet weld

Case 2.   If a weld of length c is provided along the end of the angle, the equilibrium conditions of forces and moments again lead to the following expressions:

$$l = l_1 + l_2 + c = \frac{P}{q_a} \tag{6.8}$$

$$c_1 l_1 q_a + c\left(\frac{c}{2} - c_2\right)q_a = c_2 l_2 q_a \tag{6.9}$$

Then

$$l_1 = \frac{c_2}{c}l - \frac{c}{2} \tag{6.10}$$

$$l_2 = \frac{c_1}{c}l - \frac{c}{2} \tag{6.11}$$

Concentrically loaded welded connection design will now be illustrated by means of Examples 6.5 through 6.7.

Example 6.5 *Lap Joint (Welded)*
Similar to Example 6.1, except that plates are $PL\frac{1}{2} \times 8$ loaded to a required (factored) force of 75 kips tension. Determine the transverse weld size required, using A36 steel and E70 electrodes.

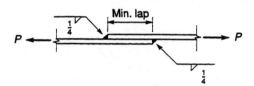

Solution

minimum lap $= 5 \times$ thickness of thinnest plate   (AISCS, Sec. J2.2b)

$$= 5 \times \frac{1}{2} = 2.5 \text{ in.}$$

total length of weld $=$ approx. $8 \times 2 = 16$ in.

Required weld unit shear force:

$$q = \frac{75}{16} = 4.69 \text{ kips/in.}$$

Required weld size:

$$N = \frac{q}{\bar{q}} = \frac{4.69}{1.39} = 3.4 \qquad w = \frac{N}{16} = \frac{3.4}{16}$$

Hence use $\frac{1}{4}$-in. fillet welds. Check the minimum permitted weld size, AISCS, Table J2.4:

$$\text{minimum weld size} = \frac{3}{.16} \text{ in.} \qquad < \frac{1}{4} \quad \text{OK}$$

Example 6.6 *Butt Joint with Groove Weld*
A $PL\frac{1}{2} \times 12$ carrying a required tensile force of 125 kips is to be spliced. Use a groove-welded butt splice, A36 steel, and E70 electrodes.

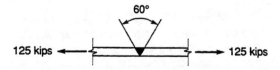

Solution   Actually, the only consideration here is to join the two pieces with a complete-penetration groove weld using electrodes that provide a weld with equal or greater tensile strength than that of the plate. A single V complete-penetration groove weld requires that the weld root on the underside must be gouged and back welded. This is only one of several groove welds that could be used. For example, if the underside is inaccessible to the welder, a backup strip must be tack welded to one of the pieces prior to fit up.

Example 6.7

Same as Example 6.3, except that fasteners are replaced by welds. The bolted tee may now be replaced by a single $\frac{5}{8}$-in. plate[*] that is welded directly to the center of the column flange using a double-bevel full-penetration groove weld. Fillet welds attaching the angles to the plate should be arranged so as to balance the forces about the neutral axis of the connection of the double angles to eliminate any eccentricities.

Solution   *Connection A:* Assume that no weld is placed along the end of angles. Use maximum allowable weld size to connect angles to bracket. Then $w = \frac{1}{2} - \frac{1}{16} = \frac{7}{16}$-in. weld (AISCS, Sec. J 2.26). The design strength of $\frac{7}{16}$-in. fillet weld (use E70 electrodes) is

$$q_a = 7 \times 1.39 = 9.73 \text{ kips/in.}$$

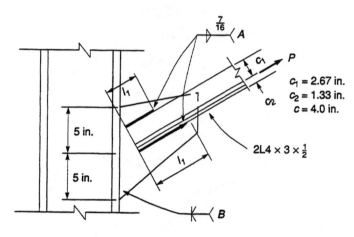

$c_1 = 2.67$ in.
$c_2 = 1.33$ in.
$c = 4.0$ in.

$2L4 \times 3 \times \frac{1}{2}$

The required total weld length is

$$l = \frac{P}{q_a} = \frac{143/2}{9.73} = 7.35$$

$$l_1 = \frac{c_2}{c}l = \frac{1.33}{4}7.35 = 2.44 \text{ in.} \qquad\qquad \text{[Eq. (6.6)]}$$

$$l_2 = \frac{c_1}{c}l = \frac{2.67}{4}7.35 = 4.91 \text{ in.} \qquad\qquad \text{[Eq. (6.7)]}$$

Use $l_1 = 2\frac{1}{2}$ in.; $l_2 = 5$ in.

Connection B: According to J2.5 of AISCS, the design stresses in tension and shear in a full penetration groove weld are the same as for the base metal. AISCS does not deal explicitly with the problem of combined shear and tension in groove welds, but an adequate design may be obtained by limiting the maximum principal tensile stress to the allowable stress in tension. As discussed later in Chapter 11, this may be accomplished by a closely approximate formula

---

[*] *Note:* As an alternative to the use of the $\frac{5}{8}$-in. plate, the WT15 × 58 could again be used, as in Example 6.3, but the simplicity inherent to welded design would then be lost. It may be noted, however, that the protruding plate attached permanently in the shop to the column is a nuisance and subject to damage in shipment.

[Eq. (10.13)], which simply reduces the nominal tension stress in the direction of applied tensile force to a value F_n:

$$F_{rt} = \left[1 - \left(\frac{f_v}{F_t}\right)^2\right] F_t$$

The tension and shear components of the applied force in the groove weld, designated, respectively, as T and V, are (Example 6.3)

$$T = 95.2 \text{ kips} \qquad V = 55 \text{ kips}$$

Try $l = 10$ in. (double-bevel full-penetration groove weld):

$$f_v = \frac{V}{tl} = \frac{55}{\frac{5}{8} \times 7.5} = 11.7 \text{ ksi} \qquad f_t = \frac{T}{tl} = \frac{95.2}{\frac{5}{8} \times 7.5} = 20.3 \text{ ksi}$$

By Eq. (10.13),

$$F_{rt} = \left[1 - \left(\frac{11.7}{0.9 \times 36}\right)^2\right] 0.9 \times 36 = 28.2 \text{ ksi} \qquad > f_t \quad \text{OK}$$

6.5 Eccentrically Loaded Connections

Previous design examples in this chapter have been limited to concentrically loaded connections. Concentricity is always desirable, but eccentrically loaded connections are sometimes required. Riveted, bolted, and welded connections will be considered.

When eccentricity of load imposes only shear force on riveted and bolted fasteners, with no change of initial tension, the assumption may be made that the eccentric load may be replaced by an equivalent force and couple acting at the centroid of the fastener group. If, on the other hand, both shear and tension are induced by load eccentricity, the design will require consideration of the following:

1. Reduced allowable tension stress in fasteners for bearing-type riveted or bolted connections, or
2. Reduced allowable shear stress in friction-type fasteners using high-strength bolts

The foregoing effects were considered in Section 6.2.

Compact filled-welded connections between heavy parts may be designed simply by providing a weld size adequate to resist the maximum resultant force per lineal inch of weld due to the combined effect of applied force and the moment induced by eccentricity.

Initial attention is now given to riveted or bolted shear connections that produce no tensile force change in the fasteners, as shown by the two-place bracket bolted to the flanges of a W column in Fig. 6.10(a). The load P may be vertical, or inclined as shown, and there is an eccentric moment of magnitude Pe acting at the centroid of the fastener group, as shown in Fig. 6.10(b). In this illustration, because of the double symmetry of the fastener pattern the centroid is readily located by inspection. It is convenient to replace P by its components P_x and P_y, as shown in Fig. 6.10(b). The problem is to determine the maximum resultant shear force on the particular most-stressed fastener, which usually can be located by elementary considerations.

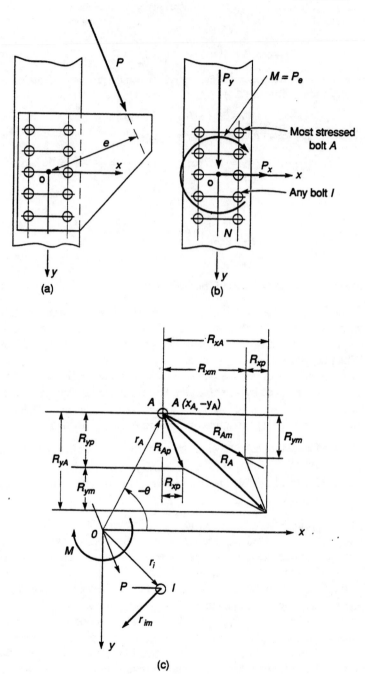

Figure 6.10 Eccentrically loaded riveted or bolted connection.

In Chapter 8 of AISCM this method of analyzing eccentric connections is called the *elastic method*, and it is presented as a conservative alternative to the preferred *instantaneous center of rotation* method, which is based on a computer implementation of an inelastic iterative analysis of the ultimate strength of the eccentric bolt group. In the inelastic method the shear force versus deformation relationship of the bolt or the weld is represented by an experimentally based nonlinear curve. The values in the load tables in Chapter 8 of AISCM for eccentric bolted and welded connections and the analysis presented here will thus not match exactly. The designer should use the load tables in AISCM where this is possible; however, the conservative method presented here is very useful for obtaining a solution for the cases not tabulated.

In Fig. 6.10(c), R_{AP} represents the resultant force due to applied load P acting on any *particular* bolt A. As assumed for concentric connections, $R_{AP} = P/n$, where n is the total number of fasteners. R_{Am} is the resultant force due to moment, assumed to be proportional to the radial distance r_A from centroid O, and to act normal to r_A. The total shear force on the fastener is shown as R_A, the resultant of R_{AP} and R_{Am}, and is shown graphically as the diagonal of the force parallelogram. To avoid the determination of angles and the use of unwieldy arithmetic, it is convenient to break down each of these resultant forces into its x and y components.

We let R_{xp} and R_{yp} represent the shear force components in a single fastener due to P, respectively, in the x and y directions. Let R_{xm} and R_{ym} be the shear force components in the same fastener due to M, respectively, in the x and y directions.

Consider P to be replaced by its components, P_x and P_y:

$$R_{xp} = \frac{P_x}{n} \quad \text{and} \quad R_{yp} = \frac{P_y}{n} \tag{6.12}$$

In Fig. 6.10(c) let I be *any* fastener at radial distance r_i from the centroid of the fastener group (O), and R_{im} the force due to the eccentric moment. The contribution of fastener I to the moment resistance is equal to $R_{im}r_i$, and the total moment is equal to the sum of these individual contributions for all n fasteners:

$$M = \sum R_{im}r_i \tag{6.13}$$

As assumed,

$$\frac{R_{Am}}{R_{im}} = \frac{r_A}{r_i} \quad \text{or} \quad R_{im} = \frac{R_{Am}r_i}{r_A}$$

Substituting this value of R_{im} into Eq. (6.13) gives

$$M = \frac{R_{Am}}{r_A} \sum r_i^2 \tag{6.14}$$

or the force R_{Am}, due to moment, on any particular fastener is equal to

$$R_{Am} = \frac{Mr_A}{\sum r_i^2} \tag{6.15}$$

Breaking R_{Am} into its x and y components, and noting, by similar triangles, that

$$\frac{R_{xm}}{R_{Am}} = -\frac{y_A}{r_A} \qquad \frac{R_{ym}}{R_{Am}} = \frac{x_A}{r_A}$$

then

$$R_{xm} = -\frac{y_A R_{Am}}{r_A} = -\frac{M y_A}{\sum r_i^2} \tag{6.16a}$$

$$R_{ym} = \frac{x_A R_{Am}}{r_A} = \frac{M x_A}{\sum r_i^2} \tag{6.16b}$$

In calculating $\sum r_i^2$, it is convenient to use x and y component distances:

$$\sum r_i^2 = \sum x_i^2 + \sum y_i^2 \tag{6.17}$$

The x and y components of total shear force on fastener A due to both applied load and moment are

$$R_{xA} = R_{xp} + R_{xm} = \frac{P_x}{n} + \frac{M(-y_A)}{\sum r_i^2} \tag{6.18a}$$

$$R_{yA} = R_{yp} + R_{ym} = \frac{P_y}{n} + \frac{M x_A}{\sum r_i^2} \tag{6.18b}$$

Finally, the resultant force on the most-stressed fastener is determined:

$$R_A = \sqrt{R_{xA}^2 + R_{yA}^2} \tag{6.19}$$

and the connection design is adequate if

$$\text{max. } R_A \leq \text{allowable shear force}$$

In Fig. 6.10 it is obvious by means of the following reasoning that the most-stressed bolt is at A:

1. R_{xp} and R_{yp} act in the positive x and y directions.
2. A is one of four locations where R_{im} is maximum.
3. A is the only bolt for which both R_{xm} and R_{ym} act in the same direction and thereby increase the magnitude of R_{xp} and R_{yp}.

Turning now from the riveted or bolted connection to the eccentrically loaded fillet-welded connection, the analysis is essentially similar and will not be developed in as much detail. In place of any single fastener, one now considers a differential length of fillet weld dl, and in place of $\sum r_i^2$ (which is the polar moment of inertia of the pattern of unit bolt areas), we substitute for a unit width of fillet-weld throat,

$$\int r_i^2 dl = I_z$$

The calculation of I_z is illustrated in Example 6.9 which makes use of the relationship $I_z = I_x + I_y$.

Figure 6.11 shows an eccentrically loaded plate bracket welded to a column flange to provide a connection similar to the bolted bracket in Fig. 6.10. The same-size fillet weld is assumed along the three edges of the bracket plate and one hidden edge

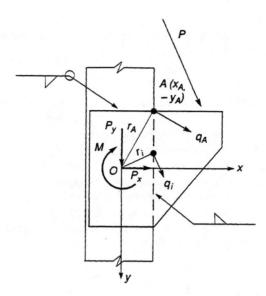

Figure 6.11 Eccentrically loaded welded connection.

of the column flange. Let the component forces in the x and y directions per unit length of weld at location A and due to applied force P be

$$q_{xp} = \frac{P_x}{l} \tag{6.20a}$$

$$q_{yp} = \frac{P_y}{l} \tag{6.20b}$$

l represents the total length of fillet weld measured at the root location.

The component shear forces per unit length at the same location due to moment can be shown to be

$$q_{xm} = -\frac{M y_A}{I_z} \tag{6.21a}$$

$$q_{ym} = \frac{M x_A}{I_z} \tag{6.21b}$$

I_z is the moment of inertia of the unit width weld about the z axis (through O in Fig. 6.11) and normal to the xy plane. As in the case of the bolted connection,

$$q_x = q_{xp} + q_{xm} = \frac{P_x}{l} + \frac{M(-y_A)}{I_z} \tag{6.22a}$$

$$q_y = q_{yp} + q_{ym} = \frac{P_y}{l} + \frac{M x_A}{I_z} \tag{6.22b}$$

Thus the resultant force per unit length of weld at point A—the determining factor in selecting the weld size—is

$$q_A = \sqrt{q_x^2 + q_y^2} \qquad (6.23)$$

The foregoing procedure may be readily extended to a general three-dimensional fillet-welded joint with the addition of a q_z force component.

Riveted and bolted connections in which eccentricity induces change in fastener tension will be considered in Section 6.7. Examples 6.8 and 6.9 illustrate the design of shear-type connections using high-strength bolts and welds, respectively.

Example 6.8 *Eccentric Loads on High-Strength Bolted Bracket*

Design a bracket connected to the faces of a W14 × 193 column to support a required girder reaction of 200 kips applied eccentrically, as shown. Using four vertical lines of bolts in each face of the column, determine the number and size of A325 high-strength bolts in friction-type connection in accordance with AISCS.

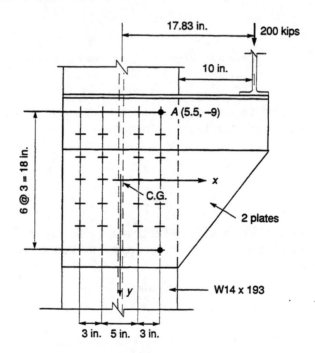

Solution   Assume seven bolts with 3-in. spacing in each vertical line. The centroid of the bolt group is at point O from symmetry.

$$P_x = 0$$

$$P_y = \frac{200}{2} = 100 \text{ kips}$$

$$M_z = 100 \times 17.83 = 1783 \text{ kip-in.} \qquad \text{(in each plate)}$$

$$n = 7 \times 4 = 28 \text{ bolts}$$

$$\sum r_i^2 = \sum (x_i^2 + y_i^2) = 14(5.5^2 + 2.5^2) + 8(9^2 + 6^2 + 3^2) = 1519 \text{ in}^2$$

By inspection, the bolt A in the right-hand upper corner (or right-hand lower corner) is the most-stressed bolt. Compute the resultant shear force R on the upper-right-corner bolt where $x = 5.5$ in. and $y = -9$ in.; then according to Eqs. (6.12), (6.16), (6.17), and (6.19),

$$R_x = \frac{P_x}{n} + \frac{M_z(-y)}{\sum r_i^2} = 0 - \frac{1783 \times (-9)}{1519} = 10.56 \text{ kips/bolt} \rightarrow (+ \text{ direction})$$

$$R_y = \frac{P_y}{n} + \frac{M_z x}{\sum r_i^2} = \frac{100}{28} + \frac{1783 \times 5.5}{1519} = 10.03 \text{ kips/bolt} \downarrow (+ \text{ direction})$$

$$R_A = \sqrt{R_x^2 + R_y^2} = \sqrt{10.56^2 + 10.03^2} = 14.56 \text{ kips/bolt}$$

Try $\frac{3}{4}$-in. bolts; the design shear force in single shear is (assuming threads are not excluded from the shear plane)

$$\phi F_v A_b = 0.75 \times 48 \times 0.442 = 15.90 \text{ kips/bolt} \qquad > R_{max} \quad \text{OK}$$

where A_b is the nominal body area of a bolt and F_v is the nominal shear stress on bolts (see AISCS, Table J3.2).

Use 28 A325 high-strength bolts of $\frac{3}{4}$-in. diameter.

Example 6.9

For a single plate bracket, welded to the face of a column, as shown in the figure, determine the size of the fillet welds required. Use E70 electrodes and A36 steel.

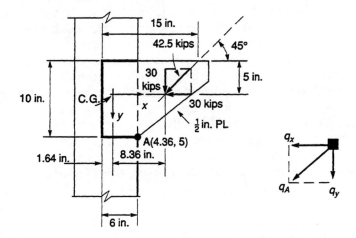

Solution   Find the center of gravity of the weld root line. Vertically, by inspection, it is 5 in. from the top or bottom of the plate, because of symmetry. Horizontally, $(2 \times 6 \times 3)/22 = 1.64$ in.

Design loads acting at c.g. of weld group:

$$P_x = -42.5 \cos 45° = -30 \text{ kips} \qquad P_y = 42.5 \sin 45° = 30 \text{ kips} \qquad P_z = 0$$

$$M_x = M_y = 0 \qquad M_z = 30 \times 8.36 = 251 \text{ kip-in.}$$

Moment of inertia:

$$I_x = 2 \times 6 \times 5^2 + \frac{1}{12} \times 10^3 = 384 \text{ in}^4/\text{in.}$$

$$I_y = 10 \times 1.64^2 + 2\left[\frac{1}{3} \times (1.64^3 + 4.36^3)\right] = 85 \text{ in}^4/\text{in.}$$

$$I_z = I_x + I_y = 384 + 85 = 469 \text{ in}^4/\text{in.}$$

Note the shearing stresses on the sketch, which indicate that the lower right edge (where $x = 4.36$, $y = -5$ and all stress components are additive) is the most highly stressed:

$$q_x = \frac{P_x}{l} + \frac{M_z(-y)}{I_z} = -\frac{30}{22} - \frac{251 \times 5}{469} = -1.36 - 2.68 = -4.04 \text{ kips/in.} \leftarrow$$

$$q_y = \frac{P_y}{l} + \frac{M_z x}{I_z} = \frac{30}{22} + \frac{251 \times 4.36}{469} = 1.36 + 2.33 = 3.69 \text{ kips/in.} \downarrow$$

$$q_A = \sqrt{q_x^2 + q_y^2} = \sqrt{4.04^2 + 3.69^2} = 5.47 \text{ kips/in.}$$

From Table 6.1, E70 weld value $\bar{q} = 1.39$ kips/in. Weld size required is $\frac{1}{16}s = N = 5.47/1.39 = 3.9$. Use $\frac{1}{4}$ in.

6.6 Shear Connections for Building Frames

A variety of beam-to-column or beam-to-girder connections is available to support simple beam reactions. They are purposely made flexible with regard to rotation between the ends of the beam and the column or girder. These are designated by AISCS as simple connections and are used in structures for which lateral forces do not need to be considered, or where other bents in the building resist wind and seismic forces by frame action, truss framing, or shear walls. Flexible connections for reactions may involve attachment only to the beam web, as shown in Fig. 6.12, or may consist of top and bottom angles, respectively, designated by AICS as "framed beam connections" and "seated beam connections." They may include rivets, bolts, or welds, alone or in combination. The AISCM provides descriptive and tabular design information covering the most commonly used types. These connections do develop a certain amount of moment, which may amount to 10% of the full fixed end moment or even more. However, these moments are disregarded in design. A nominal end

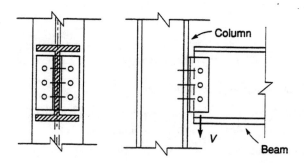

Figure 6.12   Flexible framing angle connection.

clearance or "setback" between beam end and column of $\frac{1}{2}$-in. is assumed, but connections are designed for a $\frac{3}{4}$-in. setback to allow for possible underrun in beam length.

In the case of the riveted or bolted stiffened seat angle, the fasteners that attach the angles to the column or girder are in combined shear and tension due to the eccentricity of the applied load. The bending moment is transmitted through the connection by tension in the upper fasteners and by bearing pressure in the lower part between the angles and column, as shown by the shaded area in Fig. 6.13. The equivalent effective

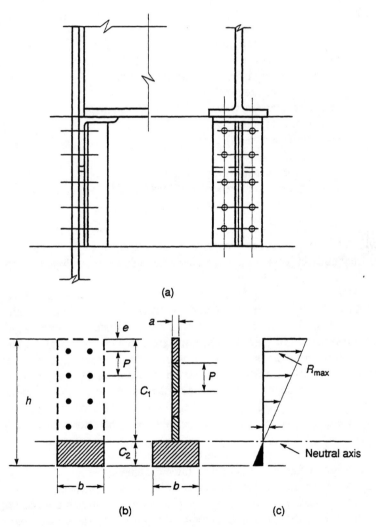

(a)

(b)   (c)

Figure 6.13 Seat angle connection with fasteners subjected to tension: (a) actual stress portions; (b) equivalent cross section; (c) stress distribution.

area in transmitting the bending moment is shown in Fig. 6.13(b) in which area ac_1 equals the area of the eight fasteners and whose stress distribution, due to moment, is shown in Fig. 6.13(c). Then

$$a = \frac{mA}{p} \tag{6.24}$$

$$c_1 = \frac{\sqrt{b}}{\sqrt{a} + \sqrt{b}} h \tag{6.25}$$

$$R_{max} = \frac{M}{I}(c_1 - e)A \tag{6.26}$$

where a = width of the equivalent fastener area
　　p = pitch of the fasteners
　　m = number of the fasteners per horizontal row
　　A = cross-sectional area of a fastener
　c_1, c_2 = distance between neutral axis and extreme fibers
　　b = width of the framing angles or T
　　h = height of the framing angles or T
　R_{max} = maximum tensile load in a fastener
　　I = moment of inertia = $(ac_1^3 + bc_2^3)/3$
　　e = edge distance between the maximum tensile load of the fastener and the edge of the framing angles

The procedure illustrated in Fig. 6.13 applies only to the specialized case in which the vertical spacing of the fasteners is uniform. However, this happens very frequently in practice, and the equations that were developed are very simple to use.

If the vertical spacing of the fasteners varies, we would have a varying thickness (a) that could not be used in these equations. Therefore, we must use another method, in which we consider the moments of the individual fasteners about the neutral axis. The neutral axis usually lies between one-sixth to one-seventh of the length of the connection (h) from the bottom of the connection. This assumption is only used to estimate the number of fasteners in tension (above the neutral axis). An equation is then written in which the moment of the compression area about the neutral axis is equated to the moment of the areas of the tension fasteners about the neutral axis, since these must be equal. The resulting answer will indicate whether we have made a correct assumption.

Once the neutral axis is located, we can calculate the moment of inertia. Next, using the flexure formula, we can determine the tensile stress in the critical fasteners and the compression stress on the extreme fiber of the connection.

An application of the preceding procedure is illustrated in Example 6.11. Other connection design principles required in the examples of beam shear connections that follow are based on procedures that have been covered previously in this chapter.

Example 6.10   *Seated Beam Connection (Bolted or Riveted)*

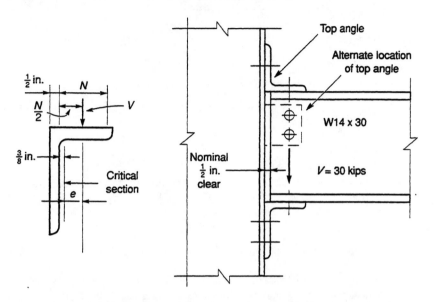

Solution   For W14 × 30, $b_f = 6.73$ in., $t_w = 0.27$ in., and $k = 0.875$. The force $V = 30$ kips is composed of a dead load of 20 kips and a live load of 10 kips. The factored ultimate load is $1.2 \times 20 = 1.6 \times 10 = 40$ kips. The length of bearing required is

a. *Local web yielding* (AISCS, Sec. K1.3):

$$\phi = 1 \qquad R_n = (2.5k + N)F_y t_w = V = 40 \text{ kips}$$

$$N \geq \frac{40}{36 \times 0.27} - 2.5 \times \frac{15}{16} = 1.77 \text{ in.}$$

b. *Web Crippling* (AISCS, Sec. K1.4): Assume that $N/d > 0.2$ [Eq. (K1-5b)]

$$\phi = 0.75 \qquad R_n = 68 t_w^2 \left[1 + \left(\frac{4N}{d} - 0.2 \right) \left(\frac{t_w}{t_f} \right)^{1.5} \right] \sqrt{\frac{F_y t_f}{t_w}}$$

Solving for N after setting $R_n = 40/0.75 = 53.33$ kips; we get $N = 3.65$ in.
$N/d = 3.65/13.84 = 0.26 > 0.2$. Equation (K1-5b) was the correct equation to use.

Use a 4-in. leg, 10 in. long. Assume $\frac{7}{8}$-in. angle thickness (governed by bending); the distance from the back of the angle to the critical section in bending is $\frac{7}{8} + \frac{3}{8} = 1.25$ in. It is usually considered that the load is concentrated at the center of the required bearing (3.65 in. in this example). Allow 0.25 in. underrun.

$$\text{eccentricity} = 0.5 + 0.25 + 3.65/2 - 1.25 = 1.325 \text{ in.}$$
$$M = 40 \times 1.325 = 53.0 \text{ kip-in.} \qquad \phi = 0.9$$

Plastic moment of a rectangular section:

$$M_p = \frac{bt^2 F_y}{4}$$

Solving for the angle thickness yields

$$t \geq \sqrt{\frac{4M}{\phi b F_y}} = \sqrt{\frac{4 \times 53}{0.9 \times 10 \times 36}} = 0.809$$

Since $t = 0.875$ in. the bending criterion is satisfied.
Use $L6 \times 4 \times \frac{7}{8} \times 10$ in. long.

Alternative procedures for attachment to the column are:

a. *Riveted:* $\frac{3}{4}$-in. A502 grade 1 rivets good for $0.75 \times 25 \times 0.4418 = 8.28$ kips (AISCS, Table J3.2):

$$n = \frac{40}{8.28} = 4.84$$

Use six rivets.

b. *Bolted:* $\frac{3}{4}$-in. A325 bolts in friction-type connection with threads not excluded from shear plane, good for $1.0 \times 17 \times 0.4418 = 7.51$ kips (AISCS, Table J3.6):

$$n = \frac{30}{7.51} = 3.99$$

Use four bolts. Note that the unfactored load of 30 kips was used since connection slip is a serviceability condition.

Connections of the beam to the seat angle, the beam to the top angle, and the top angle to the column can be made using two $\frac{3}{4}$-in.-diameter unfinished bolts. The top angle simply holds the beam in a vertical position and must be flexible to permit simple beam end rotation. Use $L4 \times 4 \times \frac{1}{4} \times 8$ in. long.

Example 6.11

Design a riveted stiffened seat connection for a W18 × 46 and an unfactored end reaction of 40 kips (10 kips dead load and 30 kips live load). All material is A36 steel. Refer to the figure at top of next page. Sketches (b) and (c) give additional details of the seat, and sketch (d) is for use in calculating rivet stress.

Solution   For W18 × 46, $b_f = 6.06$ in., $t_w = 0.360$ in., and $k = 1.25$ in.
The required reaction is $R_u = 1.2 \times 10 + 1.6 \times 30 = 60$ kips. Required bearing length N:

a. *Web yielding* (AISCS, Sec. K1.3, $\phi = 1.0$):

$$N \geq \frac{R_n/\phi}{F_y t_w} - 2.5\,k = \frac{60/1.0}{36 \times 0.36} - 2.5 \times 1.25 = 1.505 \text{ in.}$$

b. *Web crippling* (AISCS, Sec. K1.4, $\phi = 0.75$): Try Eq. (K1-5a), assuming that $N/d < 0.2$.

$$N \geq \frac{d}{3(t_w/t_f)^{1.5}} \left[\frac{R_n/\phi}{68 t_w^2 \sqrt{F_y t_f/t_w}} - 1 \right]$$

$$= \frac{18.06}{3 \times (0.36/0.605)^{1.5}} \left[\frac{60/0.75}{68/0.36^2 \times \sqrt{(36 \times 0.605)/0.36}} - 1 \right] = 2.19 \text{ in.}$$

$N = 3.25$ in. > 2.19 in.   OK

$$\frac{N}{d} = \frac{2.19}{18.06} = 0.12 \qquad < 0.2$$

Equation (K1-5a) was the correct equation.

Assume a nominal setback of 0.5 in. plus 0.25 in. allowance for a beam length underrun. Eccentricities of load for the stiffened seats will be assumed as 0.75 in. plus half the remaining seat length.

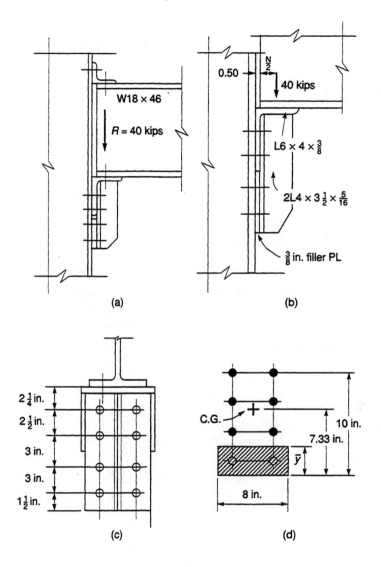

(a)   (b)

(c)   (d)

Assumed eccentricity of end reaction:

$$e = 0.75 + \frac{3.25}{2} = 2.38 \text{ in.}$$

$$\text{moment} = 2.38 \times 60 = 142.8 \text{ kip-in.}$$

Try $\frac{3}{4}$-in. A325 slip-critical bolts with standard holes.

Assume the neutral axis to be above the two lowest rivets and locate the center of gravity (c.g.) of the upper six rivets. From the bottom of the vertical angles to the c.g. the distance is

$$\frac{2 \times 4.5 + 2 \times 7.5 + 2 \times 10}{6} = 7.33 \text{ in.}$$

Locate the neutral axis [see sketch (d) and the discussion of Fig. 6.13].

$$8\bar{y}\frac{\bar{y}}{2} = 6 \times 0.442(7.33 - \bar{y})$$

$$4\bar{y}^2 + 2.65\bar{y} - 19.4 = 0$$

Solving,

$$\bar{y} = 1.90 \text{ in.}$$

$$I = 8 \times \frac{1.90^3}{3} + 2 \times 0.442(2.6^2 + 5.6^2 + 8.1^2) = 110 \text{ in}^4$$

The tensile force in the top bolt is

$$T = 142.8 \times \frac{10 - 1.90}{110} \times 0.442 = 4.65 \text{ kips}$$

The minimum bolt tension is $T_b = 28$ kips (AISCS, Table J3.1). The tensile force reduction factor (AISCS, Sec. J3.9A) is

$$1 - \frac{T}{T_b} = 1 - \frac{4.65}{28} = 0.834$$

The shear stress is calculated for the unfactored service loads for slip-critical joints.

$$f_v = \frac{40}{8 \times 0.442} = 11.3 \text{ ksi}$$

The design shear strength is

$$\phi F_v\left(1 - \frac{T}{T_b}\right) = 1.0 \times 17 \times 0.834 = 14.2 \text{ ksi} \qquad > 11.3 \text{ ksi} \quad \text{OK}$$

The required bearing area is (AISCS, Sec. J8)

$$A_{pb} \geq \frac{R_u}{\phi \times 1.8F_y} = \frac{60}{0.75 \times 1.8 \times 36} = 1.235 \text{ in}^2$$

Example 6.12
Redesign the stiffened seat of Example 6.11 as all welded, using A36 steel and E70 electrodes. Use a structural tee section, 4 in. in length, as shown.

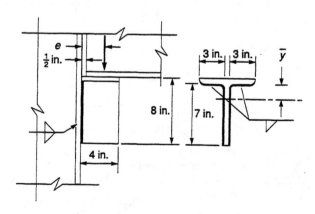

Solution   End eccentricity and moment are the same as for Example 6.11. Try WT8× 28.5 (which has a wider flange than the beam, to permit downhand fillet welds between beam and seat). Assume effective weld lengths as shown in the sketch, and locate the c.g. of the weld pattern (assume unit width of weld):

$$\bar{y} = \frac{3.5 \times 14.0}{20} = 2.45 \text{ in.}$$

$$I_x = 6 \times 2.45^2 + \frac{2}{3}(2.45^3 + 4.55^3) = 108.6 \text{ in}^4/\text{in.}$$

Assume bearing of stem on column at bottom, making the resultant weld stress at the top the critical value. As a result of the eccentric moment,

$$q_H = \frac{142.8 \times 2.45}{108.6} = 3.22 \text{ kips/in.}$$

$$q_v = \frac{60}{20} = 3.0 \text{ kips/in.}$$

$$\text{resultant } q_R = \sqrt{3.22^2 + 3.0^2} = 4.40 \text{ kips/in.}$$

$$N = \frac{q_R}{q} = \text{weld size in } \frac{1}{16}s = \frac{4.40}{1.39} = 3.16 \quad \text{(Table 6.1)}$$

Use $\frac{1}{4}$-in. fillet welds, which is the smallest size that meets the requirements of AISCS, Table J2.4. Still greater size would be required if the column flange thickness exceeds $\frac{3}{4}$ in.

6.7 Moment-Resisting Connections

The two types of moment-resisting connections classified in Sec. A2.2 of AISCS as "rigid" (type FR, fully restrained) and "semirigid" (type PR, partially restrained) were discussed in Section 6.1.

Rigid connections are used in continuous-frame construction that resists lateral forces caused or induced by wind or earthquake. Rigid connections are also a requirement for frames that are proportioned according to plastic design, in which case the connections must be strong enough to develop complete yield moment at adjacent *plastic hinges*. Rigid connections are always advantageous if a building is loaded accidentally by explosive blast, earthquake, or high wind beyond its intended normal-use load. In such cases continuous-frame behavior, whether or not proportioned by plastic design, will provide a residual strength against ultimate collapse that may be a life-saving feature.

Semirigid connections are used in semicontinuous-frame construction primarily in office or apartment buildings of moderate height. The concept is intended to provide an economical balance between simple beam design, for which the maximum bending moment of $0.125wL^2$ is at the center, and fully continuous construction, for which the maximum moment of $0.083wL^2$ is at the ends in the fully fixed end condition. In a beam with semirigid connections, the end and center moments for uniform gravity load could, ideally, be balanced at $0.0625wL^2$, thus achieving savings in the weight of the beam.

The use of structures with semirigid joints is gaining popularity because extensive research in the 1980s and 1990s has given the designers experimentally verified analytical/computerized tools to deal not only with the determination of the strength of these connections but also with the nonlinear flexibilities that need to be included in a correct structural analysis. This analysis is considerably more complicated than what can be profitably treated in this book, and the reader is referred to the specialized

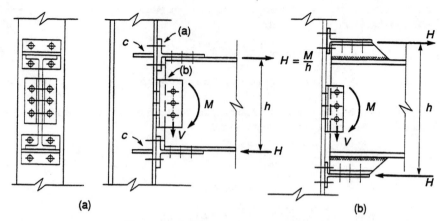

Figure 6.14   Moment and shear connection.

literature in this field. For example, the text by Chen and Lui (Ref. 6.1) has several chapters dealing with this subject. In Example 6.13 we illustrate the design of a semirigid connection, but we will only show how its *strength* is to be determined. If it were necessary also to calculate to deformations of the frame in which this connection occurs, a more involved analysis would need to be performed.

In current practice it is customary to label a connection as "semirigid" if it is designed for some specific moment capacity at any level less than that required to develop full continuity.

Figure 6.14 shows a riveted or bolted moment and shear connection that may be designed as either semirigid or rigid. It is a type that introduces the problem of prying action that increases bolt or rivet force. Another topic that is covered briefly concerns the required strenghthening of the column web against local deformation adjacent to the beam flanges.

In the connection shown in Fig. 6.14(a) the bending moment (M) is transmitted to the column flange by two tees (a) on the top and bottom flanges of the beam by tension and compression loads (H), respectively. The shear load (V) is transmitted to the columns by the two angles (b) connected to the beam web. Then

$$H = \frac{M}{h}$$

where h is the depth of the beam. It should be noted that the moment capacity of this connection can be increased by increasing the distance between the tees, as shown in Fig. 6.14(b). This greater distance will decrease the magnitude of force (H).

The top-flange moment connection shown in Fig. 6.14(a) transmits the applied moment to the column by means of fasteners acting in tension. Consequently, the connecting element (the tee section) is subjected to a bending stress and deforms as shown in Fig. 6.15, thereby creating a prying action. This prying action, in turn, creates a pressure (prying force ΔH) at the outer edge of the T and thus adds directly to the total tension in the fasteners.

Research on the behavior of bolts in T connections has led to the development of empirical formulas that approximate the prying force on the basis of simplifying

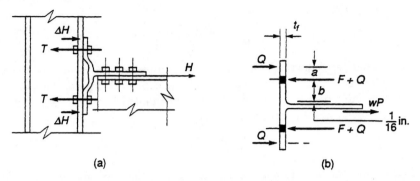

Figure 6.15   Prying action in connection.

and conservative assumptions. The prying force will be minimized and the desirable stiffness of the T flange will be enhanced if the dimension b in Fig. 6.15(b) is kept as small as erection clearance requirements will permit.

The determination of the prying force has undergone a number of evolutionary changes since the problem was first recognized, and the present method, which is used in Sec. 11 of AISCM, is conservative but less so than the methods used previously. AISCM bases its semiempirical approach on the recommendations contained in the book by Kulak, Fisher, and Struik (Ref. 6.2). Following is an outline of the steps in the prying force analysis given in AISCM.

Let $a' = a + d/2$ and $b' = b - d/2$, where a and b are defined in Fig. 6.15; a should not be taken greater than $1.25b$, and d is the nominal bolt diameter. Also, $p = b'/a'$. Next, a coefficient β is defined by the following formula:

$$\beta = \frac{1}{\rho}\left(\frac{\phi r_n}{r_{ut}} - 1\right) \tag{6.27}$$

where ϕr_n is the design strength of one bolt as determined from Table J3.2 of AISCS and r_{ut} equals the required force per bolt. If $\beta \geq 1$, set $\alpha' = 1$, and if $\beta < 1$, then $\alpha' = 1/\delta[\beta/(1 - \beta)]$, where $\delta = 1 - d'/p$. The symbol d' denotes the hole diameter $(d + \frac{1}{16}$ in.$)$ and p is the distance between holes perpendicular to the beam flange. The required thickness of the T section to account for both the strength of the flange of the T and the strength of the bolts is

$$t_{req} = \sqrt{\frac{4.44 r_{ut} b'}{p F_y (1 + \delta \alpha')}} \tag{6.28}$$

The factored prying force per bolt is determined from the following equations:

$$\alpha = \frac{1}{\delta}\left[\frac{r_{ut}}{\phi r_n}\left(\frac{t}{t_c}\right)^2 - 1\right] \geq 0 \tag{6.29}$$

$$q_u = \phi r_n \left[\delta \alpha \rho \left(\frac{t}{t_c}\right)^2\right] \tag{6.30}$$

where t_c is the flange thickness required to develop a design strength so that prying action is negligible.

$$t_c = \sqrt{\frac{4.44\phi r_n b'}{pF_y}} \tag{6.31}$$

When the flanges are thick and the gage lines are closely spaced, the prying action will be very small. In the preliminary selection of flange thickness it may be assumed that a point of inflection exists at the midlength of dimension b, in which case the flange moment per lineal inch is

$$M = \frac{Pb}{4} \tag{6.32}$$

In the final check of flange bending adequacy, the maximum moment may be assumed to be the grater of two values: at the fastener line, the total moment per fastener,

$$M_2 = Qa \tag{6.33}$$

and $\frac{1}{16}$ in. from the face of the T stem,

$$M_1 = (F+Q)b - Q(a+b) = Fb - Qa \tag{6.34}$$

The foregoing nomenclature corresponds to that used in AISCM.

The design of column web stiffeners [plates (c) in Fig. 6.14] is covered by Sec. K1 of AISCS. Such stiffeners may be needed to prevent excessive local deformation in the column web and flange at locations near the top and bottom of the beam flanges where the moment resisting plates or tees are attached to the column flange.

Example 6.13 will illustrate the design of a semirigid beam-to-column connection with a moment resistance less than that required for full continuity.

Example 6.13 *Semirigid High-Strength Bolted Beam-to-Column Connection*
The connection is to be designed for a shear of 100 kips and a bending moment of 175 kip-feet using a design similar to that shown in Fig. 6.14(a) (see also the accompanying sketch). Forces are assumed to be due to maximum combined unfactored dead, live, and wind loads equal to 80 kips. Use A325 high-strength bolts and A36 steel.

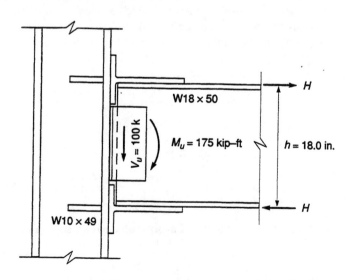

Solution   *Moment connectors:* The tension and compression forces transmitted by the tees, due to moment, at the top and bottom of the beam flanges are

$$H = \frac{M}{h} = \frac{175 \times 12}{18} = 117 \text{ kips}$$

1. *Required thickness of T flanges:* Flange width of the W18 × 50 beam is 7.5 in. Use the same length of T section for the top and bottom connectors, selected initially on the basis of required flange thickness.

   Prying action will be neglected in the initial selection of the tees. Assuming a bolt gage of $g = 4$ in., we determine the tentative flange thickness on the assumption of full bending fixity at the bolt line. Bending moments in the flange are approximated as shown in the accompanying sketch.

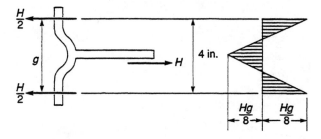

   Moment in the T flange:

$$m = \frac{Hg}{8} = \frac{117 \times 4}{8} = 58.3 \text{ kip-in.}$$

   Required thickness:

$$t = \sqrt{\frac{4m}{b\phi F_y}} = \sqrt{\frac{4 \times 58.3}{7.5 \times 0.9 \times 36}} = 0.98 \text{ in.}$$

   where b is the length of T. Try a WT9 × 59.5:

$$t_f = 1.06 \text{ in.} \qquad > 0.98$$
$$t_w = 0.655 \text{ in.} \qquad b_f = 11.265 \text{ in.}$$

2. *T web to beam flange** (use $\frac{7}{8}$-in. A325 bolts in a bearing-type connection, threads excluded from shear plane):

$$\phi = 0.75 \qquad F_v = 60 \text{ ksi} \quad \text{(AISCS, Table J3.2)}$$
$$\text{number of bolts required} = \frac{117}{0.75 \times 60 \times 0.601} = 4.33$$

   Use six bolts, in two rows.

3. *Check the tensile capacity of the T web* :

   a. *Gross section:*

$$H = 7.5 \times 0.655 \times 0.9 \times 36 = 159 \text{ kips} \qquad > 117 \text{ kips} \quad \text{OK}$$

   b. *Net section:*

$$H = (7.5 - 2.0) \times 0.655 \times 0.75 \times 58 = 157 \text{ kips} \qquad > 117 \text{ kips} \quad \text{OK}$$

* Alternatively, shop welding could have been used.

4. *Check the bolt bearing stress:*

$$N\phi R_n = 6 \times 0.75 \times 2.4 \times d \times t \times F_u = 6 \times 0.75 \times 2.4 \times 0.875 \times 0.655 \times 58$$
$$= 359 \text{ kips} \quad > 117 \text{ kips} \quad \text{OK} \quad \text{(AISCS, Sec. J3.10)}$$

5. *Bolts connecting T to column flange* (neglect prying action in initial trial selection):

nominal tensile stress, A325 bolts, $F_t = 90$ ksi   (AISCS, Table J3.2)

$$\text{trial number of bolts required} = \frac{117}{0.75 \times 90 \times 0.601} = 2.88$$

Try four $\frac{7}{8}$-in.-diameter A325 bolts. Refer to Fig. 6.15.

Determine the prying force.

$$b = \frac{g - t_w}{2} - \frac{1}{16} = \frac{4 - 0.655}{2} - \frac{1}{16} = 1.61 \text{ in.}$$

$$a = \frac{b_f - g}{2} = \frac{11.265 - 4}{2} = 3.63 \text{ in.} \qquad > 1.25b = 2.01 \text{ in.; use } a = 2.01 \text{ in.}$$

$$b' = b - \frac{d}{2} = 1.61 - \frac{0.875}{2} = 1.17 \text{ in.}$$

$$a' = a1 + \frac{d}{2} = 2.01 + \frac{0.875}{2} = 2.45 \text{ in.}$$

$$\rho = \frac{b'}{a'} = 0.479$$

$$\phi r_n = 0.75 \times 90 \times 0.601 = 40.57 \text{ kips/bolt}$$

$$r_{ut} = \frac{117}{4} = 29.25 \text{ kips/bolt}$$

$$\beta = \frac{1}{\rho}\left(\frac{\phi r_n}{r_{ut}} - 1\right) = \frac{1}{0.479}\left(\frac{40.57}{29.25} - 1\right) = 0.808 \qquad < 1$$

$$\delta = 1 - \frac{d'}{p} = 1 - \frac{0.875 + 1/16}{4} = 0.766$$

$$\alpha' = \frac{1}{\delta}\frac{\beta}{1 - \beta} = \frac{1}{0.766}\left(\frac{0.808}{1 - 0.808}\right) = 5.49 \qquad > 1 = 1.0$$

$$t_{req} = \sqrt{\frac{4.44 r_{ut} b'}{p F_y (1 + \delta\alpha')}} = \sqrt{\frac{4.44 \times 29.25 \times 1.17}{4 \times 36 (1 + 0.766 \times 1)}} = 0.773 \text{ in.}$$

$$< t_f = 1.06 \text{ in.} \quad \text{OK}$$

Since the actual flange thickness is less than the required thickness, both the flange and the bolts are satisfactory for resisting the prying force.

Shear connectors:

1. *Angles to beam web** (use $\frac{7}{8}$-in.-diameter A325 bolts in friction-type connection). The allowable double shear strength per bolt is

$$2\phi F_v A_b = 2 \times 1.0 \times 17 \times 0.601 = 20.43 \text{ kips}$$

$$\text{number of bolts required} = \frac{80}{20.43} = 3.92$$

---

* Alternatively, shop welding could have been used.

Use four $\frac{7}{8}$-in.-diameter A325 bolts, angles to beam web. Use eight $\frac{7}{8}$-in.-diameter A325 bolts, angles to column flange. (The single shear value of these is one-half double shear.)

2. *Required thickness of the shear connector angles* (try 9-in. length): The web angles must be checked for the shear-rupture strength along the line of the three bolts as per Sec. J4.1 of AISCS. The spacing between the bolts is 3 in., and the edge distances are 1.5 in. The capacity is determined from ϕR_n, where $\phi = 0.75$ and $R_n = 0.6F_u A_{nv}$. The shear-rupture area is

$$A_{nv} = t\,[9-3(\tfrac{7}{8}-\tfrac{1}{16}-\tfrac{1}{16})] = 6t \qquad t_{req} = \frac{100}{2 \times 0.75 \times 0.6 \times 58 \times 6} = 0.319 \text{ in.}$$

Here b is the length of the angles. Use two angles, $4 \times 3\frac{1}{2} \times \frac{3}{8} \times 9$ in. web angles.

Design column web stiffeners, if needed: Local web yielding (AISCS, Sec. K1.3) and web crippling (AISCS, Sec. K1.4) must be checked to determine if it is necessary to place column web stiffeners. First we consider local web yielding. In AISCS Eq. (K1-2) the $5k$ term may be modified by adding $2t_f$ to it to allow for the additional spread of force 1:1 slope through the flange of the T section. N in the equation becomes the web thickness of the T connector. The web thickness of the W10×49 column is 0.34 in. and its $k = 1.875$ in. $\phi = 1.0$ and $R_n = (5k + 2 \times t_f + N)\,t_w F_y$.

$$\phi R_n = (5 \times 1.1875 + 2 \times 1.06 + 0.655) \times 0.34 \times 36 = 107 \text{ kips} \qquad < 117 \text{ kips}$$

Column web stiffeners are needed. These should be placed as shown on the sketch up to the center of the column web. The stiffeners should have about the same thickness as the thickness of the stem of the T connector and they should be close to the width of the column flanges. Therefore, select $\frac{3}{4}$ in. × 6 in. × 7 in. plates. The width thickness ratio is 6/0.75 = 8.8, which is less than $95/\sqrt{F_y} = 15.8$ (AISCS, Table B5).

Further study of moment-resisting connections designed for full continuity of frame action will be made in Chapter 8.

6.8 Bolted End-Plate Connections

The bolted end-plate connection, illustrated in Fig. 6.16, has become increasingly popular. Shop-welded under controlled conditions and field-bolted, it can be designed for full end restraint and in many areas is the most economical type to use if continuity of frame action is utilized.

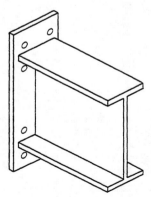

Figure 6.16   Bolted end-plate connection.

Design practice had been patterned after the procedures described in Example 6.13 on the assumption that the upper and lower portions of the end plates are essentially similar in behavior to equivalent structural tees. However, the continuity of the full plate, its attachment to the beam web, and the fact that the design of the connection with structural tees is already based on overly conservative assumptions make use of this procedure grossly overconservative for end-plate design. The correctness of this statement has been demonstrated by recent extensive research sponsored in part by AISC. These studies have resulted in empirical equations which are used as the basis for the development of the connection strength tables in Chapter 10 of AISCM.

6.9 Concluding Remarks Concerning Connections

As has been stated, connections are extremely vital elements in a structure. Only a few standard types have been treated herein, and no textbook can really cover the subject adequately. The complexity and variety of connection details is infinite, and their design can only be partially covered by any specification. In special situations connection design requires the application of engineering judgment and experience to a much greater degree than does the design of simple beams and columns. Consider, for example, Fig. 6.17, showing a complex intersection of columns and diagonal elements, shop-welded and preassembled for proper fit up of high-strength bolted field connections. The weldment shown is part of a 70-ton assembly to support a 350-ft TV antenna on top of the north tower of the World Trade Center. The assembly provides the transition from the structural framing of the building to the eight bearing plates for

Figure 6.17 Shop assembly of welded and bolted connection for World Trade Center. (Courtesy of Montague-Betts Company, Lynchburg, Virginia.)

the antenna 12 ft above roof level. It was shop-assembled in an inverted position (as pictured) and all field connections were match-drilled.

REFERENCES

6.1. W. F. CHEN AND E. M. LUI, *Stability Design of Steel Frames*, CRC Press, Boca Raton, FL, 1991.

6.2. G. L. KULAK, J. W. FISHER, AND J. H. A. STRUIK, *Guide to Design Criteria for Bolted and Riveted Joints*, 2nd ed., Wiley, New York, 1987.

PROBLEMS

6.1. Design a butt splice similar to that illustrated in Example 6.2, using $\frac{7}{8}$ -in.-diameter A325 bolts in a friction-type connection, for A36 steel. Select plate sizes adequate to supply net section to transmit a load of 210 kips in tension and determine the number of bolts required.

6.2. Design a bracket connection similar to that of Example 6.3. Select two angles adequate for a tensile force of 210 kips, deducting for a single row of $\frac{7}{8}$ -in-diameter A490 bolts that connect the angles to the web of the WT. Design the connection between flanges of the tee and column using A490 friction-type bolts. Omit design for prying action. Use A36 steel.

6.3. Rework Example 6.4, with a horizontal force component of 200 kips and a vertical force component of 150 kips.

6.4. Redesign the bracket connection, Problem 6.2, using welds made with E70 electrodes. Follow the general pattern of Example 6.7, but include selection of angle sizes based on the tensile requirement for the gross section.

6.5. Assume in Example 6.4 that the horizontal component of force is transmitted to the pin plate and channel web in proportion to their respective thicknesses. Select a suitable fillet-weld size for the attachment of pin plate to channel web. Use E70 electrodes.

6.6. Design the eccentrically loaded bracket shown. Assume that there are two plates, one hidden from view on the far side of the column flanges, so that the 150-kip load is divided equally to the two plates. Use A490 $\frac{7}{8}$ -in. friction-type high-strength bolts in single shear. Determine the required number. Refer to Example 6.8.

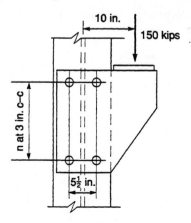

6.7. Redesign as welded the eccentrically loaded bracket of Problem 6.6 using the dimensions indicated, A36 steel, and E70 electrodes.

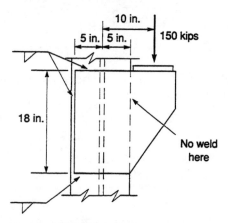

6.8. Rework Problem 6.7, but now incline the 150-kip load outward and down at 60° from the horizontal.

6.9. Design an HS bolted seat and top angle beam-to-column connection for a reaction of 50 kips. The beam is a W16 × 36 with $F_y = 36$ ksi. Use $\frac{7}{8}$-in. friction-type high-strength bolts.

6.10. Design a stiffened seat for a reaction of 100 kips using HS friction-type bolts, a W16× 36 beam, and $F_y = 36$ ksi.

6.11. Rework Problem 6.10, but now do an all-welded design using E70 electrodes.

6.12. Following the pattern in the moment connection section of AISCM, design a shop-welded and field-bolted moment connection for a W21×62 beam framed to a W14×61 column. The required moment is 300 kip-ft and the end reaction is 75 kips. All materials are ASTM A36 steel. Use A325 bearing-type bolts and E70 electrodes.

6.13. Using Example 6.13 as a guide, design a bolted semirigid beam-to-column connection, using top and bottom tees. A W24 × 76 beam is connected to a W14 × 68 column for a required reaction of 100 kips and a required bending moment of 150 kip-ft. All steel is type A36. Check for prying action.

Note: A large number of additional problems may be developed simply by requiring a check verification of any of the tabular shear and/or moment values of connections described in the connection section of AISCM.

7

Plate Girders

7.1 Introduction

Plate girders are built-up steel beams that require a section modulus greater than any rolled beam can offer. The most common form consists of two heavy flange plates between which is welded a relatively thin web plate. Girder depths range up to 20 ft or more and spans of several hundred feet are not unusual. At points of concentrated load or reaction the girder webs usually must be reinforced by *bearing stiffeners* to distribute the concentrated local forces into the web. *Intermediate* and/or *longitudinal stiffeners* may be added to serve in a quite different role—primarily that of increasing the buckling strength and thereby improving web effectiveness in resisting shear, moment, or combined stresses. Longitudinal stiffeners permit greatly increased web depths and corresponding increased spans while maintaining small web thickness. Reduction in required web thickness has also resulted from the use of the tension-field concept that permits utilization of the postbuckling strength of the girder web.

Plate girders are particularly favored for highway bridges; they permit unlimited vision and minimize clearance problems in traffic interchanges and complex multilevel overpasses. Plate girders are also frequently used in various types of buildings and industrial plants to support heavy loads. They are often used, for example, to provide a large space with no interfering columns on a lower floor of a high-rise building, as shown in Fig. 7.1.

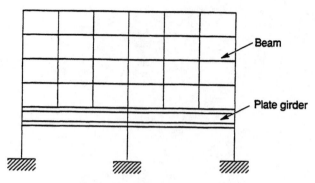

Figure 7.1 Plate girders in building.

Plate girders may be built up with bolts or rivets, as shown in Fig. 7.2(a) and (c), or welded, as shown in Fig. 7.2(b), (d), and (e). In situations where lateral support of the compression flange cannot be provided, box girders [Fig. 7.2(c) and (d)] are especially recommended because of their superior effectiveness against lateral-torsional

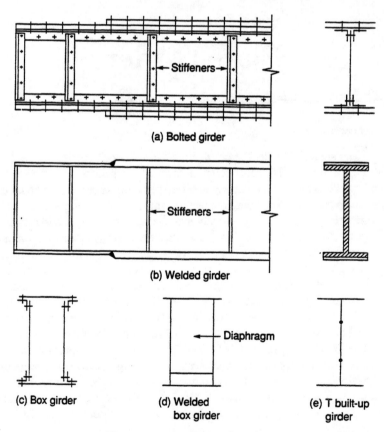

Figure 7.2 Typical types of plate girders.

buckling and in resisting lateral loads. This results from their greater strength and stiffness in torsion and in bending about the weak axis. The design of box members in torsion as well as bending is treated in Chapter 11.

Figure 7.3 shows a long cantilever section of a plate girder being lifted into position for field splicing during the construction of an aircraft hangar.

Figure 7.3 Plate girders supporting roof of United Airlines hangar in San Francisco, consisting of center span with 142-ft cantilevers. (Courtesy of American Institute of Steel Construction.)

7.2 Selection of Girder Web Plate

The selection of the web plate as shown in Fig. 7.4 involves the following steps:

1. Choose a web depth in relation to the span.
2. Choose the minimum thickness in terms of the permissible depth/thickness ratio.

7.2.1 Choose Web Depth in Relation to Span

The depth of girders ranges from approximately one-twentieth to one-sixth of the length, depending on span and load requirements. The shallower girders are desirable if the service loads are light; deeper girders will be needed if the loads are heavy or if it is desired to keep deflections at a minimum. It may be desirable to make several preliminary designs with corresponding cost estimates to achieve an optimum depth.

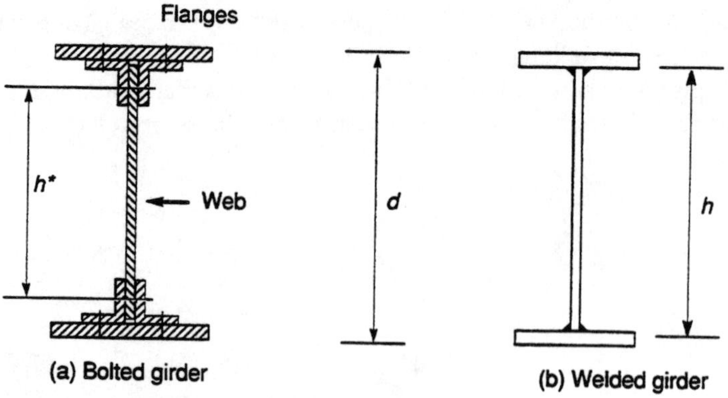

Flanges

h^* ← Web d h

(a) Bolted girder   (b) Welded girder

h^* as shown to be used as clear depth in
calculating web h/t ratios, but h = full web
depth in calculating web shear stress

Figure 7.4   Typical plate girder.

7.2.2 Choose Web Thickness in Terms of Permissible Depth/Thickness Ratio

The AISCS for plate girders with intermediate stiffeners permits the girder web at ultimate loads to go into the postbuckling range to develop tension-field action. After a stiffened thin web panel buckles in shear, it can continue to resist increasing load. The buckled web can resist diagonal tension (left half of Fig. 7.5) much as the diagonals (right half of Fig. 7.5) perform their function as tension members in a truss. The diagonal web tensions create compressive forces in the intermediate stiffeners, and these vertical stiffeners must be designed to meet this added requirement, thus acting as indicated in a manner analogous to the vertical members of the truss. After the initial buckling of a plate girder web, the girder stiffness decreases, and the girder deflection may reach a value greater than predicted by ordinary bending theory.

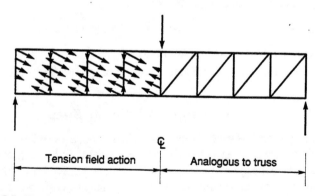

Tension field action   Analogous to truss

Figure 7.5   Tension-field action in plate girder web analogous to truss with tension diagonals.

When the postbuckling strength of the web plate is utilized, the criterion for permissible depth/thickness ratio of web is still determined by buckling considerations, but these arise from the fact that under stress the curvature of a plate girder creates vertical compression in the web due to a downward component of flange stress in a curved length of girder on the compression side and an upward component on the tension side, as shown in Fig. 7.6.

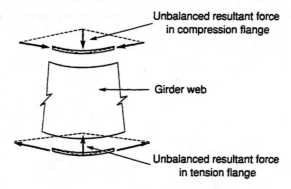

Figure 7.6   Vertical web compression due to unbalanced force in flanges.

The vertical buckling strength of the web plate must be sufficient to withstand this squeezing action, and this is taken care of if the depth/thickness ratio of the web meets the requirement of Sec. A-G1 of AISCS, that the ratio of the clear distance between flanges to the web thickness shall not exceed

$$\frac{h}{t} \leq \frac{14,000}{\sqrt{F_y(F_y + 16.5)}} \qquad \text{(see Table 7.1)} \qquad (7.1)$$

where h is the clear distance between flanges (see the following note) and t is the web thickness.

When the foregoing limitation is met, intermediate stiffener spacing as determined by the actual h/t and required shear force V_u is allowed to run as high as $3.0h$. However, if intermediate stiffeners are held to an upper spacing limit of $1.5d$, where d is the girder depth, the maximum permissible h/t ratio is greater:

$$\frac{h}{t} \leq \frac{2000}{\sqrt{F_y}} \qquad \text{(see Table 7.1)} \qquad (7.2)$$

TABLE 7.1   Maximum Ratio of the Clear Distance between Flanges to Web Thickness

	F_y								
	36	42	46	50	55	60	65	90	100
Eq. (7.1)	322	282	261	243	223	207	192	143	130
Eq. (7.2)	333	309	295	283	270	258	248	211	200

These upper limits on h/t are listed for various yield stress levels in Table 7.1, where it will be noted that the advantage of the $1.5d$ limit becomes more pronounced as the yield stress increases.

After selection of the web plate, the maximum required shear force V_u must be determined from the applicable factored loads acting on the girder by performing a structural analysis to obtain the shear diagram. The condition to be checked is $\phi V_n \geq V_u$, where the resistance factor $\phi = 0.9$ and V_n is the nominal shear capacity of the web panel. If V_u is less than ϕV_n for an unstiffened web, and if h/t is less than 260 as well as the limit provided by Eq. (7.1), no intermediate stiffeners are needed. The applicable equations for this shear capacity are given in Sec. F2 of AISCS as well as in Flowchart 7.1.

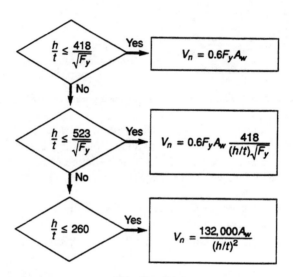

Flowchart 7.1

For example, what is the design strength in shear of a web without stiffeners if $F_y = 50$ ksi, $t = \frac{3}{8}$ in., and $h = 56$ in.?

Area of web: $A_w = 0.75 \times 56 \times 0.375 = 21$ in^2

$$\frac{h}{t} = \frac{56}{0.375} = 149.3 \qquad \leq 260$$

$$\frac{418}{\sqrt{F_y}} = \frac{418}{\sqrt{50}} = 59.1 \qquad \leq 149.3 \qquad \frac{523}{\sqrt{F_y}} = \frac{523}{\sqrt{50}} = 74.0 \qquad \leq 149.3$$

Therefore, from Flowchart 7.1,

$$\phi V_n = \phi \times \frac{132,000 A_w}{(h/t)^2} = \frac{0.9 \times 132,000 \times 21}{(149.3)^2} = 112 \text{ kips}$$

7.3 Selection of Girder Flanges

After the girder web is tentatively selected, the next step is to determine the sizes of the girder flanges. The steps in selecting girder flanges are:

1. Make the preliminary flange selection.
2. Select a trial flange plate and check its width/thickness ratio and its laterally unbraced length.
3. Determine the reduced design flexural strength in flanges.
4. Select reduced-size flanges for use away from maximum moment and determine the location of flange area transitions.

7.3.1 Make Preliminary Flange Selection

As stated in Chapter 3, the required section modulus for a beam is

$$(S_x)_{req} = \frac{M_x}{\phi F_{cr}} \tag{7.3}$$

The bending strength of a plate girder equals the sum of the bending strengths of the girder web and flanges. An approximate evaluation of the girder's bending strength can be obtained as follows:

$$S_x = \frac{I_x}{d/2} \tag{7.4}$$

$$I_x = (I_x)_{web} + (I_x)_{flanges} \tag{7.5}$$

or approximately

$$I_x \approx \frac{th^3}{12} + 2A_f\left(\frac{h}{2}\right)^2 \tag{7.6}$$

where t = thickness of girder web
h = depth of girder web
d = depth of girder
A_f = area of one girder flange

Then the section modulus of the girder for $h/d \approx 1$ becomes approximately

$$S_x = \frac{th^2}{6} + A_f h \tag{7.7}$$

Combining Eqs. (7.3) and (7.7), the following expression for the required girder flange area is obtained:

$$A_f = \frac{M_x}{\phi F_{cr}} - \frac{th}{6} \tag{7.8}$$

The first term on the right side of Eq. (7.8) represents the flange area that would be needed to resist the bending moment, M_x, without help from the web. The next term, $th/6$, is the *equivalent* flange area contributed by the girder web. Equation (7.8) provides a tentative trial selection that is subject to later verification by the moment-of-inertia method. An approximate lower bound of the flange area is obtained by setting $\phi = 0.9$ and $F_{cr} = F_y$.

7.3.2 Select Trial Flange Plate and Calculate Critical Stress

After the tentative flange area is determined, a trial flange plate is selected. The critical flange stress is determined next. This stress is either (1) the yield stress F_y, (2) the lateral-torsional buckling stress of the compression flange (LTB), or (3) the local buckling stress of the compression flange (FLB). Web buckling is not prevented in this design method since the web has a large postbuckling strength. The applicable equations from Appendix G, Sec. A-G2 of AISCS, are presented in Flowchart 7.2 for nonhybrid symmetric I girders of uniform depth. AISCS gives the necessary formulas for unsymmetrical, hybrid, and web-tapered girders in Appendices B5 and G2. In Flowchart 7.2 the parameter r_T is the y-axis radius of gyration of the compression

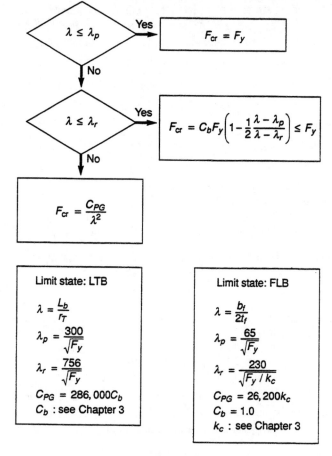

Flowchart 7.2

flange plus one-sixth of the adjacent web, and L_b is the laterally unbraced length of the compression flange. The terms C_b and k_c are defined and discussed in Chapter 3. Both limit states (LTB and FLB) need to be checked.

7.3.3 Determine Design Bending Moment

In tension-field design, the web is allowed to buckle, as previously discussed, and the bending stress on the compression side is no longer proportional to the distance from the neutral axis. As shown in Fig. 7.7, the stress at the extremity of the compression flange is slightly greater than given by the beam formula. To compensate for this in building design, a reduction in the critical flange stress is made according to Sec. A-G2 of AISCS.

When the web depth/thickness ratio

$$\frac{h}{t} > \frac{970}{\sqrt{F_y}} \tag{7.9}$$

the design capacity is ϕM_n, where the resistance factor $\phi = 0.9$, and for a nonhybrid symmetric I girder

$$M_n = S_x R_{PG} F_{cr}$$

The parameter S_x is the elastic section modulus about the x axis, F_{cr} is the critical stress calculated from the appropriate formulas in Flowchart 7.2, and R_{PG} is the reduction factor that accounts for the stress redistribution from the buckled web to the compression flange.

$$R_{PG} = 1 - \frac{a_r}{1200 + 300a_r}\left(\frac{h}{t} - \frac{970}{\sqrt{F_{cr}}}\right) \le 1.0 \tag{7.10}$$

In this equation $a_r = A_w/A_{fc}$, where A_w is the web area and A_{fc} is the area of the compression flange of the girder.

After the trial flange selection is made, the moment of inertia and section modulus are calculated on the basis of actual dimensions and properties. The required moment is then calculated and checked against the design value.

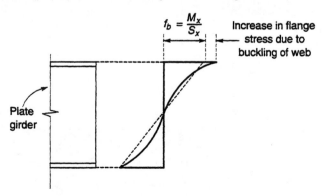

$$f_b = \frac{M_x}{S_x}$$ Increase in flange stress due to buckling of web

Plate girder

Figure 7.7 Stress distribution after buckling of web.

7.3.4 Select Reduced-Size Flanges for Use Away from Maximum Moment and Determine Location of Flange Area Transitions

For least weight the flange area should be in proportion to the bending moment. The size of the flange plates can be conveniently reduced in regions where the bending moments have decreased appreciably below the maximum values. For light loadings it is reasonable to use a constant cover plate to run the full girder length. In the case of welded girders it is preferable to use a single flange plate, without flange angles or cover plates. The flange plate thickness may be reduced at appropriate intervals, keeping in mind that such reductions should be made only if the saving in the cost of the flange material more than offsets the added expense of introducing the butt welds at thickness transition locations.

Cover plates in bolted girders (Fig. 7.8) can be cut off or flange plate thickness in welded girders reduced where the bending moments have dropped appreciably, as shown in Fig. 7.9. The resisting moments of the various girder sections where plates are cut off or reduced in thickness can be obtained by calculating the section modulus of the girder at each location [Eq. (7.3)]. One can locate the thickness transition point (or cutoff point) graphically by plotting horizontal lines that indicate the magnitude of the various resisting moments between transition points. The intersections of the moment diagram and the various horizontal lines, as shown in Fig. 7.9, determine the transition points. In the case of a uniform load the transitions can be determined mathematically from the equation for bending moment as follows:

$$\frac{x^2}{(l/2)^2} = \frac{M - \phi M_n}{M} \quad \text{and} \quad \frac{M - \phi M_n}{M} = \frac{A_c}{A}$$

where x = distance from the maximum moment in the girder to the theoretical thickness transition point (or cutoff point)

l = span length of simply supported girder

M = maximum moment in the girder due to uniform load

ϕM_n = resisting moment at theoretical thickness transition point

A_c = area of plate to be cut off or difference of flange area between two flange plates at a transition point

A = flange area plus web equivalent

$\quad = \dfrac{M}{\phi F_{cr}}$ [see Eq. (7.8)]

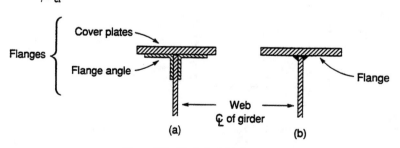

Figure 7.8 Typical girder flanges.

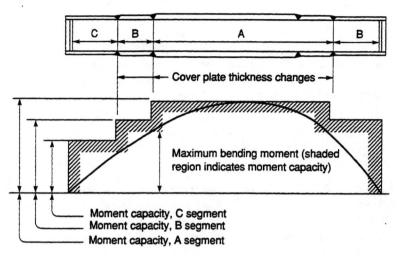

Figure 7.9   Thickness change locations in plate girder flanges.

Then the theoretical thickness transition point can be found:

$$x = \frac{l}{2}\sqrt{\frac{A_c}{A}} \tag{7.11}$$

If cover plates are used in riveted, bolted, or welded girders, they should be extended beyond the theoretical transition points with enough fasteners or welds to develop the plate stress at the theoretical cutoff points.

In welded design with a single flange plate, transitions in section are best achieved by changing the flange plate thickness (and width, if desired) with the ends of the two flange plates joined by a full-penetration groove butt weld. The weld should have a concave contour that is smoothly tangent with the surface of the thinner plate.

If repeated load is a design requirement, the choice of details may be a critical factor. Referring to Appendix K of AISCS, the following comparisons are tabulated for loading condition 3, Table A-K3.3, for the range 500,000 to 2,000,000 loading cycles:

	Illustrative Case (AISCS, Fig. A-K3.1)	Stress Category (AISCS, A-K3.2)	Allowable Stress Range (ksi), Load Condition 3
Welded girder without stiffeners with no cover plates or splices	4	B	18
Welded groove weld transition (full penetration)	12	C	13
Welded girder with intermediate stiffeners	7	C	13
Welded girder with cover plates (stress at cutoff location)	5	E	8

It is readily seen that the use of cutoff welded cover plates may introduce a severe penalty under the conditions stipulated by the foregoing tabulation. If butt joints are used at transition points with full-penetration groove welds, the section may be changed where the reduced section stress would have a maximum stress *range* of no more than 13 ksi, in which case there might not be any penalty due to repeated load.

7.4 Intermediate Stiffeners

When transverse intermediate stiffeners are provided, the shear strength is increased and it is equal to the sum of the buckled web plate strength and the vertical component of the diagonal tension that is induced in the buckled web. The intermediate stiffeners then play the dual role of improving buckling resistance and acting as compression struts as in truss action.

The design of intermediate stiffeners, when such stiffeners are required, consists of:

1. Plotting a maximum shear force diagram
2. Locating first stiffeners away from each end
3. Locating the remaining intermediate stiffeners
4. Selecting the size of the intermediate stiffeners
5. Checking interaction between flexure and shear in the web

We consider these now in greater detail.

7.4.1 Plot Maximum Shear Force Diagram

Stiffener spacing is a function of shear force in the particular panel under consideration. It is also a function of the h/t thickness ratio, which has been established at the outset. The factored shear capacity is ϕV_n, where $\phi = 0.9$ and V_n is the nominal shear strength defined in Sec. A-G3 in AISCS. The shear strength equations are given in Flowcharts 7.3 and 7.4. The term in brackets in the lower equation in Flowchart 7.4 when multiplied by the expression in front of the term in brackets is the buckling strength. When the second term is included (i.e., the entire equation is used), the post-buckling strength (or tension-field action) is accounted for.

7.4.2 Locate First Stiffeners Away from Each End

Tension-field action may not be utilized in the end panels of plate girders, nor in all panels of hybrid or tapered girders. Thus only the first part of the lower equation in Flowchart 7.4 may be used (i.e., $V_n = 0.6A_w F_y C_v$). For steels with a yield stress of 36 or 50 ksi, Tables 9-36 and 9-50 in AISCS may be used in lieu of the formulas in Flowcharts 7.3 and 7.4.

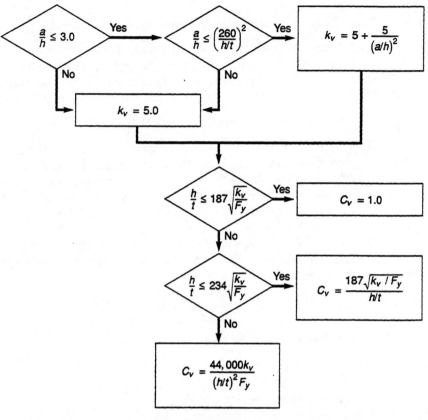

Flowchart 7.3

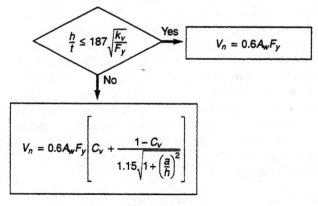

Flowchart 7.4

7.4.3 Locate Remaining Intermediate Stiffeners

For internal panels in the plate girder, tension-field action is permitted if the girder is not a hybrid girder and it is not tapered. This means that the complete equation in Flowchart 7.4 is to be used. For steels with a yield stress of 36 or 50 ksi, Tables 10-36 and 10-50 in AISCS may be used in lieu of the formulas in Flowcharts 7.3 and 7.4.

7.4.4 Select Size of Intermediate Stiffeners

The role of intermediate transverse stiffeners is twofold: They must be stiff enough to force a node in the shear buckling deformation pattern, the *moment of inertia* requirement, and they must be strong enough to be an anchor for the force component of the diagonal tension-field force in the web, the *area* requirement. The moment-of-inertia requirement is prescribed in Sec. A-F2.3 of AISCS, and the area requirement is found in Sec. A-G4.

The moment of inertia is with respect to an axis in the plane of the web. The stiffener moment of inertia I_s must be larger than $at^3 j$, where a is the panel spacing, t is the web thickness, and

$$j = \frac{2.5}{(a/h)^2} - 2 \geq 0.5$$

The stiffener area must not be less than

$$\frac{(F_y)_{web}}{(F_y)_{stiffener}} \left[0.15 D h t (1 - C_v) \frac{V_u}{\phi V_n} - 18 t^2 \right] \geq 0$$

where $D = 1.0$ for stiffeners furnished in pairs
$D = 1.8$ for single angle stiffeners
$D = 2.4$ for single plate stiffeners

Solutions for required area are provided by the italicized listings in AISCS Tables 10-36 and 10-50, for $F_y = 36$ ksi and $F_y = 50$ ksi, respectively.

It is noted that the intermediate stiffeners may be stopped short of the tension flange, provided that bearing is not needed to transmit a concentrated load or reaction. Near the tension flange the welds by which intermediate stiffeners are attached to the web shall be terminated not closer than four times the web thickness nor more than six times the web thickness from the near toe of the web to the flange weld. When single stiffeners are used, they shall be attached to the compression flange, if it consists of a rectangular plate, to resist any uplift tendency due to twist of the flange. Lateral bracing may be attached to the intermediate stiffeners, and these shall then be connected to the compression flange to transmit 1% of the total flange stress, unless the flange is composed only of angles.

7.4.5 Check Interaction between Bending and Shear in Girder Web

Plate girder webs, which depend upon tension-field action shall be so designed that the interaction between the bending and shear actions obeys the following

condition from Sec. A-G4 of AISCS if the webs are designed for tension-field action
and $0.6\phi V_n \leq V_u \leq \phi V_n$ ($\phi = 0.9$) and $0.75\phi M_n \leq M_u \leq M_n$:

$$\frac{M_u}{\phi M_n} + 0.625\frac{V_u}{\phi V_n} \leq 1.375$$

The use of this formula bypasses the more complex calculation of maximum
direct tensile stress, which acts at an angle to the girder axis.

7.5 Bearing Stiffeners

Bearing stiffeners serve three interrelated functions, which are illustrated in Fig. 7.10.

1. They distribute the transfer of local reactive forces to web shear, as illustrated in Fig. 7.10(a).
2. They prevent local yielding or crippling in the web immediately adjacent to concentrated reactions or loads. This type of failure is illustrated in Fig. 7.10(b). If no bearing stiffeners are used, the local compressive stress in the web must be checked by AISCS formulas (K1-2) and (K1-4) for interior loads, or (K1-3) and (K1-5) for end loads or reactions. This topic has been treated previously in Section 3.9.
3. Finally, bearing stiffeners prevent a more general vertical buckling of the web, of the type illustrated in Fig. 7.10(c). In this connection the design strengths are specified by AISCS formulas (K1-6) and (K1-7), the choice being determined by whether or not the top flange is restrained against rotation. In these formulas the vertical force is assumed to be distributed over a length of web equal to the girder depth or the length of the stiffened panel in which the load is placed—whichever dimension is the lesser.

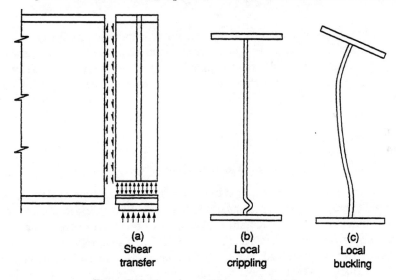

(a)
Shear
transfer

(b)
Local
crippling

(c)
Local
buckling

Figure 7.10   Support conditions at end of girder.

Bearing stiffeners should have close contact with the flanges adjacent to application points of applied or reactive loads and should extend approximately to the edges of the flanges, as shown in Fig. 7.11. According to Sec. K9 of AISCS, the stiffeners are to be designed as columns, assuming the column section to comprise the pair of stiffeners and a centrally located strip of the web whose width is equal to not more than 25 times its thickness at interior stiffeners, or a width equal to not more than 12 times its thickness when the stiffeners are located at the end of the web. The effective length shall be taken as not less than three-fourths the length of the stiffeners in computing the slenderness ratio l/r. The stiffeners shall also be checked for local bearing pressure. Only that portion of the stiffeners outside the flange angle fillet or the flange-to-web welds, as shown in Fig. 7.11, shall be considered effective in bearing and the bearing stress shall not exceed the design value of ϕR_n, where $\phi = 0.75$ and $R_n = 1.8F_y A_{pb}$. A_{pb} is the projected bearing area (AISCS, Sec. J8).

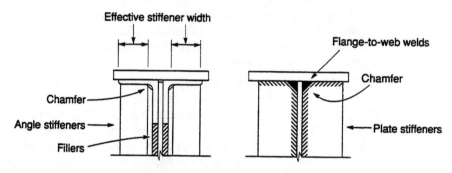

Figure 7.11   Typical stiffeners.

Local detail is important in the transfer of a load concentration into bearing stiffeners. For example, if a very heavy column introduces a load or acts as a support, two pairs of stiffeners are desirable so as to introduce the column flange bearing stress directly into the stiffeners without local bending of the girder flanges, as shown in Fig. 7.12.

7.6 Connections of Girder Elements

7.6.1 Flange-to-Web Connection

Bolts or welds connecting flange to web, or cover plate to flange, shall be designed to resist the horizontal shear resulting from the bending forces on the girder, as shown in Fig. 7.13. The longitudinal spacing of bolts shall not exceed the maximum permissible provided in Sec. J3.5 of AISCS. The flange-to-web connection shall also transmit any direct loads that are applied unless bearing stiffeners are provided.

The flange-to-web connection shall be designed at all locations to transmit the shear force between flange and web due to the moment variation.

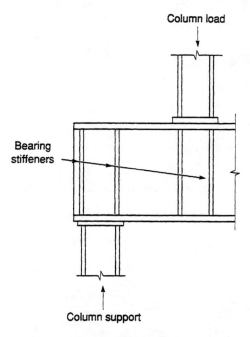

Figure 7.12   Bearing stiffeners.

1. In the case of a bolted girder, as shown in Fig. 7.13(a), let

   P_1, P_2 = flange force due to moments M_1, M_2 at any two adjacent
   cross sections, respectively

   I = girder moment of inertia

   p = pitch of bolts

   R_x, R_y = force components of resultant shear force (R) on bolt

   V = shear force

   $\bar{y}$ = distance between flange centroid and neutral axis of girder
   section

   w_y = vertical applied load per unit length

   Then

   $$R_x = P_2 - P_1 = \frac{M_2 - M_1}{I}\bar{y}A_f = \frac{Vp\bar{y}A_f}{I} \qquad R_y = w_y p$$

   $$R = \sqrt{R_x^2 + R_y^2}$$

2. In the case of a welded girder shown in Fig. 7.13(b), let

   q_x, q_y = components of stress resultant loads per unit length of
   fillet weld

   q_r = resultant shear between web and flange

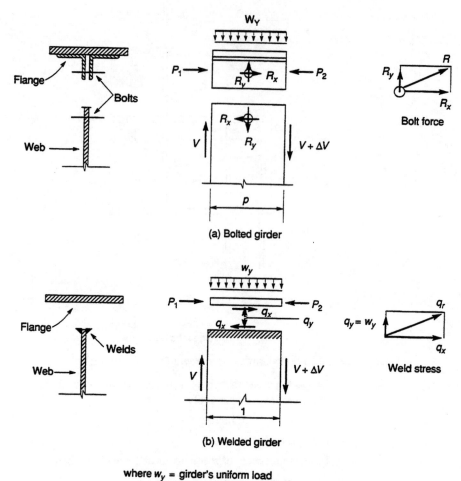

where w_y = girder's uniform load
P_1, P_2 = flange forces at cross sections
R_x, R_y = bolt's resisting components of its resultant R
q_x = stress intensity in horizontal direction
q_r = resultant shear between web and flange
p = bolt pitch

Figure 7.13 Stress transmittance from girder flange to web.

Then

$$q_x = P_2 - P_1 = \frac{M_2 - M_1}{I}\bar{y}A_f = \frac{V\bar{y}A_f}{I}$$

$$q_r = \sqrt{q_x^2 + w_y^2}$$

7.6.2 Stiffener Connections

The welds, rivets, or bolts that attach bearing stiffeners to the girder web are designed to transmit the total reactive or applied load into the web. Similarly, the intermediate stiffeners in a tension-field panel must transfer the vertical component of the total tensile force into the stiffener. Such stiffeners are to be connected to transmit at least

$$F_{vs} = \frac{5}{3}h\left(\frac{F_y}{340}\right)^{3/2} \text{ kips per linear inch}$$

where h is as defined previously and F_y is the yield stress of web steel.

According to Sec. F2.3 of AISCS, the rivets connecting stiffeners to the girder web shall be spaced not more than 12 in. on center. For intermittent fillet welds connecting stiffeners to the girder web, the clear distance between welds shall not be more than 16 times the web thickness nor more than 10 in.

7.6.3 Web Splices

Splices for girders offer no problem in welded girders, but are expensive and should be avoided whenever possible in riveted or bolted girders. The need for girder splices is dictated by erection requirements and shipping limitations. It may be desirable to locate the flange splice and web splice in different locations. The girder web primarily transmits shear force, and therefore web splices are most economically located at places where the shear force is small. Riveted or bolted splices should provide a net section area through the splice plates sufficient to resist the shear force and bending moment carried by the girder web at the splice location. Tests have shown the simple butt splice illustrated in Fig. 7.14 to be quite adequate.

In designing the web splice for moment, the web moment M_w, which is the portion of the total moment (M) carried by the girder web at the splice, can be obtained approximately from the proportion of the equivalent web area to flange area, which has been introduced in Section 7.3; then

$$M_w = \frac{th/6}{(th/6) + A_f}M$$

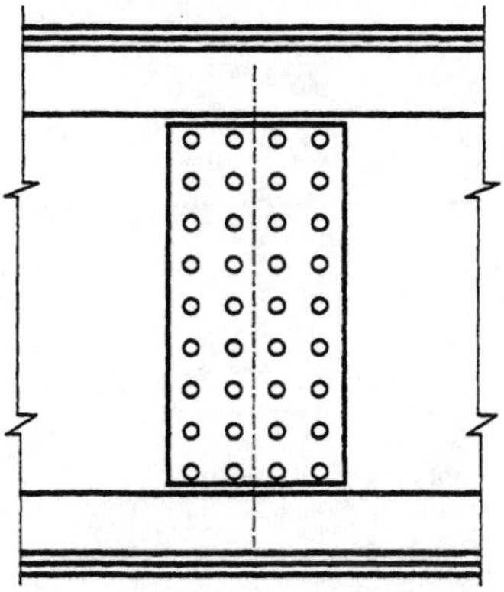

Figure 7.14   Web splice for riveted or bolted girder.

After design shears and web moments are determined, the splice design for either bolted or welded connections follows standard procedures for connections, as covered in Chapter 6. When fillers are needed in the splices, fillers should be designed according to Sec. J6 of AISCS. In welded girders, splices present no special problem; complete-penetration groove welds shall be provided to develop the full strength of the smaller spliced section.

7.7 Illustrative Examples

The information relative to plate girder design in AISCM should be studied in conjunction with the following design example, which is similar to one provided in AISCM.

Example 7.1

Design a plate girder for the loads shown.

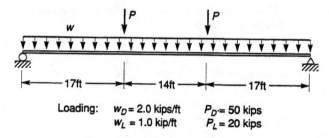

Loading: $w_D = 2.0$ kips/ft $P_D = 50$ kips
           $w_L = 1.0$ kip/ft $P_L = 20$ kips

Factored loads:

$$P = 1.2 \times 50 + 1.6 \times 20 = 92 \text{ kips}$$
$$w = 1.2 \times 2 + 1.6 \times 1 = 4 \text{ kips/ft}$$

The material is A36 steel:

$$F_y = 36 \text{ ksi} \qquad E = 29,000 \text{ ksi}$$

Use an E70 electrode, $F_{Exx} = 70$ ksi. The maximum depth is 72 in.

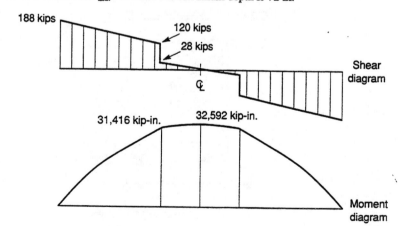

Geometry:

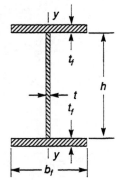

$$r_T = \frac{b_f}{\sqrt{12}}\sqrt{\frac{1}{1+(1/6)(ht/b_ft_f)}}$$

Radius of gyration of the flange plus one-sixth of the web about the y axis.

$$A_w = ht \qquad A_f = b_ft_f$$

Solution *Check a trial section*:

Web: $\frac{5}{16}$ in. × 70 in.   Flanges: $\frac{7}{8}$ in. × 15 in.

total depth 71.75 in.   < 72 in.   OK

$t = 0.3125$ in. $\dfrac{h}{t} = 224$ $A_w = ht = 21.875$ in^2 $h = 70$ in.

$b_f = 15.0$ in. $\dfrac{b_f}{2t_f} = 8.57$ $A_f = b_ft_f = 13.125$ in^2 $t_f = 0.875$ in.

$$r_T = \frac{b_f}{\sqrt{12}}\sqrt{\frac{1}{1 + A_w/6A_f}} = 3.831 \text{ in.}$$

$$I_x = \frac{b_ft_f^3}{12} \times 2 + 2A_f\left(\frac{h+t_f}{2}\right)^2 + \frac{th^3}{12} = 41{,}899 \text{ in}^4$$

$$S_x = \frac{2I_x}{h + 2t_f} = 1168 \text{ in}^3$$

$$\left(\frac{h}{t}\right)_{max} = \frac{2000}{\sqrt{F_y}} = 333 \quad \text{or} \quad \left(\frac{h}{t}\right)_{max} = \frac{14{,}000}{\sqrt{F_y(F_y + 16.5)}} = 322 \qquad > 224 \quad \text{OK}$$

Check the flexure:

$$\frac{970}{\sqrt{F_y}} = 162 \qquad < 224 \quad \text{plate girder design applies}$$

Limit-state LTB: Lateral support at the ends and under the concentrated loads (see Flowchart 7.2):

Outside span:

$$\frac{M_1}{M_2} = 0 \qquad C_b = 1.75 + 1.05\frac{M_1}{M_2} + 0.3\left(\frac{M_1}{M_2}\right)^2 = 1.75$$

$$L_b = 17 \text{ ft} = 204 \text{ in.} \qquad \frac{L_b}{r_T} = \frac{204}{3.83} = 53.25 = \lambda$$

$$\lambda_p = \frac{300}{\sqrt{F_y}} = \frac{300}{\sqrt{36}} = 50 \quad <53.25 \qquad \lambda_r = \frac{756}{\sqrt{F_y}} = 126 \quad >53.25$$

$$F_{cr} = C_bF_y\left(1 - \frac{1}{2}\frac{\lambda - \lambda_p}{\lambda_r - \lambda_p}\right) = 1.75 \times 36 \times \left[1 - \frac{1}{2}\left(\frac{53.25 - 50}{126 - 50}\right)\right] = 61.65 \text{ ksi} \qquad > 36 \text{ ksi}$$

$$a_r = \frac{A_w}{A_{fc}} = \frac{21.875}{13.125} = 1.67$$

$$R_{PG} = 1 - \frac{a_r}{12 + 300a_r}\left(\frac{h}{t} - \frac{970}{\sqrt{F_{cr}}}\right) = 1 - \frac{1.67}{1200 + 300 \times 1.67}\left(224 - \frac{970}{\sqrt{36}}\right) = 0.939$$

$\phi M_n = \phi S_x R_{PG} F_{cr} = 0.9 \times 1168 \times 0.939 \times 36 = 35{,}530$ kip-in. $> 31{,}416$ kip-in.   OK

Center span:

$$\frac{M_1}{M_2} = -1 \qquad C_b = 1.0$$

$$L_b = 14 \text{ ft} = 168 \text{ in.} \qquad \frac{L_b}{r_T} = \frac{168}{3.83} = 43.85 = \lambda$$

$$\lambda_p = \frac{300}{\sqrt{F_y}} = \frac{300}{\sqrt{36}} = 50 \; > 43.85 \qquad F_{cr} = F_y = 36 \text{ ksi}$$

$$\phi M_n = \phi S_x R_{PG} F_{cr} = 35{,}530 \text{ kip-in.} \qquad > 32{,}592 \text{ kip-in.}\quad \text{OK}$$

Limit-state FLB:

$$\lambda_b = \frac{b_f}{2t_f} = \frac{15}{2 \times 0.875} = 8.57 \qquad \lambda_{bp} = \frac{65}{\sqrt{F_y}} = 10.83 \; > (8.57) \qquad F_{cr} = F_y$$

No further check is needed. The flexure is OK.

   Check the shear: End panel: Try

$$a = 40 \text{ in.} \qquad \frac{a}{h} = \frac{40}{70} = 0.571 \qquad < 3 \quad \text{and} \; < \left(\frac{260}{h/t}\right)^2 = 1.347$$

$$\frac{h}{t} = 224$$

From Flowchart 7.3,

$$k = 5 + \frac{5}{(a/h)^2} = 20.31 \qquad \frac{187\sqrt{k_v}}{\sqrt{F_y}} = 140.61$$

$$\frac{234\sqrt{k_v}}{\sqrt{F_y}} = 176 \qquad < 224$$

$$C_v = \frac{44{,}000k_v}{F_y(h/t)^2} = 0.496 \qquad V_n = 0.6A_w F_y C_v = 234.26 \text{ kips}$$

$$\phi V_n = 0.9V_u = 210.83 \text{ kips} \qquad > 188 \text{ kips}$$

$$a = 40 \text{ in.}\quad \text{OK}$$

Next panel:

$$V = 188 - 4 \times \frac{40}{12} = 174.67 \text{ kips}$$

Try

$$a = \frac{17 \times 12 - 40}{2} = 82 \text{ in.} \qquad \frac{a}{h} = \frac{82}{70} = 1.171$$

$$k_v = 5 + \frac{5}{(a/h)^2} = 8.644 \qquad \frac{187\sqrt{k_v}}{\sqrt{F_y}} = 91.63 \qquad \frac{234\sqrt{k_v}}{\sqrt{F_y}} = 115 \; < 224$$

$$C_v = \frac{44{,}000k_v}{F_y(h/t)^2} = 0.2105$$

$$V_u = 0.6A_wF_y\left[C_v + \frac{1-C_v}{1.15\sqrt{1+(a/h)^2}}\right] = 310.1 \text{ kips} \qquad > \frac{174.67}{0.9} = 194 \text{ kips}$$

$a = 84$ in. Ok

Try

$$a = 164 \text{ in.} \qquad \frac{a}{h} = \frac{164}{70} = 2.343 \quad > \left(\frac{260}{h/t}\right)^2 = 1.347$$

$$k = 5 \qquad \frac{235\sqrt{k_v}}{\sqrt{F_y}} = 87.58$$

$$C_v = \frac{44{,}000k}{F_y(h/t)^2} = 0.1218$$

$$V_n = 0.6A_wF_yC_v = 57.55 \text{ kips} \qquad < \frac{174.67 \text{ kips}}{0.9} \quad \text{NG}$$

Use $a = 82$ in.

Stiffener spacing:

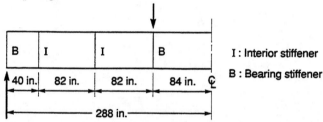

I : Interior stiffener

B : Bearing stiffener

Interaction check in panel under concentrated load (to the left of the load):

$$V_u = 188 - \left(\frac{4}{12}\right)(40 + 82) = 147.33 \text{ kips} \qquad \phi V_n = 0.9 \times 310.1 = 279.1 \text{ kips}$$

$0.6\phi V_n = 167.4$ kips > 147.3 kips

$M_u = 31{,}416$ kip-in. $\phi M_n = 35{,}530$ kip-in.

$0.75\phi M_n = 26{,}647$ kip-in. $< 31{,}416$ kip-in.

Interaction check is required.

$$\frac{M_u}{\phi M_n} + 0.625\frac{V_u}{\phi V_n} = \frac{31{,}416}{35{,}530} + \frac{0.625 \times 147.3}{279.1} = 1.214 \qquad < 1.375 \quad \text{OK}$$

Interaction check is OK. Use $\frac{5}{16}$ in. $\times$ 70 in. web and $\frac{7}{8}$ in. $\times$ 15 in. flanges.

Interior stiffener design: Single plate stiffener:

$$\frac{b_s}{t_s} \leqslant \frac{95}{\sqrt{F_y}} = 15.83$$

$$I_s = \frac{t_s(b_s + t/2)^3}{3} \qquad I_s \geq at^3 j$$

$$a = 82 \text{ in.} \qquad t = 0.3125 \text{ in.} \qquad h = 70 \text{ in.}$$

$$j = \frac{2.5}{(a/h)^2} - 2 = \frac{2.5}{(1.171)^2} - 2 = -0.18 \qquad < 0.5, \quad \text{hence } j = 0.5$$

$$I_s = \frac{t_s(b_s + t/2)^3}{3} \geq 82 \times 0.3125^3 \times 0.5 = 1.251 \text{ in}^4$$

$$A_s = t_s b_s \geq 0.15 D h t (1 - C_v) \frac{V}{\phi V_n} - 18 t^2$$

For a single plate stiffener, $D = 2.4$:

$$h = 70 \text{ in.} \qquad t = 0.3125 \text{ in.}$$

$$C_v = 0.2105 \qquad V = 174.67 \text{ kips} \qquad \phi = 0.90 \qquad V_n = 310.08 \text{ kips}$$

$$A_s \geq 0.15 \times 2.4 \times 70 \times 0.3125(1 - 0.2105)\frac{174.67}{0.9 \times 310.08} - 18 \times 0.3125^2 = 2.134 \text{ in}^2$$

Try a $\frac{1}{2}$ in. × 5 in. plate:

$$\frac{b_s}{t_s} = 10 \qquad < 15.83 \quad \text{OK}$$

$$I_s = 22.85 \text{ in}^4 \qquad > 1.25 \text{ in}^4 \quad \text{OK}$$

$$A_s = 2.5 \text{ in}^2 \qquad > 2.13 \text{ in}^2 \quad \text{OK}$$

Use $\frac{1}{2}$ in. × 5 in. single plate interior stiffeners.

Bearing stiffener design: At the reaction end (critical):

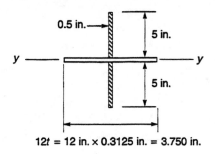

$$12t = 12 \text{ in.} \times 0.3125 \text{ in.} = 3.750 \text{ in.}$$

Try $\frac{1}{2}$ in. × 5 in. plates

$$I_y = \frac{0.5(5 \times 2 + 0.3125)^3}{12} = 45.696 \text{ in}^4$$

$$A = 2 \times 5 \times 0.5 + 3.75 \times 0.3125 = 6.172 \text{ in}^2$$

$$r = \sqrt{\frac{I_y}{A}} = 2.721 \text{ in.} \qquad \frac{KL}{r} = \frac{0.75h}{r} = \frac{0.75 \times 70}{2.721} = 19.294$$

$$\lambda = \frac{KL}{r}\frac{1}{\pi}\sqrt{\frac{F_y}{E}} = 0.216 \qquad \phi_c = 0.85 \qquad P_u = AF_{cr}$$

$$F_{cr} = 0.658^{\lambda^2}F_y \qquad F_y = 0.658^{0.216^2} \times 36 = 35.30 \text{ ksi}$$

$$\phi_c A_s F_{cr} = 0.85 \times 6.172 \times 35.30 = 185 \text{ kips}$$

$$R = 188 \text{ kips} \qquad \approx 185 \text{ kips} \quad \text{OK}$$

Use $\frac{1}{2}$ in. $\times$ 5 in. bearing stiffeners on both sides at end supports.
Design of fillet weld between flange and web:

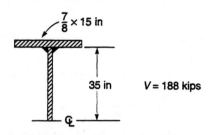

Shear flow across weld : $F_s = \dfrac{V}{I_x}\left(\dfrac{7}{8}\times 15\right)\left(35 + \dfrac{1}{2}\times\dfrac{7}{8}\right)$

$$I_x = 41,899 \text{ in}^4$$
$$F_s = 2.087 \text{ kip/in.}$$

Weld throat:

$$\frac{D}{\sqrt{2}} \qquad \text{where} \quad D = \text{weld size}$$

Weld stress:

$$f_v = \frac{2.1\sqrt{2}}{D}$$

Weld capacity:

$$\phi F_x = 0.75 \times 0.6 F_{Exx} = 0.75 \times 0.6 \times 70 = 31.5 \text{ ksi}$$

$$\frac{2.1\sqrt{2}}{D} = 31.5 \qquad D = 0.10 \text{ in.}$$

Minimum required weld size (AISCS, Table J2.4): $D = \frac{5}{16}$ in. Use $\frac{5}{16}$ -in. fillet weld.
Check if bearing stiffener is required under concentrated load:
Local web yielding (AISCS, Sec. K1.3): ·

$$92 \text{ kips} \le \phi_{bs} t (5k + N) F_y$$

$\phi_{bs} = 1.0 \qquad t = 0.3125 \text{ in.} \qquad F_y = 36 \text{ ksi} \qquad k = \dfrac{7}{8} + \dfrac{5}{16} = 1.19 \text{ in.}$

Assume that the bearing length is 4 in.:

$$\phi_{bs} t (5k + N) F_y = 112 \text{ kips} \qquad > 92 \text{ kips} \quad \text{OK}$$

Web crippling (AISCS Sec. K1.4): Interior concentrated load; therefore, use Eq. (K1-4) in AISCS.

$\phi = 0.75 \qquad t_f = 0.875 \text{ in.} \qquad t_w = 0.3125 \text{ in.} \qquad F_y = 36 \text{ ksi} \qquad N = 4 \text{ in.} \qquad d = 71.75 \text{ in.}$

$$R_n = 135 t_w^2 \left[1 + 3\left(\frac{N}{d}\right)\left(\frac{t_w}{t_f}\right)^{1.5}\right]\sqrt{\frac{F_y t_f}{t_w}} = 137.1 \text{ kips}$$

$$\phi R_n = 0.75 \times 137.1 = 102.8 \text{ kips} \qquad > 92 \text{ kips} \quad \text{OK}$$

No bearing stiffener is needed.

Example 7.2

Design a plate girder for flexure and shear, using SI units. The general loading as well as the shear and bending moment diagrams are shown below.

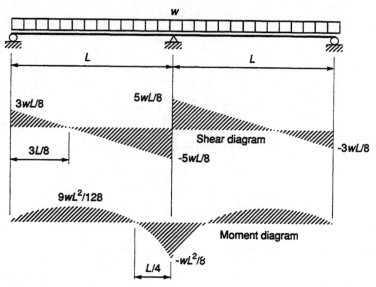

Given: Dead load = 60 kN/m

Assumed girder weight: 6 kN/m (force) or 612 kg/m (mass)

Live load = 85 kN/m

Span L = 18 m

Yield stress = 345 MPa

Solution

$$w_D = 66 \text{ kN/m} \qquad w_L = 85 \text{ kN/m} \qquad L = 18 \text{ m}$$
$$w_u = 1.2w_D + 1.6w_L = 215.2 \text{ kN/m}$$

Required moment:

$$M_u = \frac{w_u L^2}{8} = 8715.6 \text{ kN·'m}$$

Flexural design: Unbraced length: The girder is laterally braced along its length at the top chord.

Cross section:

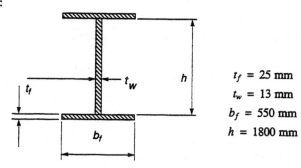

$t_f = 25$ mm

$t_w = 13$ mm

$b_f = 550$ mm

$h = 1800$ mm

Material properties:

$$F_y = 345 \text{ MPa} \qquad E = 200,000 \text{ MPa}$$

Calculated properties:
  Web area:

$$A_w = ht_w = 23,400 \text{ mm}^2$$

  Flange area:

$$A_f = b_f t_f = 13,750 \text{ mm}^2$$

  Web slenderness:

$$\lambda_w = \frac{h}{t_w} = 138.462$$

  Flange slenderness:

$$\lambda_f = \frac{b_f}{2t_f} = 11$$

Maximum permissible web slenderness (the slenderness limit formulas from Appendix G in AISCS were taken from the draft of AISC metric LRFD Specification):

$$(\lambda_w)_{max} = \frac{0.48E}{\sqrt{F_y(F_y - 115 \text{ MPa})}} = 340.799 \qquad > \lambda_w = 138.462 \quad \text{OK}$$

Moment of inertia about x axis:

$$I_x = 2b_f\frac{t_f^3}{12} + 2A_f\left(\frac{h}{2} + \frac{t_f}{2}\right)^2 + \frac{h^3 t_w}{12} = 2.922 \times 10^{10} \text{ mm}^4$$

Elastic section modulus:

$$S_x = \frac{I_x}{h/2 + t_f} = 3.159 \times 10^7 \text{ mm}^3$$

Calculation of critical stress:
  (a) Lateral buckling: Not needed because girders are laterally braced.
  (b) Flange local buckling:

$$\lambda_p = 0.38\sqrt{\frac{E}{F_y}} = 9.149 \qquad < \lambda_f = 11$$

$$k_c = \text{if}\left[\frac{4}{\sqrt{\lambda_w}} < 0.35, 0.35, \text{if}\left(\frac{4}{\sqrt{\lambda_w}} > 0.763, 0.763, \frac{4}{\sqrt{\lambda_w}}\right)\right] = 0.35$$

$$\lambda_r = 1.35\sqrt{\frac{Ek_c}{F_y}} = 19.23 \qquad > \lambda_f = 11$$

$$F_{cr} = F_y\left(1 - \frac{1}{2}\frac{\lambda_f - \lambda_p}{\lambda_r - \lambda_p}\right) = 313.331 \text{ MPa}$$

Calculation of design capacity:

$$a_r = \frac{A_w}{A_f}$$

$$R_{PG} = 1 - \frac{a_r}{1200 + 300a_r}\left(\lambda_w - 5.7\sqrt{\frac{E}{F_{cr}}}\right) = 1.006$$

Use $R_{PG} = 1.00$.

$$M_n = S_x R_{PG} F_{cr}$$

Resistance factor:

$\phi = 0.9$ $\phi M_n = 8956.46$ kN·m $> M_u = 8716$ kN·'m   OK

Use 25 mm × 550 mm flanges and 13 mm × 1800 mm web.

    Check if lateral bracing is needed in the negative moment region. The unbraced length is $L/4 = 18/4 = 4.5$ m. The required unbraced length is calculated as follows (AISCS, Appendix G):

$$r_T = \sqrt{\frac{b_f^3 t_f}{12(A_f + A_w/6)}} = 140.136 \text{ mm}$$

$L_p = 1.76\sqrt{\dfrac{E}{F_y}}r_T = 5.938$ m > 4.5 m;   no bracing needed in negative moment region

    Check if we can use steel with $F_y = 250$ MPa in the positive moment region. Local buckling of flange:

$$F_y = 250 \text{ MPa}$$

$$\lambda_p = 0.38\sqrt{\frac{E}{F_y}} = 10.748 \qquad < \lambda_f = 11.00$$

$$\lambda_r = 1.35\sqrt{\frac{k_c E}{F_y}} = 22.59 \qquad > \lambda_f = 11.00$$

$$F_{cr} = F_y\left(1 - \frac{1}{2}\frac{\lambda_f - \lambda_p}{\lambda_r - \lambda_p}\right) = 247.34 \text{ MPa}$$

$$M_n = S_x R_{PG} F_{cr} \qquad \phi M_n = 7070.146 \text{ kN·m}$$

From the moment diagram:

$$w_u = 215.2 \text{ kN/'m} \qquad L = 18 \text{ m}$$

$$M_u = \frac{9 w_u L^2}{128} = 4902.525 \text{ kN·'m} \qquad < 7070 \text{ kN·m} \quad OK$$

Material distribution in the girder:

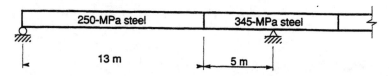

    Shear design—end panel at exterior support: This is a steel with $F_y = 250$ MPa. Because it is an end panel, we cannot use the tension-field action contribution to the web panel strength.

Stiffener spacing $a = 2.5$ m.

Cross section:

$t_f = 25$ mm $t_w = 13$ mm $b_f = 550$ mm $h = 1800$ mm

Material properties:

$$F_y = 250 \text{ MPa} \qquad E = 200,000 \text{ MPa}$$

Calculated properties:
   Web area:

$$A_w = ht_w = 23,400 \text{ mm}^2$$

Web slenderness:

$$\lambda_w = \frac{h}{t_w} = 138.462$$

Calculation of C_v:

$$k_v = 5 + \frac{5}{(a/h)^2} = 7.592 \qquad \lambda_p = 1.10\sqrt{\frac{k_v E}{F_y}} = 85.727 \qquad \lambda_r = 1.37\sqrt{\frac{k_v E}{F_y}} = 106.769$$

$$C_v = \text{if}\left[\lambda_w \le \lambda_p, 1, \text{if}\left(\lambda_w \le \lambda_r, \frac{1.1\sqrt{k_v E/F_y}}{\lambda_w}, \frac{1.52 k_v E}{\lambda_w^2 F_y}\right)\right] = 0.482$$

Design strength when tension-field action is not permitted: $\phi = 0.9$

$$(V_n)_{ntf} = 0.6 A_w F_y C_v$$
$$(\phi V_n)_{ntf} = 1521.182 \text{ kN} \qquad > 3w_u L/8 = 1453 \text{ kN}$$

Shear design—panel to each side of the center support: This is an interior panel and therefore it is permissible to utilize the tension-field action. The yield strength of the material is 345 MPa.

Stiffener spacing $a = 3.5$ m.

Cross section:

$$t_f = 25 \text{ mm} \qquad t_w = 13 \text{ mm} \qquad b_f = 550 \text{ mm} \qquad h = 1800 \text{ mm}$$

Material properties:

$$F_y = 345 \text{ MPa} \qquad E = 200,000 \text{ MPa}$$

Calculated properties:
   Web area:

$$A_w = ht_w = 23,400 \text{ mm}^2$$

Web slenderness:

$$\lambda_w = \frac{h}{t_w} = 138.462$$

Calculation of C_v:

$$k_v = 5 + \frac{5}{(a/h)^2} = 6.322 \qquad \lambda_p = 1.10\sqrt{\frac{k_v E}{F_y}} = 66.595 \qquad \lambda_r = 1.37\sqrt{\frac{k_v E}{F_y}} = 82.941$$

$$C_v = \text{if}\left[\lambda_w \le \lambda_p, 1, \text{if}\left(\lambda_w \le \lambda_r, \frac{1.1\sqrt{k_v E/F_y}}{\lambda_w}, \frac{1.52 k_v E}{\lambda_w^2 F_y}\right)\right] = 0.291$$

Design strength when tension-field action is permitted:

$$V_{n\_tf} = \text{if}\left\{\lambda_w \le 1.1\sqrt{\frac{k_vE}{F_y}},\ 0.6A_wF_y,\ 0.6A_wF_y\left[C_v + \frac{1-C_v}{1.15\sqrt{1+(a/h)^2}}\right]\right\}$$

$$\phi = 0.9 \qquad \phi V_{n\_tf} = 2496.72 \text{ kN} \qquad > 5w_uL/8 = 2421 \text{ kN} \quad \text{OK}$$

Moment of inertia requirement:

$$j = \text{if}\left[\frac{2.5}{(a/h)^2} - 2 \ge 0.5,\ \frac{2.5}{(a/h)^2} - 2,\ 0.5\right] \qquad I_{st\_req} = at_w^3 j = 9.237 \text{ in}^4$$

Stiffener area requirement (only if tension field action is utilized):

$D = 1$ for stiffeners in pairs.

$D = 1.8$ for single angle stiffener.

$D = 2.4$ for single plate stiffener.

$$D = 2.4 \qquad F_{yst} = 250 \text{ MPa} \qquad V_u = 2421 \text{ kN}$$

$$A_{st\_req} = \frac{F_y}{F_{yst}}\left[15Dht_w(1-C_v)\frac{V_u}{\phi V_{n\_tf}} - 18t_w^2\right] = 3798.897 \text{ mm}^2$$

The student may now design the size of the stiffeners for these requirements.

Final placement of stiffeners:

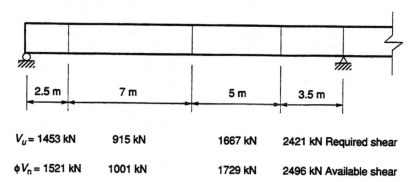

| 2.5 m | 7 m | 5 m | 3.5 m |

$V_u = 1453$ kN   915 kN   1667 kN   2421 kN Required shear

$\phi V_n = 1521$ kN   1001 kN   1729 kN   2496 kN Available shear

PROBLEMS*

7.1. For a welded girder cross section made up of single $\frac{15}{16}$ in. × 24 in. flange plates, top and bottom, connected by a $\frac{7}{16}$ in. × 100 in. web plate, determine:

(a) The approximate resisting moment based on the flange area method.

(b) The resisting moment by the moment of inertia method.

(c) The required stiffener spacing at a location where the shear is 400 kips.

7.2. The girder cross section shown is made up of an MC18×58 channel attached to the top flange of a W36 × 135 section by means of friction-type A325 high-strength bolts of $\frac{7}{8}$-in. diameter. What bolt pitch is required at a location where the required shear on the girder is 400 kips?

* Assume use of A36 steel in all problems.

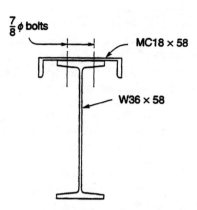

$\frac{7}{8}\phi$ bolts

MC18 × 58

W36 × 58

7.3. On the basis of the structural details illustrated below, using welding electrodes, determine the factored nominal end reaction R (same as end shear V), in each of the following five different ways:

(a) On the basis of the fillet weld size and weld spacing at the end of the girder, attaching web to flange plates.

(b) On the basis of the local contact compressive bearing stress at the bottom ends of the bearing stiffeners.

(c) On the basis of the size and arrangement of bearing stiffeners. (Assume that welds between bearing stiffeners and web are adequate.)

(d) On the basis of the first space (36 in.) from the bearing stiffener to the first intermediate stiffener.

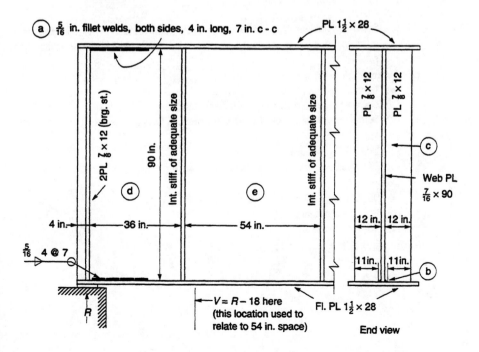

(e) On the basis of the second space (54 in.) between the first and second intermediate stiffeners. (The shear used in the design of this space is assumed to be 18 kips less than the end reaction; that is, $V = R - 18$.)

7.4. Design an all-welded plate girder for the following conditions. Include a general design drawing and detailed sketches of such portions as are judged needed to provide complete information. Span, 124 ft. Simple supports. A36 steel. Uniform live load of 3.5 kips/ft. Concentrated factored load of 910 kips, 20 ft from left support. Assume adequate lateral support. Web plates are to be as thin as possible, in commercially available thicknesses, to carry the maximum shear. Web plate thickness may be varied, if desired, to suit the high shear area at one end.

Note: This is a rather lengthy assignment and is suitable as a term project. For classroom use it is suggested that groups of three or four students each be assigned arbitrary web plate depths, between 100 and 150 in., varying in increments of 10 in. A plot of girder weight versus web plate depth can then be made as an exercise in the search for the most economical proportion.

The following is a summary of steps that may be followed in carrying out this assignment:

(1) Web plate
  (a) Choose a clear depth in relation to the span.
  (b) Choose the thickness.
  (c) Check for shear.
  (d) Check the interaction in the web.

(2) Flange plates
  (a) Make the preliminary selection by the flange-area method.
  (b) Determine the critical stress.
  (c) Check the strength.
  (d) Select reduced size flanges for use away from the maximum moment and determine the location of the flange transitions. (Check the flange width/thickness ratios.)

(3) Intermediate stiffeners
  (a) Locate the first stiffener away from each end.
  (b) Locate the remaining intermediate stiffeners.
  (c) Select the size of the intermediate stiffeners. (Check the area and I requirements.)

(4) Bearing stiffeners
  (a) Design for maximum reaction.
  (b) (For assigned problem, assume other bearing stiffeners to be identical.)

(5) Design welds:
  (a) Stress transfer for intermediate stiffeners.
  (b) Bearing stiffeners.
  (c) Web-to-flange shear transfer.

(6) Weight takeoff for complete girder.

8

Continuous Beams
and Frames

8.1 Introduction

In the design of a continuous structure, attention is turned from the individual member to the complete structure and the interrelated behavior of all its members. Continuous beams and frames in steel take advantage of the inherent continuity of all-welded construction. Weight reduction will result from the partial equalization of positive and negative moments at midspans and over supports, respectively. Deflections are reduced, the need for special bracing to resist lateral forces may be eliminated, and greater resistance to ultimate collapse due to earthquake or other shock loads is achieved.

Section A5 of AISCS states that "the required strength of structural members and connections shall be determined by structural analysis for the appropriate load combinations. . . ." In the case of statically indeterminate beams and frames, analysis of the internal forces is always permitted by *elastic analysis*. If the cross section is compact and the member is adequately braced, *plastic analysis* may be used. If the elastic analysis is used, the limiting strength of the structure is controlled by the capacity of that cross section which is subjected to the largest forces. In plastic design the limit state is the formation of a kinematic mechanism that includes enough cross sections with plastic hinges to permit the collapse of the entire structure.

In the elastic range of behavior, continuous beams and frames are statically indeterminate. The well-known method of moment distribution will be reviewed and used for analysis in some of the examples. In other cases tabular information is available in the AISCM for a limited number of span arrangements. If one is not familiar with methods for the analysis of continuous structures, the inclusion of these examples and problems is optional. Examples will omit design of details of types covered in earlier chapters, with primary attention given to member-size selection.

While the method of moment distribution is emphasized here because of its simplicity and transparency, many other methods of structural analysis could be used. In the design office the engineer will probably use computer programs specifically created to analyze continuous beams and rigid frames. From such programs the designer will know the forces in any part of the structure that are needed for checking the members or cross sections. However, the engineer should be familiar with manual methods of analysis in order to check the computer results and for preliminary design. The student wishing to become acquainted with the background of computerized structural analysis is referred to the text by Weaver and Gere (Ref. 8.1).

Coverage herein is limited to continuous beams and single-story frames. Multi-story building design would involve complexities going beyond desired basic limits. Emphasis is on design examples that are specially selected to illustrate important concepts.

8.2 Moment-Distribution Analysis: A Summary

The following condensed treatment of the *moment-distribution* method of analysis for continuous beams and frames is provided as a preliminary to the presentation of the design examples. The moment-distribution method provides a simple procedure to determine, through a converging iterative process, the bending moments at supports or joints in a continuous beam or frame. Once these "end moments" are known, the shears, reactions, and moments can be determined at any location by statics. The method is readily applied to problems involving sidesway. The procedure for frames under lateral force will be demonstrated when sidesway is induced by a lack of symmetry of applied vertical load.

The procedure is summarized as follows:

1. Assume the members to be locally locked against rotation at all support points and joints.
2. Enter the *fixed-end moments*, using available tabular information or deriving them by basic beam analysis procedures. An important feature is the systematic entry of numerical data in a standardized tabular format. In entering the fixed-end moments, a special *rotational* sign convention is usually used; that is, fixed-end moments are positive if applied to the end of a member in a clockwise sense, negative if applied counterclockwise.
3. *Unlock* the joints. If any given joint is unlocked, the ends of the members entering the joint, along with the joint, will rotate until equilibrium is established. This process introduces equilibriating moments in each member

entering the joint in proportion to their relative rotational stiffness. As the joint rotates, there is a carryover of moment to the far end of each member if that member is either temporarily locked against rotation (as may be assumed) at the far end or if the far end is actually assumed to be permanently fixed against rotation. For a member of uniform cross section it is readily shown that the *carryover factor* is equal to $\frac{1}{2}$.

4. At those joints assumed to be temporarily locked against rotation, the moments introduced by carryover from the opposite ends will usually introduce a new unbalance of overall joint moment. In this case steps (3) and (4) must be repeated, in sequence, until the residual unbalanced moments at all joints are inconsequential to the calculation accuracy desired for design purposes.

The foregoing procedure is best reviewed in connection with numerical examples, such as those in Examples 8.2 and 8.5.

The basic concepts of rotational stiffness and carryover are illustrated in Fig. 8.1(a), where a positive (clockwise) moment M_A is introduced at A, causing (by carryover) a positive moment at fixed end B of magnitude $M_A/2$. The rotational stiffness at A, or M_A/ϕ_A, is equal to $4EI/L$. If member AB is located and loaded in a frame in such a way that the relation between ϕ_A and ϕ_B is known in advance, due to symmetry, antisymmetry, or because the end at B is hinged, modified rotational stiffness values permit a reduction in the required computation because no carryover needs to be made in the particular span involved. The modified stiffness values for these special

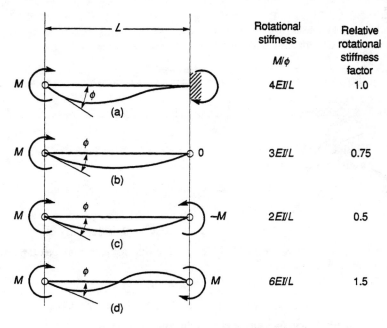

Figure 8.1   Moment rotational stiffness factors.

cases are tabulated in Fig. 8.1(b)–(d), along with their relative values normalized with respect to the basic case shown in Fig. 8.1(a). The use of these modifications to expedite the analysis is illustrated in Examples 8.2 and 8.5.

8.3 Elastic Design of Continuous Beams

In the elastic design procedure, elastic continuous beam analyses are made to determine maximum positive and negative bending moments for various critical positions of live load. If the chosen W beam section meets the compact section requirements of Sec. B5 of AISCS, and the unbraced length requirements of Sec. F1.2d of AISCS, the beam may be "proportioned for $\frac{9}{10}$ of the negative moments produced by gravity loading which are maximum at points of support, provided that, for such members, the maximum positive moment shall be increased by $\frac{1}{10}$ of the average negative moments." This adjustment in moments is justified by the fact that when yielding starts at the supports, due to negative moment, the positive moments increase at a more rapid rate, and by the time the failure load is reached the positive and negative moments are more or less equalized.

In Example 8.1 advantage is taken of the availability of tables of moments for continuous beams of two, three, or four spans of equal length in AISCM, pp. 4–202. The final result should be identical with that based on continuous beam analyses, as in examples that follow. In Example 8.6, plastic design will be used for the same problem.

Example 8.1

Select a W beam of A36 steel for a continuous beam of three equal spans of 30 ft each. A factored dead load of 1.2 kips/ft (including weight of beam) and factored live load of 3.2 kips/ft are assumed to be uniform. Use the elastic design procedure. Simple supports are assumed at extreme ends. Continuous lateral support will be provided by the floor system.

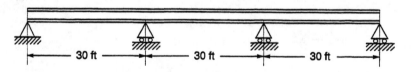

Solution   Calculate moments due to dead load (AISCM Tables, pp. 4–202). Positive moment in end span, $0.4l$ from end:

$$M = 0.080wl^2 = 0.080 \times 1.2 \times 30^2 \times 12 = 1037 \text{ kip-in.}$$

Negative moment at interior supports:

$$M = -0.100wl^2 = -0.10 \times 1.2 \times 30^2 \times 12 = 1296 \text{ kip-in.}$$

Calculate moments due to live load (AISCM Tables) with only end spans loaded for maximum positive moment in end span.

Maximum positive moment $0.45l$ from end:

$$M = 0.1013wl^2 = 0.1013 \times 3.2 \times 30^2 \times 12 = 3501 \text{ kip-in.}$$

Negative moment for same load condition (not maximum):

$$M = -0.050wl^2 = -0.05 \times 3.2 \times 30^2 \times 12 = 1728 \text{ kip-in.}$$

Maximum negative moment (AISCM Tables); one end span unloaded:

$$M = -0.1167wl^2 = -0.1167 \times 3.2 \times 30^2 \times 12 = -4033 \text{ kip-in.}$$

Make beam selection for $\frac{9}{10}$ maximum negative moment or for maximum positive moment increased by $\frac{1}{10}$ of the average negative moments (AISCS, Sec. A5.1).

Negative required moment:

$$M_u = 0.9(1296 + 4033) = 4796 \text{ kip-in.}$$

Positive required moment (maximum dead and live load moments are conservatively assumed to be at the same location):

$$M_u = (1037 + 3501) + 0.1\left(\frac{0 + 1296 + 1728}{2}\right) = 4689 \text{ kip-in.}$$

Negative moment controls beam selection. Section modulus required:

$$Z_x = \frac{M_u}{\phi F_y} = \frac{4796}{0.9 \times 36} = 148 \text{ in}^2$$

Referring to the load factor design beam selection tables, AISCM pp. 4–18, a W24 × 62 is selected. $Z = 153 \text{ in}^3$. Check the shear force. The maximum is at the end of the loaded end span when the other end span is unloaded (refer to AISCM Tables):

$$V_u = 0.6 \times 1.2 \times 30 + 0.617 \times 3.2 \times 30 = 80.8 \text{ kips}$$

Design shear strength:

$$\phi V_n = 0.9 \times 0.6 \times F_y \times A_w = 0.9 \times 0.6 \times 36 \times 23.74 \times 0.43 = 198.4 \text{ kips}$$

Since 80.8 kips < 198.4 kips, the shear is OK.

Note: If the required shear is quite large, it might be desirable to check the direct stress in the web, adjacent to the flange fillet, at the interior support. The procedure in Example 11.3 could be used, but the check is not required by AISCS and the stress is usually not critical.

The complete design would also involve a check on the need for bearing stiffeners over the supports, their design if needed, along with bearing plates and other framing details. As mentioned earlier, examples in this chapter will be concerned primarily with member-size selection, as the design of the details has been covered previously. Whether or not splices are needed depends on availability and shipping restrictions—if needed, complete continuity can be provided by a full-penetration butt weld, and the splice or splices should be located in a region of low bending moment.

In a second example the same overall length and loads will be designated, but the interior supports will be moved toward the ends by 1 ft 6 in., making the center span 33 ft. Since the AISCM tabular information is no longer applicable, the required analyses will be made by moment distribution, as summarized in Section 8.2. Only two analyses will be required, (1) load on one end span, and (2) load on the center span; all load combinations thereby follow by superposing results of these solutions and by taking advantage of the overall symmetry of the structure.

Example 8.2

Rework Example 8.1, but now for new interior support locations, as shown in the sketch. Determine the bending moments at the interior supports with the left span loaded with 1 kip/ft.

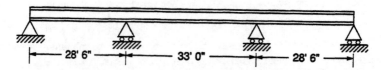

Solution Since the moment of inertia is constant over all three spans, only relative values of the beam stiffness factors (I/L) are needed. Choose any convenient value for I, 1000, for example. Since the end spans are hinged at their extremities, the stiffness modification factor of $\frac{3}{4}$, as explained in connection with Fig. 8.1(b), will be introduced. The distribution factors, in fractions or decimal parts of unity, are entered at each support, and the FEM (fixed-end moments) are calculated and entered. The FEM at the hinged supports are balanced and half of the balancing moments are carried over to the interior supports, thenceforth all carryover and balancing of moments is confined to the center span. The tabulation of the foregoing operations follows:

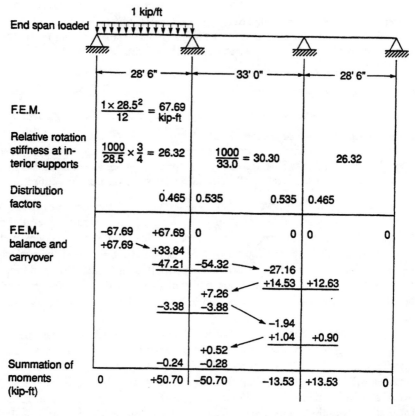

We next apply uniform load of 1 kip/ft to the center span, in which case complete symmetry of load and structure permit use of modified stiffness factors for all spans. Final moments are obtained in one balancing operation with no carryover, since there are no unbalanced moments.

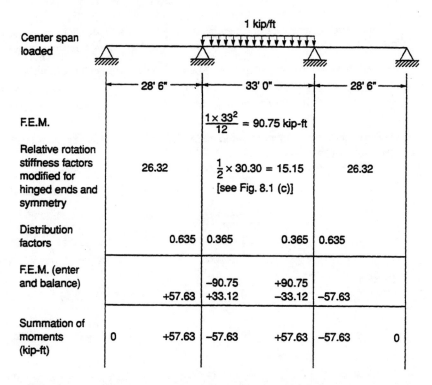

Center span loaded		1 kip/ft		
	28' 6"	33' 0"	28' 6"	
F.E.M.		$\frac{1 \times 33^2}{12}$ = 90.75 kip-ft		
Relative rotation stiffness factors modified for hinged ends and symmetry	26.32	$\frac{1}{2}$ × 30.30 = 15.15 [see Fig. 8.1 (c)]	26.32	
Distribution factors	0.635	0.365	0.365	0.635
F.E.M. (enter and balance)	+57.63	−90.75 +33.12	+90.75 −33.12	−57.63
Summation of moments (kip-ft)	0 +57.63	−57.63	+57.63	−57.63 0

The moments at the supports due to the dead load alone will now be determined as the summation of (1) the left end span loaded, (2) the center span loaded, and (3) the right end span loaded, obtained by reversing, end for end, the results for the left end span loaded.

Moments at supports due to the dead load alone are as follows:

Loaded span		1 kip/ft		
Left end span	0	+50.70 −50.70	−13.53 +13.53	0
Center span	0	+57.63 −57.63	+57.63 −57.63	0
Right end span	0	−13.53 +13.53	+50.70 −50.70	0
Summation of dead load moments (kip-ft)	0	+94.80 −94.80	+94.80 −94.80	0

The maximum moment that determines the required beam size will be the greatest of the following, after modification as allowed by Sec. A5.1 of AISCS, for compact sections.

1. *Maximum positive moment in end span.* Put the live load on both end spans.
2. *Maximum positive moment in center span.* Put the live load on the center span only.
3. *Maximum negative moment at support.* Put the live load on one end span and the center span.

The moments at the supports due to the live load of 3.2 kips/ft are obtained by direct proportion from the previous solutions.

1. *Maximum positive moment in the end span:*

DL+LL = 1.2 + 3.2 = 4.4 kips/ft   1.2 kips/ft   4.4 kips/ft

	A	B				
Dead-load moments	0	+113.76	−113.76	+113.76	−113.76	0
Live load, left span	0	+162.24	−162.24	−43.30	+43.30	0
Live load, right span	0	−43.30	+43.30	+162.24	−162.24	0
Summation of moments (kip-ft)	0	+232.70	−232.70	+232.70	−232.70	0

The maximum positive moment in the end span will be at the location of zero shear. Determine the left reaction. Consider the left span as isolated and take moments about the right end:

$$R_A \times 28.5 + 232.70 - 4.4 \times 28.5 \times 14.25 = 0$$

Solving for R_A,

$$R_A = 54.54 \text{ kips}$$

Let x be the distance from A to the location of zero shear:

$$R_A - 4.4x = 0 \qquad x = 12.40 \text{ ft}$$

Maximum moment:

$$M_{max} = 54.54 \times 12.40 - 4.4 \times \frac{12.40^2}{2} = 338 \text{ kip-ft}$$

Maximum moment increased for design (AISCS, Sec. A5.1):

$$M_u = 338 + \frac{0 + 232.70}{2 \times 10} = 349.6 \text{ kip-ft}$$

2. *Maximum positive moment in the center span:*

1.2 kips/ft	4.4 kips/ft	1.2 kips/ft			

Dead-load moments	0	+113.76	−113.76	+113.76	−113.76	0
Live load, center span	0	+184.42	−184.42	+184.42	−184.42	0
Summation of moments (kip-ft)	0	+298.18	−298.18	+298.18	−298.18	0

The maximum positive moment is at the center of the center span and is equal to the simple beam moment: $wL^2/8$ less 298.18 kip-ft.

Maximum moment:

$$M_{max} = \frac{4.4 \times 33^2}{8} - 298.18 = 300.77 \text{ kip-ft}$$

Maximum moment increased for design (AISCS, Sec. A5.1):

$$M_u = 300.77 + \frac{298.18}{10} = 330.6 \text{ kip-ft}$$

3. *Maximum negative moment at the interior support:*

4.4 kips/ft	1.2 kips/ft			

Dead-load moments	0	+113.76	−113.76	+113.76	−113.76	0
Live load, left span	0	+162.24	−162.24	−43.30	+43.30	0
Live load, center span	0	+184.42	−184.42	+184.42	−184.42	0
Summation of moments (kip-ft)	0	+460.42	−460.42	+254.88	−254.88	0

The maximum negative moment reduced for design (AISCS, Sec. A5.1) is

$$M_u = 0.9 \times 460.42 = 414.38 \text{ kip-ft} \quad \text{(controls selection)}$$

The required section modulus is

$$Z = \frac{414.38 \times 12}{0.9 \times 36} = 153.5 \text{ in}^3$$

Use a W24 × 62, $Z = 153$ in³.

The design of the support details is omitted. It may be noted that the movement of supports from the positions in Example 8.1 has had very little effect on the maximum design moment and that the same beam selection is made as before.

As an alternative to a continuous beam, an articulated series of beams with alternate spans cantilevering over the interior supports may be considered. Such a structure has the advantage of being statically determinate and is unaffected by settlement of supports, a contributing factor in its widespread selection for highway overpass bridges. Hinges are logically placed near locations that would have zero moment if the same structure were continuous. Example 8.3 provides a comparison with Examples 8.1 and 8.2 by using the same overall length and load conditions.

Example 8.3

Here we are dealing with a three-span structure using articulated beams, 90 ft 0 in. overall. Use A36 steel, 3.2 kip/ft live load, 1.2 kip/ft dead load.

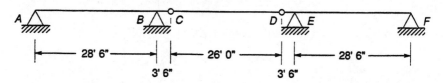

Solution For maximum positive moment in span AB, put the live load on segment AB and take moments about support B, noting that the dead load on CD causes a concentrated reactive load of 15.6 kips at C:

$$R_A \times 28.5 + 15.6 \times 3.5 + \frac{1.2 \times 3.5^2}{2} - \frac{4.4 \times 28.5^2}{2} = 0$$

Solving,

$$R_A = 60.53 \text{ kips}$$

Let x be the distance from A to the location of zero shear.

$$60.53 - 4.4x = 0$$
$$x = 13.76 \text{ ft}$$
$$M_{max} = 60.53 \times 13.76 - 4.4 \times \frac{13.76^2}{2} = 416.35 \text{ kip-ft}$$

The maximum positive moment in span CD,

$$M_{max} = \frac{4.4 \times 26^2}{8} = 371.8 \text{ kip-ft}$$

The maximum negative moment is at B or E. The entire length may be loaded with live load. Take moments about B of cantilever segment BC, noting the reactive force of 57.2 kips at C.

$$M_{max} = 57.2 \times 3.5 + \frac{4.4 \times 3.5^2}{2} = 227.2 \text{ kip-ft} \quad \text{(does not control)}$$

Beam size required for end spans AC and DF:

$$Z = \frac{416.35 \times 12}{0.9 \times 36} = 154.2 \text{ in}^3$$

Use a W21 × 68, $Z = 160 \text{ in}^3$.

Beam size required for center cantilever span CD:

$$Z = \frac{371.8 \times 12}{0.9 \times 36} = 137.7 \text{ in}^3$$

Use a W21 × 62, $Z = 144 \text{ in}^3$.

As a fourth design alternative, consider the use of three simply supported beams, 30 ft each in span.

$$M_{max} = \frac{4.4 \times 30^2}{8} = 495 \text{ kip-ft}$$

$$Z = \frac{495 \times 12}{0.9 \times 36} = 183.3 \text{ in}^3$$

Use W24 × 76 beams, $Z = 200 \text{ in}^3$.

Thus an appreciable saving in weight is achieved in comparison with simple beam design, by use of either continuity or cantilevered articulation. Moreover, especially in the case of the continuous beam, the maximum deflections will be appreciably reduced as compared with simple beam design.

8.4 Elastic Design of Continuous Frames

Attention will be given primarily to one-story frames such as are widely used in shops, supermarkets, and industrial plants. Similar design considerations are required in the case of multistory frames, but that specialized field is beyond the scope of this introduction to basic design.

The details of the elastic structural analysis will not be shown in the following design example of a continuous rigid frame. The given moments, shears, and axial forces were determined by using an elastic frame analysis program for the personal computer. These values could just as well have been determined by manual methods, such as the moment-distribution or slope-deflection methods. Many of the design examples will also be treated subsequently by the alternative plastic-design procedure.

Special attention will now be given to columns that are simply roof support elements and are lacking in continuity with the main frame. Initially, to illustrate the problem, we consider the simplest possible situation of the column that must depend on its neighboring framed column or columns for lateral support at the top.

Referring to Fig. 8.2(a), column 1 is shown in a buckled configuration caused by vertical load P_1. At the instant of buckling the external bending moment $P_1\Delta$ at the base is in exact equilibrium with the internal resisting moment developed at the base by the member. Now consider column 2, in Fig. 8.2(c), hinged at both top and base. If it is not supported laterally at the top, it will collapse as a mechanism under any load. The force required to provide lateral support, as shown at the top, must keep it in static equilibrium and is determined by taking moments about the lower hinge. It is equal to $P_2\Delta$.

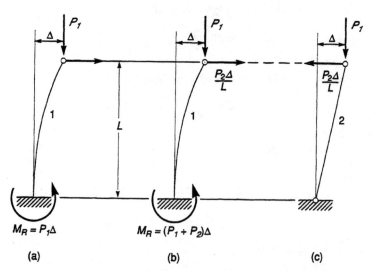

Figure 8.2 Lateral support requirements for a hinged-end column supported by a neighboring column.

Now assume that column 1 is attached at the top by a horizontal member to column 2 as shown in Fig. 8.2(b) and (c). In this case it must provide the lateral restraining force necessary to maintain equilibrium, and the required resisting moment at the base, as shown in Fig. 8.2(b), must be equal to

$$M_R = (P_1 + P_2)\Delta$$

This moment is exactly equal to the resisting moment that would be needed if column 1 were itself loaded with both loads P_1 and P_2. This fact leads to the following design procedure, which has been used in practice when hinged-end columns must be supported by their neighbors:

1. Design the hinged-end columns for their own applied loads with an effective length factor $K = 1$.

2. Framed or fixed-base columns that are connected at the top to hinged-end columns, so as to have the same lateral deflection at the top, shall be designed for their own column load plus the load, or share of the load, apportioned from otherwise unsupported hinged-end columns.

The foregoing procedure is quite accurate in the elastic or buckling range of behavior. In the inelastic range it is conservative but is recommended as a safe design procedure.

Appendix C2 in AISCS gives further detailed procedures for dealing with frames where part of the columns in the story under investigation are framed so that they provide stability for themselves and for the other columns that have pinned ends. Besides more accurate and more complicated methods, a simple modified effective length factor K'_i to account for the pinned-end "leaner" columns is defined as

$$K'_i = K_i\sqrt{N}$$

where K_i is the effective length of the column under investigation as determined by the use of the nomographs of Table 4.4b, and

$$N = \frac{\Sigma P_u}{\Sigma P_u - \Sigma P_{u'o}}$$

ΣP_u is the total factored gravity load and ΣP_{uo} is the portion of the gravity load supported directly by the pinned-end columns in the story under investigation. For the case of the simple examples presented here, the answer is identical. The AISCS method is more general in that it can handle more complex cases.

The following numerical example will further explain the discussion of Fig 8.2.

Example 8.4

The structure shown is supported by the wall and roof bracing system to prevent lateral movement at the top of the columns out of the plane of the sketch. Select column sizes for the 140-kip center load using A36 steel W shapes.

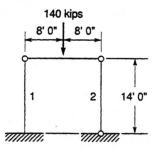

Solution   *Select the size for column 1.* Each column carries an actual load of 70 kips. Column 1 provides lateral support at the top for column 2, so it will be designed for 140 kips. Referring to Table 4.1, for buckling as a cantilever column, as shown, $K = 2.0$ in theory, but 2.1 is recommended for design to allow for some base rotation. Normal to the sketch, with lateral support at top, $K = 0.7$ in theory, with 0.8 recommended for design. Try a W8 × 28:

$$r_x = 3.45 \text{ in.} \qquad r_y = 1.62 \text{ in.}$$

$$A = 8.25 \text{ in}^2$$

$$\frac{KL}{r_x} = \frac{2.1 \times 14 \times 12}{3.45} = 102.3 \quad \text{(governs)}$$

$$\frac{KL}{r_y} = \frac{0.8 \times 14 \times 12}{1.62} = 83.0$$

Design strength:

$$\phi P_n = 17.70 \text{ ksi} \times 8.25 \text{ in}^2 = 146 \text{ kips} \qquad > 140 \text{ kips} \quad \text{OK} \quad \text{(AISCS, Table 3-36)}$$

Select the size for column 2. $K = 1$ for buckling in either direction. Design for an actual load of 70 kips. Try a W8 × 21:

$$A = 6.16 \text{ in}^2 \qquad r_y = 1.26 \text{ in.}$$

$$\frac{KL}{r_y} = \frac{1.0 \times 14 \times 12}{1.26} = 133.3 \quad \text{(governs)}$$

Design strength:

$$\phi P_n = 11.88 \text{ ksi} \times 6.16 \text{ in}^2 = 73 \text{ kips} \qquad > 70 \text{ kips} \quad \text{OK} \quad \text{(AISCS, Table 3-36)}$$

In Example 8.5 a three-span single-story frame will be designed for an assumed dead load of 1 kip/ft and a live load of 2 kips/ft. As in the case of Example 8.2, a complete study would include a check on the maximum positive moments in the center and end spans and the negative moment at an interior support. The negative moment does, in fact, control the design, and the computations for maximum positive moment will, therefore, be omitted. For a maximum negative moment at an interior support, the live load is placed on the adjacent end and center spans, as in Example 8.2. This loading, being unsymmetrical, will cause sidesway.

If a frame experiences lateral translation (sidesway) between story levels, additional story moments are introduced. These are equal to the gravity load times the lateral deflection. Such moments are called *second-order moments*, and their effect can be illustrated by the simple example of Fig. 8.3. In this figure is shown a vertical cantilever column of height L and stiffness EI which is subjected to a lateral force H and a vertical axial force P. The deflected shape is illustrated in Fig. 8.3(b). The deflection at the top is Δ_2, where the subscript 2 signifies second order; that is, it is caused by both the horizontal force H and the moment $P\Delta_2$. The bending moment at the base of the cantilever column is $M = HL + P\Delta_2$. In Fig. 8.3(c) it is shown that the same bending moment results if the horizontal force H at the top is replaced by the force $H = P_2\Delta/L$. The second-order bending moment can be expressed as

$$M_2 = HL + P\Delta_2 = HL\left(1 + \frac{P\Delta_2}{HL}\right) = M_1\left(1 + \frac{P\Delta_2}{HL}\right) = M_1 B_2$$

where $M_1 = HL$ is the *first-order moment*, which is calculated by the usual methods of elastic first-order structural analysis, and B_2 is the sway moment *amplification factor*. For the cantilever of Fig. 8.3,

$$\Delta_2 = \left(H + \frac{P\Delta_2}{L}\right)\frac{L^3}{3EI} = \frac{HL^3}{3EI}\left(1 + \frac{P\Delta_2}{HL}\right) = \Delta_1\left(1 + \frac{P\Delta_2}{HL}\right)$$

where Δ_1 is the first-order lateral deflection. Solving the equation above for Δ_2 yields

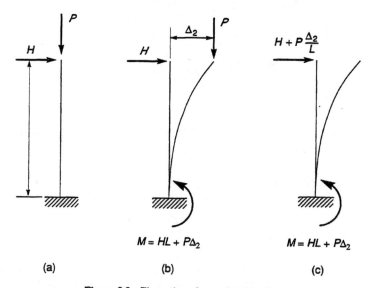

$$M = HL + P\Delta_2 \qquad\qquad M = HL + P\Delta_2$$

(a)   (b)   (c)

Figure 8.3 Illustration of second-order effect.

$$\Delta_2 = \frac{\Delta_1}{1 - P\Delta_1/HL}$$

and substituting into the expression for B_2, we obtain

$$B_2 = 1 + \frac{P\Delta_2}{HL} = \frac{1}{1 - P\Delta_1/HL}$$

The amplification factor B_2 for a more complex multistory frame is determined as shown in Fig. 8.4. The equation for B_2 is essentially the same except that ΣP_u, the sum of the factored gravity loads at the level of the story under consideration, and ΣH_u, the story shear, is used.

The B_2 method described above is one of three possible ways to calculate the effects of second-order moments. Section C1 of AISCS gives another formula for approximating B_2 [Eq. (C1-5)]. AISCS also suggests that preferably a full second-order elastic analysis may be performed in lieu of the approximate B_2 amplification factor. There are a variety of computer software packages that contain such a capability. However, for the purposes of this basic course we use the simple manual procedure discussed above.

The required design moments to be used in checking an assumed beam or beam-column size are determined by the following procedure.

$$M_u = B_1 M_{NT} + B_2 M_{LT}$$

The term B_1 is the member moment amplification factor [Eq. (5.7)]:

$$B_1 = \frac{C_m}{1 - P_u/P_e}$$

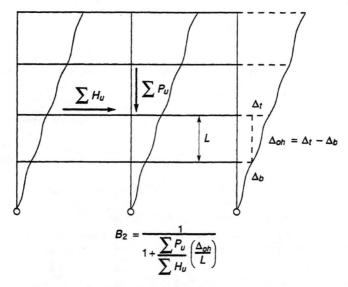

$$B_2 = \frac{1}{1 + \dfrac{\sum P_u}{\sum H_u}\left(\dfrac{\Delta_{oh}}{L}\right)}$$

Figure 8.4 Amplification factor.

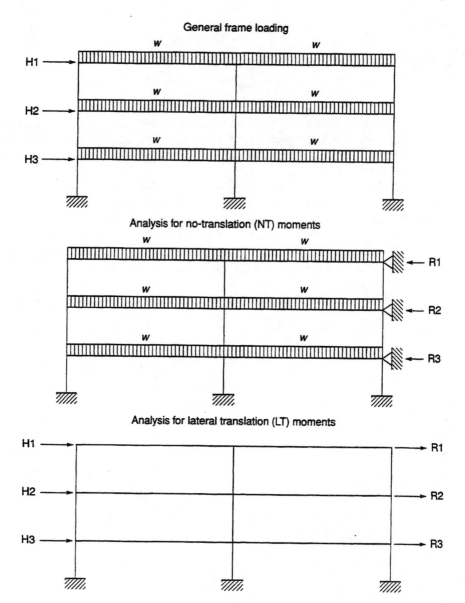

Figure 8.5 Procedure for approximate second-order analysis.

which modifies the moments M_{NT} obtained from an elastic first-order analysis that does not permit story sway, as shown in Fig. 8.5 for the *no-translation* (NT) moments. The term B_2 amplifies the moments M_{LT} obtained from a first-order elastic analysis of the frame when all lateral loads, including the restraining forces from the NT analysis, are applied to the frame. This is shown in Fig. 8.5 as the *lateral-translation* (LT) moments. For any frame that is laterally unbraced, two analyses must therefore be performed if this method of approximate second-order analysis is used: the NT analysis and the LT analysis. The following example illustrates this approach.

Example 8.5

Select column and beam member sizes for the frame shown for a live load of 2 kips/ft and an assumed dead load of 1 kip/ft. Use A36 steel.

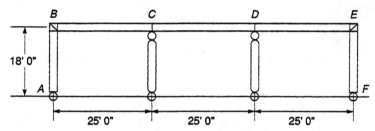

Solution   The critical load case is obtained from the three possible conditions shown below:

1. All three spans are fully loaded.
2. One span is fully loaded while the other two carry only dead load.
3. Two spans are fully loaded.

Analyses were made for all three cases, and the critical case was case 3. The factored dead load is $1.2 \times 1.0 = 1.2$ kips/ft and the factored total load is $1.2 \times 1.0 + 1.6 \times 2.0 = 4.4$ kips/ft.

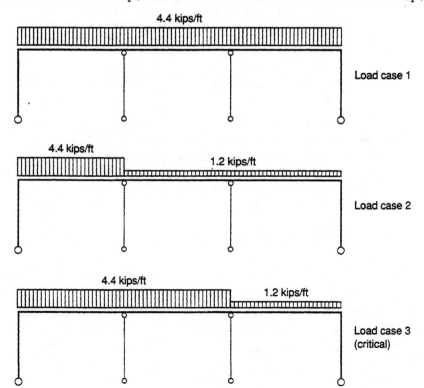

The models for the structural analysis for the no translation (NT) and the lateral translation (LT) cases are indicated on the following two sketches:

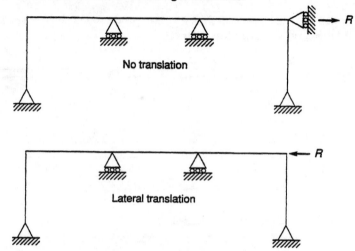

No translation

Lateral translation

The calculated moments for load case 3 at the joints are as follows:

Type of Analysis	M_B (kip-in.)	M_C (kip-in.)	M_D (kip-in.)	M_E (kip-in.)	R (kips)
NT	−1427	−3329	−1757	−144	5.94 to the left
LT	641	−213	213	−641	5.94 to the right

The positive NT moments between the joints are less than the negative joint moments. The lateral deflection at point E equals 0.744 in. for an assumed moment of inertia of 1000 in⁴.

For preliminary design assume that $\Delta_{oh}/L = 0.003$. If the actual deflection is different, calculations will be made with the computed deflection for load case 3,

$$\sum P_u = 2 \times 25 \times 4.4 + 1 \times 25 \times 1.2 = 250 \text{ kips} \quad \text{and} \quad \sum H_u = 5.94 \text{ kips}$$

$$B_2 = \frac{1}{1 - \left(\sum P_u / \sum H_u\right)\left(\Delta_{oh}/L\right)} = \frac{1}{1 - (250/5.94) \times 0.003} = 1.15$$

The maximum moment occurs at location C:

$$M_u = M_{NT} \times B_1 + M_{LT} \times B_2$$

Since at C the axial load is zero, $B_1 = 1.0$, and therefore $B_1 M_{NT} = 1.0 \times (-3329) = -3329$ kip-in.

$B_2 M_{LT} = 1.15 \times (-213) = -244$ kip-in. $M_u = -3329 - 244 = -3573$ kip-in.

The required plastic section modulus is

$$Z_{req} = \frac{3573}{0.9 \times 36} = 110 \text{ in}^3$$

Try a W21 × 50 section, $Z_x = 110$ in^3: $I_x = 984$ in^4. Calculate a new B_2:

$$\Delta_{oh} = 0.744 \times \frac{984}{1000} = 0.732 \text{ in.} \qquad \frac{\Delta_{oh}}{L} = \frac{0.732}{18 \times 12} = 0.00339$$

$$B_2 = \frac{1}{1 - (250/5.94) \times 0.00339} = 1.17 \qquad M_u = 3329 + 1.17 \times 213 = 3577 \text{ in.-kips}$$

This moment is within less than 1% of the previous value; hence use the W12 × 50 size for *the beams and the exterior columns.*

The interior pinned-base columns have an unbraced length of 18 ft in both the x and y directions and the maximum axial force as obtained from the structural analysis equals

$$P_u = P_{NT} + P_{LT} = 121.6 + 4.3 = 126 \text{ kips}$$

From the tables in AISCM select a W8 × 31, which has $\phi_c P_n = 153$ kips.

The member size of W21 × 50 must now be checked for the exterior columns in the rigid frame. The relevant data for this shape are as follows:

$$F_y = 36 \text{ ksi} \qquad A = 14.7 \text{ in}^2 \qquad r_y = 1.30 \text{ in.} \qquad r_x = 8.18 \text{ in.}$$

$$\frac{b_f}{2t_f} = 6.1 < \frac{65}{\sqrt{F_y}} = 10.83 \quad \text{flanges are compact}$$

Examination of the results of the computer analysis indicates that the critical load case is load case 1 (i.e., all spans are loaded by the load of 4.4 kips/ft). From this analysis the following forces were found:

$$P_u = 50.1 \text{ kips} \qquad M_{NT} = 1523 \text{ kip-in.} \qquad M_{LT} = 0$$

Check compactness of web according to Table B5.1 of AISCS:

$$\frac{P_u}{\phi_b P_y} = \frac{50.1}{0.9 \times 14.7 \times 36} = 0.105$$

$$\lambda_p = \frac{640}{\sqrt{F_y}}\left(1 - \frac{2.75P_u}{\phi_b P_y}\right) = 75.8 \qquad > 49.4 \quad \text{web is compact}$$

Calculate B_1. For a beam-column with one end moment $C_m = 0.6$:

$$\lambda_x = \frac{L_x}{\pi r_x}\sqrt{\frac{F_y}{E}} = \frac{18 \times 12}{\pi \times 8.18}\sqrt{\frac{36}{29,000}} = 0.296$$

$$P_{e1} = \frac{AF_y}{\lambda_x^2} = \frac{14.7 \times 36}{0.296^2} = 6040 \text{ kips}$$

$$B_1 = \frac{C_m}{1 - P_u/P_{e1}} = \frac{0.6}{1 - 50.1/6040} = 0.605 \qquad < 1.0, \quad \text{hence } B_1 = 1.0$$

Calculate M_u:

$$M_u = B_1 M_{NT} + B_2 M_{LT} = 1.0 \times 1523 + 1.17 \times 0 = 1523 \text{ kip-in.}$$

Effective length factor:

$$G_A = \frac{I/18}{I/25} = 1.39 \qquad G_B = 10 \qquad K = 1.95 \quad \text{from sway-permitted nomograph}$$

To account for the two leaner columns, multiply K by $\sqrt{N}$, where

$$N = \frac{\sum P_u}{\sum P_u - \sum P_{u'o}} = \frac{3 \times 25 \times 4.4}{25 \times 4.4 \times (3-2)} = 3 \qquad K' = K\sqrt{N} = 3.38$$

Axial capacity:

x-axis length: $L_x = 18$ ft

y-axis length: $L_y = \frac{18}{3} = 6$ ft, assuming lateral braces at the third points

$$\lambda_x = \frac{K'L_x}{\pi r_x}\sqrt{\frac{F_y}{E}} = 1.00 \qquad < 1.5 \quad \text{controls}$$

$$\lambda_y = \frac{L_y}{\pi r_y}\sqrt{\frac{F_y}{E}} = 0.621$$

$$F_{cr} = F_y(0.658^{\lambda_c^2}) = 23.69 \text{ ksi} \qquad \phi_c AF_{cr} = \phi_c P_n = 0.85 \times 14.7 \times 23.69 = 296 \text{ kips}$$

Flexural capacity: From AISCM: $L_p = 5.4$ ft and $L_r = 16.2$ ft; $\phi_b M_p = 297$ ft-kip and $\phi_b M_r = 184$ ft-kip; $C_b = 1.75$ (moment on one end only); $L_b = 6$ ft.

$$\phi M_n = C_b\left[\phi_b M_p - (\phi_b M_p - \phi_b M_r)\frac{L_b - L_p}{L_r - L_p}\right] \le \phi_b M_p = 297 \text{ ft-kip} = 3564 \text{ in.-kip}$$

Check interaction equation (AISCS, Sec. H1-1)

$$\frac{P_u}{\phi_c P_n} = \frac{50.1}{296} = 0.169 \qquad < 0.2, \quad \text{use Eq. (b)}$$

$$\frac{P_u}{\phi_c P_n} \times \frac{1}{2} + \frac{M_u}{\phi_b M_n} = \frac{0.169}{2} + \frac{1523}{3564} = 0.512 \qquad < 1.0 \quad \text{OK}$$

The W21 × 50 shape is thus the size selected for the rigid frame.

Attention is now turned to plastic design, with application to a number of the same design problems that have been treated elastically in the foregoing sections.

8.5 Introduction to Plastic Design

Before engaging in the study of plastic design for continuous beams and frames, one should review earlier material that is especially relevant. The whole concept of plastic design is dependent on the great ductility of structural steel and on its unique yield property as exhibited by the plastic portion of the stress–strain diagram shown in Fig. 1.2, the discussion of which in Section 1.3 should be reread at this time. In *elastic design* of continuous beams and frames the basis for determining the capacity of the structure is the strength of the cross section at the location of highest moment, which is calculated by assuming elastic behavior. In *plastic design* the capacity is determined on the basis of the strength of the entire structure. This plastic collapse load is the *ultimate strength* of the structure: no further increase of load is possible. When the ultimate strength is reached, various localized parts of the structure will have yielded to a varying degree and finally, when collapse is imminent, the structure is said to have become a plastic mechanism. In a true mechanism (such as one of the interior columns of Example 8.5) there is no strength whatever unless lateral support is provided.

In a plastic mechanism there is resistance to collapse, but deformation proceeds with no increase in load, just as it does in the simple tension test in the plastic range of Fig. 1.2.

Members used in plastic design are primarily those wide-flange W shapes that meet the requirements for a compact section of AISCS, Table B5.1. These sections are compact to a degree that will permit bending deformation well into the plastic range without loss of compressive strength due to plastic buckling. This added compactness gives the member a quality known as plastic *rotation capacity*. The bending behavior of the W shape in the plastic range was discussed in detail in Section 3.3 in connection with the plastic design of simple beams that are statically determinate in both the elastic and plastic stages of their load-deflection history. While there can be no objection to the plastic design of simple beams, neither is there any particular advantage. In the simple beam the relative distribution of moment throughout the beam is the same in the elastic and plastic stages. Not so for the continuous beam or frame! After yielding is initiated at the location of maximum elastic-range moment, the relative distribution of moment along the continuous beam or frame starts to change. The total load on the structure continues to increase, although the deflection for a given increment of load becomes progressively greater as the plastic mechanism or collapse load is approached. It is this redistribution of moment that provides the potential for greater economy when plastic design is used.

The specification requirements for plastic design are covered in the following sections of AISCS: Sec. A5.1 limits the yield strength of steel that is permissible in plastic design to 65 ksi; Sec. B5.2 defines compactness limits of the cross section; Sec. C2 prescribes the maximum axial force that must not be exceeded in a plastically designed beam-column; Sec. E1.2 gives the limit of beam-column slenderness ratio; and Sec. F1.2b defines the maximum permissible unbraced length of a plastically designed beam near a plastic hinge. Of these limitations the most stringent is the limit on the unbraced length. The member must not only be of compact cross section but must have a close spacing of the lateral braces in order to develop the rotation capacity required for the formation of a plastic hinge.

Plastic design should not be used if the number of cycles of repeated load (AISCS, Appendix K3) reaches a number requiring a limit on the maximum stress range. Plastic design ensures an optimum of structural integrity, or "toughness," against failure—a desirable feature in earthquake- or blast-resistant structures. Moreover, especially in the case of continuous beams, the structural analysis for plastic design is simplified because it is *statically determinate*. A complete and authoritative commentary on plastic design is provided in Ref. 8.2.

Before considering the alternative plastic design procedure applied to the conditions of Example 8.1, the ultimate strength moment–load relationships for an end span and for an interior span of a continuous beam will be examined. Assume that the M versus ϕ curve for the W section shown in Fig. 3.9 may be replaced by a two-straight-line approximation, as shown in Fig. 8.6.

For the end span shown in Fig. 8.7, elastic analysis shows that M_p would first be reached at the support where the maximum moment occurs. According to Fig. 8.6,

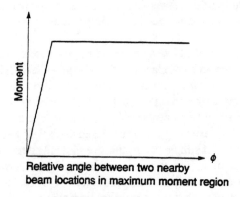

Figure 8.6 Assumed beam behavior in plastic design analysis.

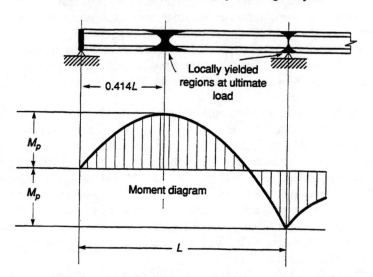

Figure 8.7 End-span moments at ultimate plastic load.

this moment would not change with increasing load, and the maximum load condition would be reached when a positive moment of M_p was also reached, thus making this beam segment a plastic mechanism. It can be shown that the maximum positive moment is $0.414L$ from the simply supported end and that, for an end span,

$$M_p = 0.086wL^2 \qquad (8.1)$$

For an interior span, M_p would be reached simultaneously at both ends, and after additional increase in load would also be reached at the center. Since the total range of moment is the same as the center moment in a simple beam, it is readily seen (Fig. 8.8) that

$$M_p = \frac{wL^2}{16} \qquad (8.2)$$

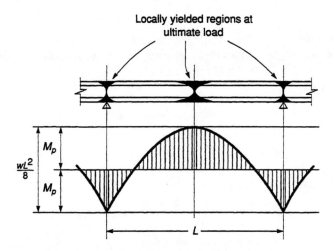

Figure 8.8 Interior span moments at ultimate plastic load.

Example 8.6
Rework Example 8.1 using plastic design. The total factored load on any one of the spans will be

$$1.2 + 3.2 = 4.4 \text{ kips/ft}$$

Solution Since the interior and end spans are of the same length, the beam selection will be determined by Eq. (8.1), which requires a greater M_p than Eq. (8.2). The required M_p, by Eq. (8.1), is

$$M_p = 0.086 \times 4.4 \times 30^2 \times 12 = 4087 \text{ kip-in.}$$

The required plastic modulus is

$$Z_{\text{req}} = \frac{4087}{0.9 \times 36} = 126.1 \text{ in}^3$$

Referring to the AISCM, try a W24 × 55, with $Z = 134 \text{ in}^3$. Check the width/thickness ratios for requirements of ASICS, Table B5.1:

$$\frac{b}{2t_f} = 6.94 \qquad < \frac{65}{\sqrt{F_y}} = 10.83 \quad \text{OK}$$

$$\frac{h}{t_w} = 54.6 \qquad < \frac{640}{\sqrt{F_y}} = \frac{640}{\sqrt{36}} = 106.7 \quad \text{OK}$$

Check the shear capacity. In the end span, adjacent to the interior support, maximum shear occurs equal to

$$V_{\text{max}} = \frac{wL}{2} + \frac{M_p}{L} = (0.50 + 0.086)wL = 0.586wL$$

$$= 0.586 \times 4.4 \times 30 = 77.35 \text{ kips}$$

Shear capacity (AISCS, Formula F2-1):

$$\phi V_n = \phi 0.6 F_y A_w = 0.9 \times 0.6 \times 36 \times 0.395 \times 23.57 = 181.0 \text{ kips} \qquad > 77.35 \text{ kips} \quad \text{OK}$$

Example 8.7

Rework Example 8.6 with supports relocated to reduce end spans to an optimum condition that would make the required M_p for the end spans the same as that for the center span.

    Solution   Let x = length of end span. Then $90 - 2x$ = length of center span. By Eqs. (8.1) and (8.2),

$$0.086wx^2 = \frac{1}{16}w(90-2x)^2$$

This reduces to

$$x^2 - 137x + 3087 = 0$$

Solving the quadratic,

$$x = \frac{137}{2} \pm \frac{1}{2}\sqrt{137^2 - 4 \times 3087} = 68.5 \pm 40.1 = 28.4$$

Try two end spans of a 28 ft 6 in. span, and a center span of 33 ft 0 in. The factored load is 4.4 kips/ft. The required M_p for the end span is

$$M_p = 0.086 \times 4.4 \times 28.5^2 \times 12 = 3688 \text{ kip-in.}$$

As a check, calculate M_p for the center span:

$$M_p = \frac{1}{16} \times 4.4 \times 33^2 \times 12 = 3594$$

(Slightly less, as an exact optimum was not chosen.) The required plastic modulus is

$$Z_{req} = \frac{3688}{0.9 \times 36} = 113.8 \text{ in}^3$$

A beam size W24 × 55, with $Z = 134 \text{ in}^3$, is still the least-weight selection (AISCM). Although, in comparison with Example 8.6, the shifting of supports has appreciably reduced the required Z from 126.1 to 113.8 in^3, no weight saving results. This, of course, is due to the discontinuous nature of available W shapes in relation to their strength properties. In comparison with Example 8.2, plastic design has effected a savings of 11.3% in weight over the elastic design. Moreover, the required computational work has been greatly reduced and simplified. No general conclusions should be drawn from these comparisons.

8.6 Plastic Design of Frames

In Examples 8.6 and 8.7 the determination of the required plastic modulus (Z) was relatively simple. There was only one distribution of moment possible at failure for each span, and the span requiring the greatest Z determined the beam selection. The solution was easily made by simple statics.

    In a frame it is usually more convenient to make the analysis by the mechanism method, and there are more ways than one in which a possible failure mechanism can develop. Each possible mechanism will, in general, account for a different load at failure, only the lowest of which is correct and which thus provides the criterion for member selection. One recalls that "a chain is no stronger than its weakest link." One way to determine the correct mechanism is to try all possibilities, but as the number of members in a frame increases, this procedure gets increasingly cumbersome. As an alternative, after some experience with similar cases, if one can guess the correct mechanism in advance, its correctness can be verified by applying equations of static equilibrium to the elements between hinges as a preliminary to the construction of the bending moment diagram for the complete frame. If the correct mechanism has been chosen, the moment at each plastic hinge will be exactly equal to the plastic moment of the corresponding member. If the mechanism is incorrect, too great a load will have been calculated, and the bending moment at one or more plastic hinge locations will exceed the plastic moment capacity of the member. In any case, whether one tries all mechanisms or guesses the correct one, the foregoing moment check should be made to ensure the correctness of the solution. In very complex cases the aid of a computer can be sought, programmed to search out the lowest load corresponding to all possible mechanism solutions.

In applying the mechanism method, the method of virtual displacements is especially appropriate. If a system is in equilibrium, the total work done during an incremental displacement is zero; that is, the positive (external) work done by the applied forces is equal in magnitude to the negative (internal) work done. In the case of a plastic mechanism at failure load, the internal work is equal to the sum of the products of each plastic moment multiplied by the rotation at the corresponding plastic hinge.

To illustrate the mechanism-virtual-displacement method, we redo the case shown in Fig. 8.8, the interior span of a uniformly loaded continuous beam. It is convenient to show the incremental displacement as if it originated from the undeformed structure, as shown in Fig. 8.9.

The external work is equal to the total uniform load times the average deflection, which is $\theta L/4$:

$$\text{external work} = wL \times \frac{\theta L}{4}$$

The internal work is equal to the sum of the products of the plastic hinge moments multiplied by the incremental hinge rotation θ at the two ends and by 2θ at the center.

$$\text{internal work} = M_p(\theta + 2\theta + \theta) = 4M_p\theta$$

Equating the external and internal work we obtain, as before, the same Eq. (8.2):

$$M_p = \frac{wL^2}{16}$$

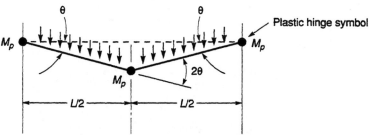

Figure 8.9 Beam mechanism for interior span of a continous beam.

We now apply the same procedure to the simple hinged-base frame shown in Fig. 8.10, loaded horizontally at the top by force $P/3$ and vertically midspan by P. There are three possible mechanisms illustrated in Fig. 8.10: beam (a), panel (b), and a combination of the two (c).

For the beam mechanism [Fig. 8.10(a)] equating external and internal work,

$$\frac{PL\theta}{2} = 4M_p\theta \quad \text{or} \quad M_p = \frac{PL}{8}$$

The mechanism shown in Fig. 8.10(b) is for failure by sidesway, usually termed a panel mechanism. In this case,

$$\frac{PL\theta}{3} = 2M_p\theta \quad \text{and} \quad M_p = \frac{PL}{6}$$

For the combined mechanism [Fig. 8.10(c)], both the horizontal and vertical external forces do work and

$$\frac{PL\theta}{3} + \frac{PL\theta}{2} = M_p(2\theta + 2\theta)$$

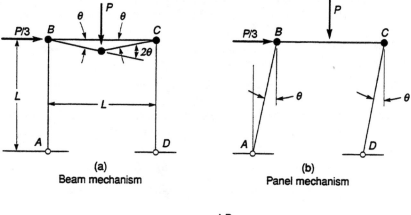

(a)
Beam mechanism

(b)
Panel mechanism

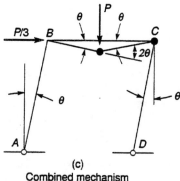

(c)
Combined mechanism

Figure 8.10   Possible failure mechanisms for a single-span hinged-base bent.

whence

$$M_p = \frac{5}{24}PL$$

The largest value of M_p corresponds to the least load for a given M_p; hence the combined mechanism is the predicted mode of failure.

We now make a moment diagram check. The vertical force at D is obtained by taking moments about A for the entire frame. The horizontal shear at D is obtained by taking moments about C for member CD; that is, the base shear at D is equal to M_p/L, or $5P/24$. Thus the base shear at A must be $P/8$, and the moment at B is now determined as $PL/8$. The complete moment diagram can now be constructed as shown in Fig. 8.11, and the moment at the center of the beam is found by statics to be $5PL/24$, which is also the previously predicted value of M_p obtained by the combined mechanism shown in Fig. 8.10(c); thus the correctness of this solution is confirmed. Had this solution been incorrect, the moment at the center of the beam would have turned out to be greater than $5PL/24$.

Example 8.8

In Fig. 8.11, assume that $L = 14$ ft 0 in. and that the factored vertical load $P_{uv} = 120$ kips and the factored horizontal load equals $P_{uh} = 40$ kips. Thus $P = 120$ kips for comparison with the diagrams of Fig. 8.11. Only this load combination will be checked here. Load combinations that involve only factored vertical loads must also be checked because the load factors are different. This will not be done for this problem. The combined mechanism controls.

Solution   Calculate the required M_p:

$$M_p = \frac{5 \times 120 \times 14}{24} = 350 \text{ kip-ft}$$

The required plastic modulus is

$$Z = \frac{350 \times 12}{0.9 \times 36} = 129.6 \text{ in}^3$$

Try a W24 × 55:

$$Z = 134 \text{ in}^3 \qquad A = 16.2 \text{ in}^2$$

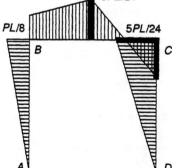

Figure 8.11   Moment-diagram check on mechanism solution.

Check the width/thickness ratio:

$$\frac{b_f}{2t_f} = 6.9 \qquad < 10.83 \quad \text{OK} \quad \text{(AISCS, Table B5.1)}$$

Check the depth/thickness ratio. Determine the vertical reaction at D by taking moments of forces on the complete structure about A.

$$V_D = \frac{40 \times 14 + 120 \times 7}{14} = 100 \text{ kips}$$

$$\text{max. column load } P_u = 100 \text{ kips}$$

$$\frac{P_u}{\phi_b P_y} = \frac{100}{0.9 \times 36 \times 16.2} = 0.191 \qquad > 0.125 \qquad \text{(AISCS, Table B5.1)}$$

$$\text{limiting } \left(\frac{h}{t_w}\right)_p = \frac{191}{\sqrt{F_y}}\left(2.33 - \frac{P_u}{\phi_b P_y}\right) = \frac{191}{\sqrt{36}}(2.33 - 0.191) = 68.1$$

For the W24 × 55, $h/t_w = 54.6$, < 68.1, thus OK; the section is compact.

Check for the combined column load and bending moment (AISCS formula (H1-1). To determine P_e, K (effective length factor) must be evaluated (refer to Section 4.3). Figure 4.5 will be used. At the column top ($I_c/L_c = I_g/L_g$), $G = 1.0$; at the column bottom, for a hinged base, assuming that $G = 10.0$, then

$$K = 1.87 \qquad \text{(by Fig. 4.5)}$$

Determine column capacity:

$$\lambda_c = \frac{KL}{\pi r_x}\sqrt{\frac{F_y}{E}} = \frac{1.87 \times 14 \times 12}{\pi \times 9.11}\sqrt{\frac{36}{29,000}} = 0.387 \qquad > 1.5$$

$$F_{cr} = \left(0.658^{\lambda_c^2}\right)F_y = 33.82 \text{ ksi}$$

$$\phi_c P_n = 0.85 \times 16.2 \times 33.82 = 465.6 \text{ kips}$$

Determine B_1.

$$B_1 = \frac{C_m}{1 - P_u/P_{e1}}$$

$$C_m = 0.6 \quad \text{no moment at base of beam-column}$$

$$P_{e1} = \frac{AF_y}{\lambda_{c1}} \qquad \text{calculated for an effective length factor of 1.0}$$

$$\lambda_c = \frac{KL}{\pi r_x}\sqrt{\frac{F_y}{E}} = \frac{1.0 \times 14 \times 12}{\pi \times 9.11}\sqrt{\frac{36}{29,000}} = 0.207$$

$$P_{e1} = \frac{16.2 \times 36}{0.207^2} = 13,634 \text{ kips}$$

$$B_1 = \frac{0.6}{1 - 100/13,634} = 0.604 \qquad < 1.0, \quad \text{hence } B_1 = 1.0$$

For one-story frames of the type considered here, the second-order frame effects can be neglected because the columns are subject mainly to flexure. Therefore, the required moment is $M_u = B_1 (M_p)_{req} = 1.0 \times 350 = 350$ kip-in. The flexural capacity of a laterally braced compact member is

$$\phi_b M_n = \phi_b M_p = 0.9 \times 134 \times 36 = 4342 \text{ kip-in.} = 361.8 \text{ kip-ft}$$

Check the interaction equations [AISCS, Eq. (H1-1)]:

$$\frac{P_u}{\phi_c P_n} = \frac{100}{465.6} = 0.215 \qquad > 0.2$$

$$\frac{P_u}{\phi_c P_n} + \left(\frac{8}{9}\right)\frac{M_u}{\phi_b M_n} = 0.215 + \left(\frac{8}{9}\right)\frac{350}{361.8} = 1.075 \qquad > 1.0 \quad \text{NG}$$

The W24×55 will not be adequate for the columns. A change to a W24×62 will not change the frame strength or the required plastic moment of the beam. The student is requested to make the necessary checks for this new size, and it will be found that the interaction equation sum equals 0.944 < 1.0. Therefore, use W24×55 for the beam and W24×62 for the two columns.

Lateral bracing requirements: Plastic hinges may form at either B or C, and at the center. Assume that bracing is to be supplied at these three locations. Considering the moment diagram in Fig. 8.11, check the adequacy of this bracing.

For the beam: the unbraced length $L_b = 7 \times 12 = 84$ in. The required unbraced length on either side of a plastic hinge is stated in Sec. F1 of AISCS, Eq. (F1-17):

$$L_{pd} = \frac{3600 + 2200(M_1/M_2)}{F_y} r_y$$

For the W24×55, $r_y = 1.34$ in. and $M_1/M_2 = (PL/8)/(5PL/24) = -0.6$; hence $L_{pd} = 84.9$ in., > 84 in. The beam is thus adequately braced by a lateral support in the center and at the two corners.

For the column a lateral brace will be provided at the center of the member so that the unbraced length is also 84 in. The ratio $M_1/M_2 = -0.5$ and $r_y = 1.38$ in.,

$$L_{pd} = \frac{3600 - 2200 \times 0.5}{36} \times 1.38 = 95.8 \text{ in.} \qquad > 84 \text{ in.}$$

Note that the lateral braces must hold the members against both lateral movement and twist. Design of these details is omitted.

At the upper corner knee connections, where a plastic hinge may form, the full plastic moment creates high shear in the web at the corner. The full yield strength of the flange areas of the beam must be transmitted by shear into the column web, and vice versa (see AISCS Commentary, Sec. K1.7). If the web thickness is inadequate, a diagonal stiffener may be added, as shown. The design will be made for the full-capacity hinge moment of 36× 134 = 4824 kip-in. Calculate the shear in the web:

$$V = \frac{4824}{0.95 \times 23.57} = 215 \text{ kips} \qquad \text{(AISCS Commentary, Sec. K1.7)}$$

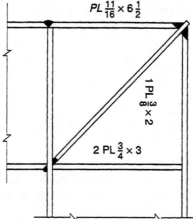

The shear capacity of the web is calculated by the provisions of Sec. K1.7 of AISCS. The axial force ratio in the column is $P_u/P_y = 100/(16.2 \times 36) = 0.171 < 0.4$ and thus Eq. (K1-9) applies:

$$R_v = 0.6F_y d_c t_{cw} = 0.6 \times 36 \times 23.74 \times 0.43 = 220.5 \text{ kips}$$

In this equation d_c is the depth of the column section and t_{cw} is the column web thickness. The design condition is

$$\phi R_v \geq V_u = 0.9 \times 220.5 = 198.4 \text{ kips} \qquad < V_u = 215.4 \text{ kips}$$

Therefore, a diagonal stiffener or a doubler plate is needed. A diagonal stiffener will be specified, as shown in the drawing of the joint. The force that must be resisted by this stiffener is equal to $F = (215 - 198)/0.707 = 24$ kips.

The stiffener area needed is $A_{st} = 24/\phi F_y = 24/(0.9 \times 36) = 0.742 \text{ in}^2$. A stiffener plate 2 in. $\times \frac{3}{8}$ in. will be specified on one side of the web only. The stiffener area is 0.75 in^2. Its width/thickness ratio is $2/0.375 = 5.33 < 65/\sqrt{F_y} = 10.83$. Use one $2 \times \frac{3}{8}$ diagonal stiffener on each corner joint.

Add end plate and vertical stiffeners to transmit column flange forces into beam web. The column flange area is $7.04 \times 0.59 = 4.15 \text{ in}^2$. Select end plate, $\frac{11}{16} \times 6\frac{1}{2}$.

$$A = 4.47 \text{ in}^2$$

The vertical stiffeners are to be two PL $\frac{3}{4} \times 3$:

$$A = 4.5 \text{ in}^2$$

Welds attaching the end plates and vertical stiffeners to the beam web should be sized so as to fully develop the web shear capacity of 198 kips. Weld design involves elementary procedures covered in Chapter 6 and will not be carried further in this example.

The plastic design of a frame having the same dimensions and load requirements of Example 8.5 will now afford a further opportunity to compare weights between a plastic and an allowable stress design. In the absence of lateral load at the top, it may be presumed that failure will be by beam mechanism.

Example 8.9

Rework Example 8.5 using plastic design.

Solution Beam selection will be on the basis illustrated in Fig. 8.5, with any one of the three beam spans requiring the same plastic moment,

$$M_p = \frac{wL^2}{16} = \frac{4.4 \times 25^2}{16} = 171.9 \text{ kip-ft}$$

The required plastic modulus is

$$Z = \frac{171.9 \times 12}{0.9 \times 36} = 63.7 \text{ in}^3$$

Try a W18 × 35, with

$$Z = 66.5 \text{ in}^3 \qquad A = 10.3 \text{ in}^2$$

Check the adequacy of the same section as column element AB or EF. As in elastic design, the framed columns must provide added stability for the interior columns, which are hinged, top and bottom, and not continuous with the main frame. End columns will be designed for half of the full load on all three spans,

$$P = \frac{4.4 \times 75}{2} = 165.0 \text{ kips}$$

Check the width/thickness ratios by Table B5.1 of AISCS:

$$\frac{b}{2t_f} = 7.1 \qquad < 10.83 \quad \text{OK}$$

$$\frac{h}{t_w} = 53.5$$

The actual factored load, $P_{u1} = 4.4 \times 12.5 = 55.0$ kips, is used as in Example 8.5.

$$\frac{P_{u1}}{\phi_b P_y} = \frac{55.0}{0.9 \times 36 \times 10.3} = 0.165 \qquad > 0.125$$

The limiting h/t_w, by AISCS, is

$$\frac{191}{\sqrt{F_y}}(2.33 - 0.165) = 68.9 \qquad > 53.5 \quad \text{OK}$$

Determine the effective length factor, K (Sec. 4.3).

$$\frac{I_c}{L_c} = \frac{510}{18 \times 12} = 2.36 \qquad \frac{I_g}{L_g} = \frac{510}{25 \times 12} = 1.70$$

$$G_A = \frac{2.36}{1.70} = 1.4 \qquad G_B = 10 \text{ (hinge)} \qquad K = 1.95 \quad \text{(from Fig. 4.5)}$$

$$\lambda_c = \frac{1.95 \times 18 \times 12}{7.04\pi}\sqrt{\frac{36}{29,000}} = 0.671 \qquad > 1.5$$

$$F_{cr} = (0.658^{0.671^2}) \times 36 = 29.82 \text{ ksi}$$

$$\phi_c P_n = 0.85 \times 10.3 \times 29.82 = 261 \text{ kips} \qquad \frac{P_u}{\phi_c P_n} = \frac{165}{261} = 0.632 \qquad > 0.2$$

$B_1 = 1.0$ based on experience with Example 8.8.

$$M_u = B_1(M_p)_{req} = 1.0 \times 171.9 = 171.9 \text{ kip-ft}$$
$$\phi_b M_n = 0.9 \times 66.5 \times 36 = 2154.6 \text{ kip-in.}$$

Interaction equation check:

$$0.632 + \frac{8}{9} \times \frac{171.9 \times 12}{2154.6} = 1.48 \qquad > 1.0 \quad \text{NG}$$

Try a new section W21 × 57.

$$K = 2.3 \quad \lambda_c = 0.667 \quad F_{cr} = 29.89 \text{ ksi} \quad \phi_c P_n = 424.32 \text{ kips} \quad \phi_b M_n = 4179.6 \text{ kip-in.}$$

$$\frac{165}{424} + \frac{8}{9} \times \frac{171.9 \times 12}{4179.6} = 0.828 \qquad < 1.0 \quad \text{OK}$$

Use W21 × 57 for the columns and W18 × 35 for the beams.

Interior column design at locations C and D: these columns are independent of the main frame and the design is essentially the same as in Example 8.5. The plastic equalization of all beam end moments would result in a slightly smaller column load at working load, but the required column size is the same.

Although heavier end columns are required in the plastic design, the overall weight of the main-frame material is 4677 lb, 873 lb less than the 5550 lb required for the elastic allowable stress design of Example 8.5. The 15.7% overall saving is appreciable, especially if a number of frames are required.

Plastic design has been used in multistory buildings, most commonly for the beams in either fully braced structures or those with relatively few floors for which lateral forces of wind or earthquake are not a primary consideration. Plastic design probably offers the greatest advantage in application to one-story frames similar to the case in Example 8.9. The greater simplicity of plastic design analysis procedures in comparison with continuous beam analysis was well demonstrated by Examples 8.5 and 8.9. For those who wish to pursue the subject further, ASCE *Manual 41* (Ref. 8.2) is recommended. An early text on the subject, *Plastic Design of Steel Frames*, (Ref. 8.3) also provides many detailed analysis and design solutions of a variety of frames.

REFERENCES

8.1. W. WEAVER, JR., and J. M. GERE, *Matrix Analysis of Framed Structures*, 3rd ed., Van Nostrand Reinhold, New York, 1990.

8.2. *Plastic Design in Steel: A Guide and Commentary*, Manual 41, American Society of Civil Engineers, New York, 1971.

8.3. L. S. BEEDLE, *Plastic Design of Steel Frames*, Wiley, New York, 1958.

PROBLEMS

8.1. Rework Example 8.1, but change to three spans of 38 ft, each designed for a dead load of 1 kip/ft and a live load of 2.5 kip/ft. Use A36 steel. Assume continuous lateral support. Use AISCM tables to calculate moments. Select member sizes. The loads are service loads.

8.2. Rework Example 8.2, but change the center span to 42 ft and the end spans to 36 ft each. The live load is 2.5 kip/ft, the dead load 1 kip/ft. Use A36 steel. The loads are service loads.

8.3. Rework Example 8.3 using end spans of 36 ft with cantilever overhang of 5 ft, leaving a simply supported suspended center beam of 32 ft. Use A36 steel. Same loads as Problems 8.1 and 8.2. Compare total weights for the three alternative solutions.

8.4. Select column and beam sizes for stress design for the two-span frame shown. Use A36 steel. The dead load is 1 kip/ft, the live load, 2.5 kip/ft. Lateral support is assumed for the end columns and the beam. Example 8.5 is similar. The loads are service loads.

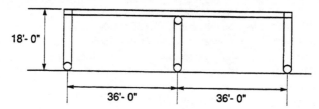

8.5. Determine the plastic shape factor for the box section shown. Check the suitability of the section for plastic design using steels with yield points of 36 and 50 ksi.

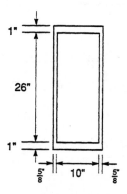

8.6. Show that $M_p = 0.086wL^2$ and that the plastic hinge for positive moment occurs at a distance $0.414L$ from A at the left end of the simple support. Same conditions as in Fig. 8.4. (*Hint:* From the equilibrium of the entire span, write an expression for R_A. Then write a general expression for M at any distance x from A. Set $dM/dx = 0$ for maximum M.)

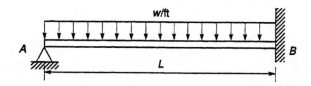

8.7. Using the mechanism method, determine an expression for M_p. Compare with results of Problem 8.6.

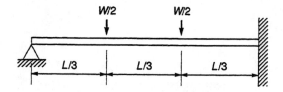

8.8. Rework Problem 8.1 by plastic design.

8.9. Rework Problem 8.2 by plastic design. Compare the total weight of the results of Problems 8.1, 8.2, 8.3, 8.8, and 8.9.

8.10. Design the frame as shown by plastic design. Use A36 steel. The loads are factored (ultimate) loads. Determine the required locations for lateral support. Design a corner knee connection.

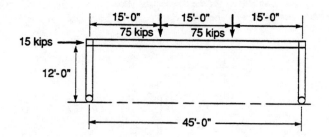

8.11. Rework Problem 8.4 by plastic design. Compare the total weight with the results of Problem 8.4.

8.12. In Example 8.9, show that if one end span carries full live load, the frame is no more likely to fail by sidesway than by the beam mechanism that was presumed to be critical.

9

Computer-Aided Technology

9.1 Introduction

In the late 1940s the electronic computer first became available as a research tool in universities and industry. Since then, as the size and price of computer components decreased and its capabilities and capacity increased, it has rapidly spread out into all fields of engineering and has produced dramatic changes in engineering practice and productivity. Processes that were impossible, overly laborious, or impractical by manual procedures are now handled rapidly by a modern design office computer facility. The computer can make layouts, isometric projections, analyses, design members and frames, make code checks, select member sizes, modify and optimize designs, and finally, prepare complete detailed design drawings.

Computer-aided technology (CAT) and its application have led rapidly to engineer and computer interaction as a design team. The engineer provides design controls and guides based on his or her professional expertise and experience and the computer performs speedy computation, checking, and iteration to provide an optimum solution. Meanwhile, the computer creates an enormous data bank of stored information which the engineer can override for partial or overall rerun for new jobs and further optimization.

9.2 Basic Flowchart Programming

The flowchart approach, introduced in earlier chapters as a guide to the use of the AISC Design Specifications, has had its primary application and origin as a basic aid in computer programming. It is a series of logical steps which direct the programmer along the flow path until the program objective is achieved. The integrated flowchart is primarily composed of the following three basic flow elements (modules): (1) sequential steps (2) decision tests, providing direction to the next step to be executed as the result of a test; and (3) conditional iteration in meeting certain requirements and/or optimization. These three basic flow elements are integrated into a logical flow system which is not only used for computer programming, but is also extremely useful in the overall design for guidance through the morass of the complex codes, preferred constraints, and procedures. The entire flow process, then, becomes more simple and systematic.

9.2.1 Sequential Steps

Logically independent sequential steps as shown in Flowchart 9.1 indicate that the control of the design process flows from box A to box B, and then on to C, and so on. The process boxes A, B, C, and so on, are all connected by one way direction flow arrows, and are explicitly representative of a set of requirements, assignments, or formulas that are to be executed by the computer subroutine (program module) with or without engineer interaction.

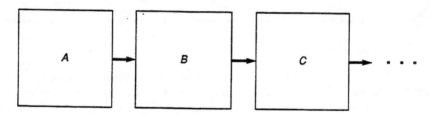

Flowchart 9.1   Sequential Steps

9.2.2 Decision Test

The *decision test symbol*, or *decision node*, indicates a test requirement as shown in Flowchart 9.2. The diamond-shaped test symbol is characterized by one flow-in arrow into the decision node and by two outflow arrows leading to the next appropriate step depending upon the result of a test requirement. The test requirement is a defined criterion to be met. Whether or not the criterion is met determines which of the two alternative paths must be followed in making an exit from that decision node.

9.2.3 Conditional Iteration

Conditional iteration is a higher-level module which is built up by sequential steps and decision test symbols. Conditional iteration provides a series of repetitive executions (loop operations) dictated by the constraint parameters required by codes

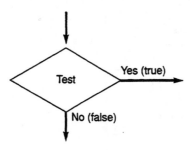

Flowchart 9.2   Decision Test

and optimization. In the conditional iteration shown in Flowchart 9.3, the flow passes through the process box to the decision node, where a test is made. If the test is no good (NG), the correction/optimization will be made in favor of the test before return- ing back to the previous process. The process is repeated until the test is met so that the flow can exit to a new process.

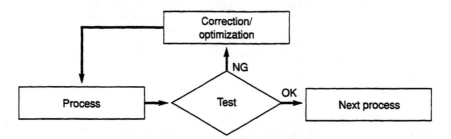

Flowchart 9.3   Conditional Iteration

9.3 Computer-Aided Design

Design is a process involving configuration, loadings, boundary conditions, material properties, optimum choices, codes, and specifications. A final design always results from a sequence of problem-solving operations coupled with various choices and optimizations. In most situations, the design problem in steel does not have any easy or direct closed-form solution at the beginning of the design. However, there are some initial and preliminary solutions to initiate any complex design problem. One of the most important of these design solutions is the use of CAT to achieve a satisfactory solution in meeting the design criteria and performance requirements. With the aid of a high-speed, electronic digital computer, the integrated design system can be exe- cuted in part or in total with great effectiveness. The design system can be used as an analytical tool to allow rapid synthesis by a number of successive iterations. The inte- grated system of computer-aided design is demonstrated in Flowchart 9.4, which con- sists of the following logical steps.

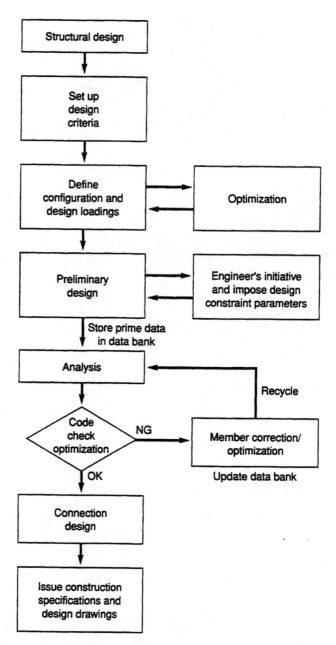

Flowchart 9.4   Computer-Aided Design System

9.3.1 Design Criteria

One must establish complete design criteria before actual design work takes place. The design criteria provide a general design guide concerning the type of structural system, material strengths and grades, structure configuration, design loads, and specifications. The design-constraint parameters should be implemented together with the code-check requirements, which will be introduced subsequently.

9.3.2 Define Configuration and Design Loadings

Define configuration.   Structural configuration can be defined, in general, by three direction parameters in space, such as two dimensions in the north-south and east-west directions in the floor plan, and one dimension in the vertical direction (elevation). It is usually defined by column and floor spacings. Then, all joint coordinates can be generated automatically by computer whenever the spacings are given. Simultaneously, all joint labels or identification names can be automatically defined by computer and composed of:

$$IJN \qquad \text{as illustrated in Fig. 9.1}$$

where $I = i$th column line (north-south), a numerical order
  $J = j$th column line (east-west), an alphabetical order
  $N = n$th floor of the joint, a numerical order

Three-dimensional geometric optimization, in general, is required to determine an optimum number of floors for a given size of building. Configuration optimization minimizes the cost function ϕ_n, which consists of the primary cost parameters such as material, land, labor, design, maintenance, and so on, as follows:

$$\phi_n = f(x_1, x_2, x_3, x_4, x_5, ..., x_m)$$

where n = number of stories
  m = number of independent primary cost and constraint parameters
  $x_1, x_2, x_3, x_4, x_5, ...$
  = independent primary cost and constraint parameters, such as material, land, labor, geometry, loading, maintenance, etc.

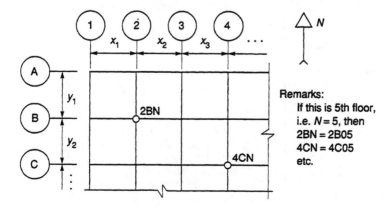

Remarks:
If this is 5th floor,
i.e. $N = 5$, then
2BN = 2B05
4CN = 4C05
etc.

Figure 9.1   Typical nth-floor plan.

To state the problem mathematically, then, the optimum conditions may be obtained by solving the following equations:

$$\frac{\partial \phi_n}{\partial x_1} = 0 \qquad \frac{\partial \phi_n}{\partial x_2} = 0$$

$$\frac{\partial \phi_n}{\partial x_3} = 0 \qquad \frac{\partial \phi_n}{\partial x_4} = 0$$

$$\frac{\partial \phi_n}{\partial x_5} = 0, \ldots, \frac{\partial \phi_n}{\partial x_m} = 0$$

In the computer-aided optimization, which will be introduced separately in Section 9.4, a simplified analytic model for a rapid configuration optimization can be developed in terms of the primary cost parameters mentioned for direct evaluation for a series of ϕ_n data, as illustrated in Fig. 9.2.

Design loadings. There are several types of primary loadings that should be considered in design; they are dead, live, wind, and seismic loads. Dead loads can be either defined manually or calculated by computer when member sizes are known. Live loads include impact, snow, thermal, and any loads that are not permanently applied to the structure. Live loads shall be considered as applied either to the entire supporting area or to a portion of the supporting area, and any probable combination of primary loads resulting in the highest stresses in the supporting member should be used as design loads. Proper provisions shall be made for dynamic loading caused by wind and earthquakes. In any case, the loads other than dead load shall be not less than those recommended in the American National Standards *Building Code Requirements for Minimum Design Loads* (ASCE 7-95). After all primary loads are defined, then, a series of possible design load combinations can be arrived at for inclusion in the structural analysis.

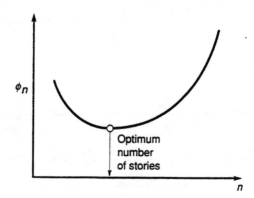

Figure 9.2   Simplified analytical model for configuration optimization.

9.3.3 Preliminary Design

During this design planning stage, the design objective is to arrive as closely as possible at an optimum design that maximizes the serviceability and reliability, and at the same time minimizes the construction cost as well as service and maintenance costs. With this goal in mind, a preliminary design strategy must be developed by an experienced structural engineer based on past experience and statistical data, which leads the design process toward an optimum solution.

In general, there are two extreme starting design strategies: (1) simple beam strategy (upper bound), and (2) fixed-beam strategy (lower bound). For the simple beam strategy, one assumes that all beams are simply supported, and the preliminary design can be started by selecting beam and column member sizes based on their simple moments and axial loads, respectively. During the iteration process, which will be discussed later, each subsequent member selection reduces the beam size and increases the column size as long as the design stresses of the beam are less than the specification allowable stresses. On the other hand, one could use fixed-beam strategy and assume that all beams are fixed-ended, except those that are actually hinged. Preliminary design, then, can be started by selecting beams based on the fixed end moments alone, and columns based on the axial loads combined with the fixed-end moments. In the iteration process, each subsequent member selection increases the beam size and reduces the column size as long as the design stresses are closed and slightly less than the specification allowables. The preliminary member sizes as selected are all stored in the computer member data bank for future update during the *code check* process.

To facilitate cost-effective construction procedures, the designer should group all structural members according to type (i.e., beams, columns, and tension members) and select one member size per group by the most critical design load combination within the group. For better utilization of the structural rolled shapes, one should establish computer data files for beam, column, and tension member properties. The first two files are available from AISCM.

9.3.4 Analysis

A complete structural analysis should be performed by computer as soon as the geometric configuration, design loadings, boundary conditions, and preliminary member shapes and sizes are defined. Such a structural analysis can best be performed using personal computers, workstations, or mainframe computers, depending on the size of the structure and the availability of the computer hardware. There are a great many computer programs that can be used for structural analysis. The students are probably already acquainted with the student versions of one or more of these software packages from their structural analysis courses, and therefore they will have no difficulty becoming proficient with the larger and more sophisticated professional versions in use in design offices.

For more complex structures, or for structures that must survive unusual loadings, an advanced analysis is performed, using programs that account for the geometric and material nonlinearities of the members and the joints. Such advanced methods provide additional data on the effects of repeated nonregular loads from earthquakes, windstorms, or explosions. Such analyses are not yet routine, but they are becoming available and are often used for investigative studies.

9.3.5 Code Check

As introduced earlier, in process 3, preliminary design, there are two starting design strategies for initial member-size selection. The code-check conditional iteration is illustrated in Flowchart 9.5 and is required to ensure that members meet the code requirements. For the correction/optimization of member selection during the iteration process, each subsequent member selection chooses smaller beams with larger columns for the simple beam strategy, or larger beams with smaller columns for the fixed-beam strategy. This iteration process revises the member group data bank until every member group meets the code requirements.

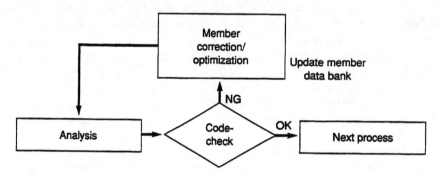

Flowchart 9.5   Code-Check Conditional Iteration

9.3.6 Connection Design

As soon as all structural members are defined, the connection detail design can begin. For simplicity, economy, and uniformity in fabrication and erection of connections in steel construction, they should be separated into two groups: (1) shop connections, and (2) field connections. The typical connections in each.group can then be designed on the basis of the critical conditions involved within the group by the aid of the AISC *Manual*, Vol. II, or by a special computer subroutine.

9.3.7 Issue Construction Specifications and Design Drawings

Construction specifications and design drawings constitute the final step in the design process. The construction specifications, in general, include all material strengths and grades, construction methods and procedures, site-testing, and inspection requirements. Design drawings are the final design products, including foundation, floor, roof and site plans, elevations, cross sections, connections, and special details.

Construction specifications can be generated by computer word processing by editing general specifications which are stored in a computer file. The editing consists of changing, adding, and deleting from the file.

High-quality design drawings can be produced rapidly by autographic systems (computer graphics) by retrieving all the stored information from the data bank. The systems consist of a computer, digitizers, cathode ray tube (CRT) terminal, input device, and output plotter or hard-copy device for copying the screen display image. The systems are also very useful for graphic verification of geometry during the computer structural analysis. They can also make various structural, perspective, or isometric views for management presentation, model studies, and as a tool for coordinating various engineering disciplines.

The digital plotter is used to translate computer data into a highly accurate graphic output. Design drawings can be made by a pen plotter or electrostatic plotter/printer, depending on the drawing requirements. The pen plotter uses a ball point, liquid ball, or nylon-tipped pen and utilizes either drum, belt bed, or flat bed technology to fit the plotting media. The electrostatic plotter/printer is more efficient for use in exceptionally high speed plotting. It produces drawings by imprinting specially treated paper with electrostatic charges which, when treated with a toner, develop into specific images.

The digitizer is used to tap the power of a computer for designing and drafting. A designer must first transmit the concept or rough idea to the computer. This can be effectively done with a digitizer which converts the points, lines, and curves of drawings or sketches to digital impulses that can be understood and acted upon by the digital computer.

9.4 Computer-Aided Optimization

Structural design optimization has been defined as the design and construction of a structure at the lowest overall lifetime cost in fulfillment of the design objectives. Its purpose is simultaneously to maximize structural appearance and space, serviceability, safety, reliability, and future adaptability, while at the same time, minimizing total cost of design, construction, maintenance, and time. With computer-aided technology one should be able to effectively vary the primary design parameters such as structural systems (e.g., geometric configuration as to column and floor systems, materials, code and specification requirements, etc.) to arrive at an optimum design. This is illustrated in Fig. 9.3 in terms of structural merit or worth versus the overall cost.

An experienced structural engineer with statistical data can simplify a complex structure to a manageable number of prime design parameters for rapid analytical processing. In order to arrive at an optimized overall solution, as illustrated in Fig. 9.4, the overall design optimization generally requires simultaneous consideration of the following primary factors: (1) structural configuration, (2) materials, (3) external loadings, and (4) applicable codes and specifications.

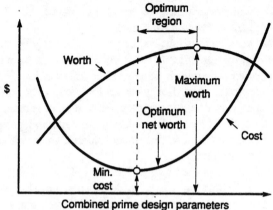

Figure 9.3 Optimum net worth.

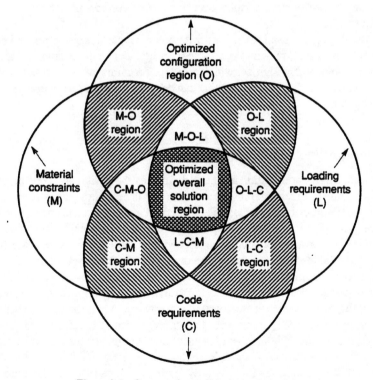

Figure 9.4 Concept of overall design optimization.

10

Composite Construction

10.1 Introduction

Steel beams frequently support concrete slabs, as in buildings and bridges. Sometimes steel beams and columns are encased in concrete for fireproofing. Even if there are no mechanical connectors between the slab and the beam, there is a certain amount of connection between the top flange of the steel beam and the bottom of the concrete slab due to bond and friction, and under a small load the steel beam deflects less than it would if it did not interact with the concrete: the two elements behave as a composite beam [compare the beams in Fig. 10.1(a) and (b)]. Friction and bond are generally not able to provide reliable composite action, except in the case of complete encasement; therefore, mechanical shear connectors are utilized to provide a reliable connection between the slab and the beam. These connectors are welded to the top flange of the steel beam and are embedded in the concrete slab, where they are kept in place by hooks or heads. Many kinds of connectors have been used (e.g., channels or angles), but the most economical and almost universally used shear connectors are round-headed studs which are electrically fused to the steel flange plate by a light, portable stud gun. Examples of a composite beam are shown in Figs. 10.2 and 10.3.

Since the slab is usually there anyway, and since the round-headed studs are economical ($1 to $2 per stud, depending on the size) and easy to install, it makes

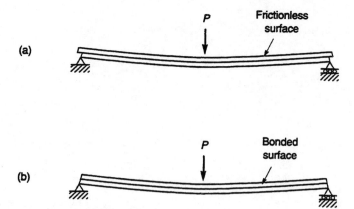

Figure 10.1 Illustration of composite and noncomposite deflection: (a) no composite action; (b) complete composite action.

good structural sense to use composite construction whenever possible. Cost savings of 10% to 20% as compared to noncomposite beams are usual.

In accordance with the AISC Specification, composite beams may be designed by elastic or plastic methods of design. When the cross section is compact, the moments may be determined by plastic analysis for statically indeterminate beams. The usual method of design is to determine the moments by elastic analysis and to proportion the cross section for its plastic capacity. The design of composite structures is defined in Sec. I in AISCS. Because the design procedures are relatively complex,

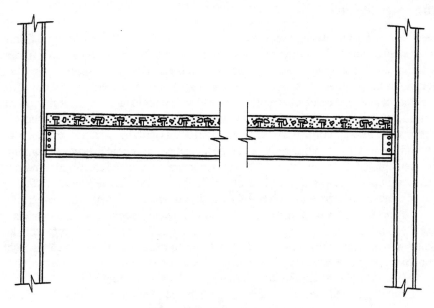

Figure 10.2 Composite beam.

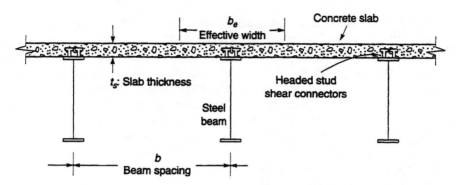

Figure 10.3   Cross section of composite floor.

AISCM provides extensive design aids for use by the designer. There are also many software packages that are available in practice. In this chapter we concentrate on the basic principles without any use of design aids. The student should be aware, however, that in routine design assignments the tedious calculations can be avoided by using tables or spreadsheets.

Basically, composite beams are primarily flexural members, and the following section will develop methods to determine their flexural capacity. No significant enhancement of the shear capacity is achieved by composite beams, and the shear force check is performed as for noncomposite beams (i.e., the web of the steel beam resists the full shear force). Composite beams are especially suited for commercial buildings and for highway bridges.

10.2 Flexural Strength of Composite Cross Section

The moment capacity of a composite beam is determined according to elastic assumptions in Sec. I3.2 of AISCS if the web is noncompact, that is, when $h/t_w < 640/\sqrt{F_y}$. No restrictions apply as to the compactness of the compression flange in the positive moment region of the beam, because the flange is attached to, and often embedded into, the concrete slab. Since all rolled wide-flange shapes have compact webs, the elastic method of determining cross-section capacity is used for thin-web built-up bridge girders. The elastic method is thus quite conservative for rolled shapes. However, elastic assumptions are needed for serviceability checks where the transformed moment of inertia is required.

10.2.1 Elastic Theory

In elastic analysis stress is proportional to strain and the limiting moment is reached when the limiting stress $\phi F_y = 0.9 F_y$ in the bottom fiber of the steel beam, or $\phi \times 0.85\, f'_c = 0.8 \times 0.85 \times f'_c$ in the top fiber of the concrete slab. For such an analysis the width of the concrete slab is transformed into an equivalent steel section by reducing the width of the slab to b_e/n, where b_e is the effective width and n, the modular ratio, is the ratio of the modulus of elasticity of steel, E_s, to the modulus of elasticity of the concrete, E_c (i.e., $n = E_s/E_c$). This ratio is approximately $n = 9$ for concrete with a crushing strength $f'_c = 3$ ksi. This value will be used in the examples throughout this chapter.

The transformed section is shown as the left sketch in Fig. 10.4(a). The effective transformed slab is $t_s \times b_e/n$. The steel wide-flange shape, attached to the concrete slab by shear connectors, has an area A_s, depth d, and moment of inertia I_s.

The strain in the composite section is linear, varying directly as the distance from the neutral axis. Two cases of stress distribution are possible: (a) when the neutral axis is in the steel beam, the concrete slab is entirely in compression, and (b) when the neutral axis is in the concrete slab, part of the slab is in tension [Fig. 10.4(a) and (b), respectively]. In the latter case the tensile part of the slab is assumed cracked and it will be assumed to be unstressed.

Neutral axis in the steel.   This case is shown in Fig. 10.4(a). The steel beam is assumed to be symmetric (i.e., both flanges have the same dimensions). The location of the neutral axis is determined by taking a statical moment about the base:

$$y\left(A_s + \frac{t_s b_e}{n}\right) = \frac{A_s d}{2} + \frac{t_s b_e}{n}\left(d + \frac{t_s}{2}\right)$$

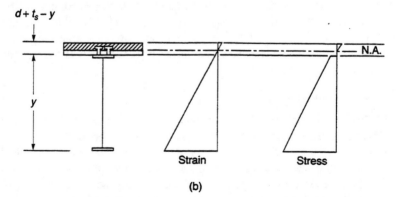

Figure 10.4   Elastic strain and stress distribution in composite beam (neutral axis in steel).

from which

$$y = \frac{0.5A_s d + (t_s b_e/n)(d + 0.5t_s)}{A_s + t_s b_e/n} \tag{10.1}$$

The neutral axis is in the steel as long as $y \leqslant d$, and this occurs if $A_s d \geq t_s^2 b_e/n$.
The transformed moment of inertia about the neutral axis is

$$I_t = I_s + A_s\left(y - \frac{d}{2}\right)^2 + \frac{b_e t_s^3}{12n} + \frac{b_e t_s}{n}\left(d + \frac{t_s}{2} - y\right)^2 \tag{10.2}$$

Neutral axis in the concrete.   This case is shown in Fig. 10.4(b). Taking a statical moment about the base,

$$y\left[A_s + \frac{b_e}{n}(d + t_s - y)\right] = \frac{A_s d}{2} + \frac{b_e}{n}(d + t_s - y)\left(d + t_s - \frac{d + t_s - y}{2}\right)$$

from which, after some algebra, we obtain

$$y = \frac{A_s n}{b_e} + d + t_s - \frac{A_s n}{b_e}\sqrt{1 + \frac{b_e(d + 2t_s)}{A_s n}} \tag{10.3}$$

The transformed moment of inertia about the neutral axis is

$$I_t = I_s + A_s\left(y - \frac{d}{2}\right)^2 + \frac{b_e(d + t_s - y)^3}{3n} \tag{10.4}$$

In either of the two cases, the maximum stress in the steel is in the bottom flange,

$$f_{su} = \frac{M_u y}{I_t} \leq \phi F_y = 0.9F_y \tag{10.5}$$

and the maximum stress in the concrete slab is at its top,

$$f_{cu} = \frac{M_u(d + t_s - y)}{nI_t} \leq \phi_c \times 0.85 f_c' = 0.8 \times 0.85 f_c' \tag{10.6}$$

The resistance factor of 0.80 is taken from the LRFD specification of the Portland Cement Association. Division by the modular ratio in Eq. (10.6) transforms the stress to the concrete properties.

Example 10.1
Select a rolled wide-flange beam to resist a factored bending moment of 30,000 kip-in. The beam is attached to a concrete slab with an effective width of 90 in. and a thickness of 6 in. The compressive strength of the concrete is 3 ksi and the yield stress of the steel is 50 ksi. Use elastic theory. This example is for choosing a rolled shape, and it is used only as an illustration of the elastic method. Normally, the plastic method would be used, since the elastic method is quite conservative for shapes with compact webs.

Solution   *Preliminary design:* Section modulus required for noncomposite beam:

$$(Z_x)_{req} = \frac{M_u}{\phi F_y} = \frac{30,000}{0.9 \times 50} = 667 \text{ in}^3$$

A W36 × 170 is needed, $Z_x = 668$ in^3.

As a trial section for the selection of the composite beam, try a section two sizes smaller from the Load Factor Design Selection tables. Try a W36 × 160.

Moment capacity of W36 × 160, composite with concrete:

$$\text{Slab: } b_e = 90 \text{ in.} \qquad \text{W36} \times 160: \quad d = 36.01 \text{ in.}$$

$$t_s = 6 \text{ in.} \qquad\qquad A_s = 47.0 \text{ in}^2$$

$$f'_c = 3 \text{ ksi} \qquad\qquad I_x = 9750 \text{ in}^4$$

$$n = 9 \qquad\qquad F_y = 50 \text{ ksi}$$

Neutral axis: If $A_s d \geq t_s^2 b_e / n$, the neutral axis is in the steel.

$$A_s d = 1692 \text{ in}^3 \qquad \frac{t_s^2 b_e}{n} = 360 \text{ in}^3 \qquad < 1692$$

Therefore, the neutral axis is in the steel; y is determined from Eq. (10.1).

$$y = \frac{0.5 A_s d + (t_s b_e / n)(d + 0.5 t_s)}{A_s + t_s b_e / n} = 29.78 \text{ in.}$$

$$I_t = I_s + A_s\left(y - \frac{d}{2}\right)^2 + \frac{b_e t^3}{12n} + \frac{b_e t_s}{n}\left(d + \frac{t_s}{2} - y\right)^2 = 21{,}558 \text{ in}^4$$

Stress in steel:

$$f_s = \frac{My}{I_t} = \frac{30{,}000 \times 29.78}{21{,}558} = 41.45 \text{ ksi} \qquad < 0.9 \times 50 = 45 \text{ ksi}$$

Stress in concrete:

$$f_c = \frac{M(d + t_s - y)}{I_t n} = 1.89 \text{ ksi} \qquad < 0.8 \times 0.85 \times 3 = 2.04$$

The beam is somewhat understressed.

Summary of trials:

Beam	W 36 × 160	W 36 × 150	W 36 × 135
y	29.78 in.	29.97 in.	30.28 in.
I_t	21,558 in^4	20,364 in^4	18,292 in^4
f_s	41.45 ksi	44.16 ksi	49.66 ksi
f_c	1.89 ksi	1.94 ksi	2.05 ksi

Use a W36 ×150.

10.2.2 Plastic Theory

Experiments performed on composite beams showed that elastic theory gives a very conservative prediction of the moment capacity. The true moment capacity can be closely approximated by assuming that the steel section is fully yielded and the compressed part of the concrete slab is everywhere stressed to $0.85 f'_c$. The effective cross section for plastic analysis consists of the steel beam and the effective slab [Fig. 10.5(a)].

Three fully plastic stress distributions are possible: the steel beam is fully yielded in tension and the tensile part of the concrete slab is ineffective [Fig. 10.5(b)];

the plastic neutral axis is in the flange of the steel beam [Fig. 10.5(c)]; and the plastic neutral axis is in the web [Fig. 10.5(d)] of the steel beam.

Steel fully yielded in tension [Fig. 10.5(b)] .   The resultant force of the concrete stress is 0.85 $f'_c b_e a$ and it acts at a distance $a/2$ from the top of the slab. The resultant force of the steel stress is $F_y A_s$ and it acts at the centroid of the steel beam. Equilibrium requires that these two forces are equal, that is,

$$F_y A_s = 0.85 f'_c b_e a$$

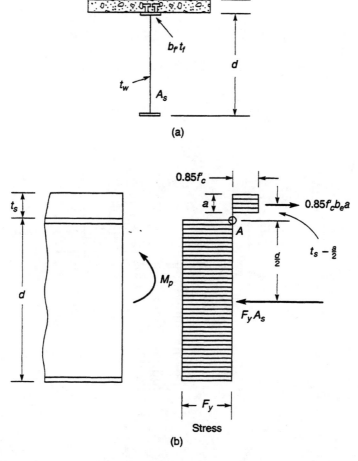

Figure 10.5   Plastic capacity of composite cross section: (a) cross section for plastic analysis; (b) steel is fully yielded in tension, concrete is cracked; (c) neutral axis in top flange; (d) neutral axis in web.

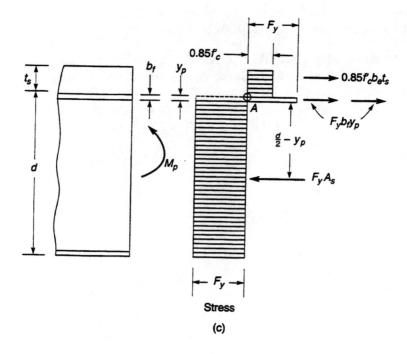

(c)

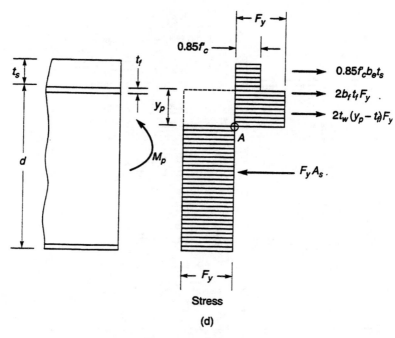

(d)

Figure 10.5 (*continued*).

from which the depth of the concrete compression zone is

$$a = \frac{F_y A_s}{0.85 f'_c b_e} \tag{10.7}$$

The neutral axis will be in the concrete if

$$F_y A_s \leqslant 0.85 f'_c b_e t_s \tag{10.8}$$

The plastic moment capacity is obtained by taking the moment of the resultant forces about the point A, which is at the top of the steel beam.

$$M_p = \frac{F_y A_s d}{2} + 0.85 f'_c b_e a \left(t_s - \frac{a}{2} \right) \tag{10.9}$$

Neutral axis in top flange [Fig. 10.5(c)].   The resultant force in the concrete is $0.85 f'_c b_e t_s$, and the corresponding quantity in the steel beam is $F_y A_s - 2F_y b_f y_p$, where y_p is the distance from the top of the flange to the neutral axis. Force equilibrium:

$$0.85 f'_c b_e t_s = F_y A_s - 2F_y b_f y_p$$

from which

$$y_p = \frac{F_y A_s - 0.85 f'_c b_e t_s}{2 F_y b_f} \tag{10.10}$$

The neutral axis will remain in the flange if $0 \leqslant y_p \leqslant t_f$, that is,

$$0.85 f'_c b_e t_s \leqslant F_y A_s \leqslant 0.85 f'_c b_e t_s + 2 F_y b_f t_f \tag{10.11}$$

Taking moments about point A:

$$M_p = 0.85 f'_c b_e t_s \left(y_p + \frac{t_s}{2} \right) + 2 F_y b_f y_p \left(\frac{y_p}{2} \right) + F_y A_s \left(\frac{d}{2} - y_p \right) \tag{10.12}$$

Neutral axis in web [Fig. 10.5(d)].   Force equilibrium:

$$0.85 f'_c b_e t_s + 2 F_y b_f t_f + 2 F_y t_w (y_p - t_f) = F_y A_s$$

from which

$$y_p = t_f + \frac{A_s}{2 t_w} - \frac{b_f t_f}{t_w} - \frac{0.85 f'_c b_e t_s}{2 F_y t_w} \tag{10.13}$$

Moment equilibrium about A:

$$M_p = 0.85 f'_c b_e t_s \left(y_p + \frac{t_s}{2} \right) + 2 F_y b_f t_f \left(y_p - \frac{t_f}{2} \right)$$
$$+ F_y t_w (y_p - t_f)^2 + F_y A_s \left(\frac{d}{2} - y_p \right) \tag{10.14}$$

Example 10.2

Select a wide-flange beam to resist an ultimate bending moment of 30,000 kip-in. Use plastic theory (this is the same problem as Example 10.1). Use $b_e = 90$ in., $t_s = 6$ in., $f_c' = 3$ ksi, and $F_y = 50$ ksi.

Solution *Preliminary design:* Plastic section modulus required for noncomposite design: W36 × 170. Since a W36 × 150 was needed for the elastic design, try a W33 × 130 for the plastic design.

Plastic moment capacity of a W33 × 130 section:

$$A_s = 38.3 \text{ in}^2 \qquad t_f = 0.855 \text{ in.} \qquad d = 33.09 \text{ in.} \qquad b_f = 11.51 \text{ in.}$$
$$A_s F_y = 38.3 \times 50 = 1915 \text{ kips} \qquad b_e = 90 \text{ in.} \qquad t_s = 6 \text{ in.} \qquad f'_c = 3 \text{ ksi}$$
$$0.85 f'_c b_e t_s = 0.85 \times 3 \times 90 \times 6 = 1377 \text{ kips}$$

Since $A_s F_y > 0.85 f'_c b_e t_s$, the neutral axis is in the steel.

$$0.85 f'_c b_e t_s + 2 F_y b_f t_f = 1377 + 2 \times 50 \times 11.51 \times 0.855 = 2361 \text{ kips} \qquad > 1915 \text{ kips}$$

Hence the neutral axis is in the flange.

From Eq. (10.10), the location of the plastic neutral axis is

$$y_p = \frac{F_y A_s - 0.85 f'_c b_e t_s}{2 F_y b_f} = \frac{1915 - 1377}{2 \times 50 \times 11.51} = 0.467 \text{ in.} \qquad < t_f$$

From Eq. (10.12), the plastic moment is

$$M_p = 0.85 f'_c b_e t_s \left(y_p + \frac{t_s}{2} \right) + 2 F_y b_f y_p \left(\frac{y_p}{2} \right) + F_y A_s \left(\frac{d}{2} - y_p \right) = 35,689 \text{ kip-in.}$$

Since $\phi M_n = \phi M_p = 0.85 \times 35,689 = 30,336$ kip-in.   > 30,000 kip-in, the W33 × 130 is OK.

Example 10.3
Determine the elastic and the plastic moment capacity of a composite beam with a cross section shown in Fig. 10.6. $F_y = 36$ ksi.

Solution *Elastic limit moment.* Properties of transformed section: Assume that the elastic neutral axis is in the steel.

$$y = \frac{6 \times (81/9)(3 + 23.73 + 1) + 20.1(23.73/2 + 1) + 8 \times 1 \times 0.5}{6(81/9) + 20.1 + 8 \times 1} = 21.44 \text{ in.}$$

Since $y = 21.44$ in. $< d + 1 = 24.73$ in., the neutral axis is in the steel; hence the whole transformed slab is active.

$$I_t = \frac{(81/9) \times 6^3}{12} + \left(\frac{81}{9} \right)(6)(24.73 + 3 - y)^2 + 1830$$

$$+ 20.1 \left(y - 1 - \frac{23.73}{2} \right)^2 + 8 \times 1 (y - 0.5)^2 + \frac{8 \times 1^3}{12} = 9115 \text{ in}^4$$

$$f_s = 0.9 F_y \geq \frac{M_u y}{I_t} \leftarrow \text{stress in steel controls}$$

$$M_u \leq \frac{0.9 \times 36 \times 9115}{21.44} = 13,775 \text{ kip-in.}$$

$$f_c = 0.8 \times 0.85 \times f'_c \geq \frac{M_u(1 + 23.73 + 6 - y)}{I_t n} \leftarrow \text{stress in concrete controls}$$

$$M_u \leq \frac{0.8 \times 0.85 \times 3 \times 9115 \times 9}{1 + 23.73 + 6 - 21.44} = 18,014 \text{ kip-in.}$$

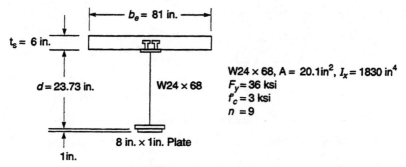

(a) Properties of cross section

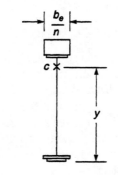

(b) Transformed section

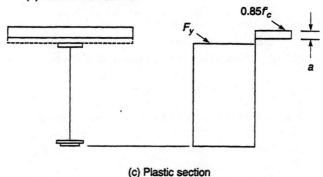

(c) Plastic section

Figure 10.6   Data for Example 10.3 .

Plastic limit moment:
Capacity of slab:

$$F_c = 0.85\, f'_c\, b_e t_s = 0.85 \times 3 \times 81 \times 6 = 1239.3 \text{ kips}$$

Capacity of steel:

$$F_s = F_y(A_{WF} + A_{CP}) = 36(20.1 + 8) = 1011.6 \text{ kips}$$

Since $F_c > F_s$, the plastic neutral axis is in the concrete slab.

Location of neutral axis:

$$F_s = 0.85 f'_c b_e a$$

$$a = \frac{1011.6}{0.85 \times 3 \times 81} = 4.90 \text{ in. below top of slab}$$

$$M_p = (0.85 f'_c b_e a)\left(t_s - \frac{a}{2}\right) + \left[20.1 \times \frac{23.73}{2} + 8(23.73 + 0.5)\right] F_y = 19{,}156 \text{ kip-in.}$$

$$M_u \le \phi M_n = \phi M_p = 0.85 \times 19{,}156 = 16{,}282 \text{ kip-in.}$$

Section 13.2 of AISCS requires a resistance factor of 0.85 for the plastic capacity of the composite cross section.

10.3 Design of Composite Beams

10.3.1 Shored and Unshored Construction

During construction the steel beam alone must carry the weight of the wet concrete, the formwork, the construction crew, the construction equipment, and its own weight. If the design of the composite beam is based on the *plastic strength* of the cross section, the steel beam alone must carry the construction loads when temporary shores are not provided. If the design is based on the *elastic strength*, the stresses must be adjusted to account for the construction sequence. After the concrete has hardened, the live load is then supported by the beam in composite action. The actual stress in the bottom flange of the steel beam is thus

$$f_b = \frac{M_D}{S_s} + \frac{M_L}{S_t} \tag{10.15}$$

where M_D and M_L are the moments due to the construction load and the live load, respectively, S_s is the elastic section modulus of the steel beam, and

$$S_t = \frac{I_t}{y} \tag{10.16}$$

is the elastic section modulus of the transformed composite beam. The designer must assure that the steel beam is not overstressed during construction. There are two ways to solve this problem.

The first is a constructional measure. Temporary shores are erected and kept in place until the concrete has reached 75% of its final strength. For example, with four temporary shores (Fig. 10.7) the dead-load stresses are negligible, and upon removal of the shores the composite beam carries both the dead load and the live load.

It is often inconvenient and expensive to use shores. In that case the dead-load stresses must be kept below $\phi F_y = 0.9 F_y$.

10.3.2 Effective Width

The actual stress distribution in the slab is nonuniform due to shear lag. The stress is highest over the steel beams and lowest between the beams (Fig. 10.8). To avoid complicated calculations, an equivalent uniform stress over the effective width

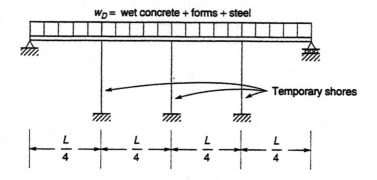

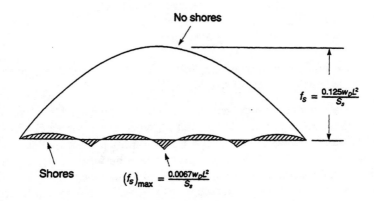

Figure 10.7   Effect of temporary shores.

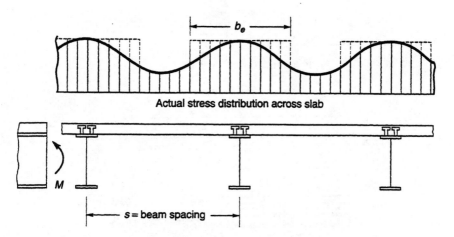

Figure 10.8   Illustration of shear lag and effective width.

b_e is used. The effective width is given in Sec. I3.1 of AISCS as follows: "The effective width of the concrete slab on each side of the centerline of the beam shall not exceed: (a) one-eighth of the beam span, center-to-center of supports; (b) one half the distance to the center-line of the adjacent beam; or (c) the distance to the edge of the slab."

10.3.3 Shear Connector Design

The strength of individual shear connectors is given by formulas in Sec. I5.3 for stud connectors and for channel connectors. The expression for individual stud shear connectors is

$$Q_n = 0.5 A_{sc}\sqrt{f_c' E_c} \leq A_{sc} F_u \tag{10.17}$$

where A_{sc} = cross-sectional area of the shear connector, in^2
f_c = crushing strength of the concrete, ksi
E_c = modulus of elasticity of concrete, ksi
$\quad = w^{1.5}\sqrt{f_c'}$
w = unit weight of concrete, lb/ft^3 (150 lb/ft^3 for normal-weight concrete)
F_u = tensile strength of stud material, ksi

The number of shear connectors required are determined by plastic design in AISCS. Shear connectors are ductile, and when one connector reaches its yield load it will continue to hold its load even while deforming plastically. Thus under static loading the shear connectors can be spaced evenly between the point of maximum moment and the point of zero moment. Under fatigue-type loading, as in a bridge, shear connectors must be spaced according to the variation of the shear diagram.

The number of shear connectors between points of maximum moment and zero moment are obtained in AISCS as follows:

$$n = \frac{F}{q_u} \tag{10.18}$$

where, conservatively (see Fig. 10.9),

$$F = \min(0.85 f_c' b_e t_s, A_s F_y) \tag{10.19}$$

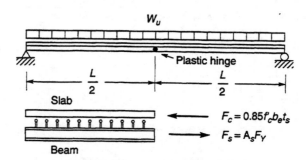

Figure 10.9 Shear connector failure.

10.3.4 Additional Design Considerations

Sec. I3 of AISCS treats the following topics, which are not covered herein but which are easy to follow from the information provided in AISCS.

1. Number of shear connectors required if there are concentrated loads between the point of maximum moment and zero moment
2. Composite beam design if fewer than the number of shear connectors required for full shear connection are present
3. Composite beam design when the shear connectors are applied through a steel deck into the top flange of the steel beam
4. Maximum and minimum spacing of connectors

10.4 Composite Columns

Composite columns can be constructed by filling a hollow steel section with concrete, or by encasing rolled or built-up steel shapes with concrete. The latter columns are especially useful because the concrete both strengthens and fireproofs the member. The design of composite columns is performed in compliance with Sec. I2 of AISCS. In Sec. I2.1 are listed a number of limitations and restrictions. The reader is urged to study these. However, they do not overly restrict the application of this type of member in design. Following are the criteria that need to be checked for the design of composite columns when the reinforcing steel is neglected. This is a conservative assumption. The basic concept in the design of such columns is to define modified yield strengths F_{my} and modified moduli of elasticity E_m, which are then used in the column formulas of Chapter 4 (also in AISCS, Sec. E2).

For concrete-filled pipes or rectangular tubes:

$$F_{my} = F_{ys} + 0.85 \frac{f_c' A_c}{A_s} \tag{10.20}$$

$$E_m = E_s + \frac{0.4 E_c A_c}{A_s} \tag{10.21}$$

$$r = r_s \tag{10.22}$$

For concrete encased shapes:

$$F_{ym} = F_{ys} + \frac{0.6 f_c' A_c}{A_s} \tag{10.23}$$

$$E_m = E_s + \frac{0.2 E_c A_c}{A_s} \tag{10.24}$$

$$r = \max(r_s, 0.3 h_2) \tag{10.25}$$

The terms in these equations are as follows:

F_{ys} = yield stress of steel, ksi; if $F_{ys} > 55$ ksi, use $F_{ys} = 55$ ksi

f'_c = concrete crushing strength, ksi

A_c = area of concrete

A_s = area of steel

E_s = modulus of elasticity of steel, $E_s = 29{,}000$ ksi

E_c = modulus of elasticity of concrete

r = radius of gyration

r_s = radius of gyration of steel section

h_2 = dimension of the rectangular concrete encasement perpendicular to the direction of buckling

Composite column design is illustrated by the following two examples.

Example 10.4

Determine the design capacity of a concrete-filled $6 \times 6 \times \frac{3}{16}$ steel tube. $F_y = 36$ ksi, $f'_c = 3$ ksi, and $L = 14$ ft 0 in.

Solution   From AISCM:

$$A_s = 4.27 \text{ in}^2 \qquad r_s = 2.36 \text{ in.}$$

$$\text{Concrete area:} \quad A_c = \left(6 - 2 \times \frac{3}{16}\right)^2 = 31.64 \text{ in}^2$$

$$E_c = w^{1.5}\sqrt{f'_c} = 150^{1.5}\sqrt{3} = 3182 \text{ ksi}$$

Eq. (10.21): $$F_{my} = 36 + 0.85 \times 3 \times \frac{31.64}{4.27} = 54.9 \text{ ksi}$$

Eq. (10.22): $$E_m = 29{,}000 + 0.4 \times 3182 \times \frac{31.64}{4.27} = 38{,}431 \text{ ksi}$$

$$\lambda_c = \frac{KL}{\pi r_s}\sqrt{\frac{F_{my}}{E_m}} = \frac{14 \times 12}{2.36\pi}\sqrt{\frac{54.9}{38{,}431}} = 0.856 \qquad < 1.5$$

$$F_{cr} = 0.658^{\lambda_c^2} F_y = 0.658^{0.856^2} \times 54.9 = 40.39 \text{ ksi}$$

$$\phi_c P_u = \phi_c A_s F_{cr} = 0.85 \times 4.27 \times 40.39 = 146.6 \text{ kips}$$

The design capacity is 147 kips.

Example 10.5

Determine the capacity of a W8 × 31 column; minor axis buckling, with an effective length of $0.85L$. Concrete encasement 10 in. × 10 in., $F_y = 50$ ksi, $f'_c = 4$ ksi, and $L = 15$ ft 0 in.

Solution   From AISCM:

$$A_s = 9.13 \text{ in}^2 \qquad r_s = 2.02 \text{ in.}$$

$$\text{Concrete area:} \quad A_c = 10 \times 10 - 9.13 = 90.87 \text{ in}^2$$

$$E_c = w^{1.5}\sqrt{f'_c} = 150^{1.5}\sqrt{4} = 3674 \text{ ksi}$$

Eq. (10.23): $$F_{my} = 50 + 0.6 \times 4 \times \frac{90.87}{9.13} = 73.9 \text{ ksi}$$

Eq. (10.24): $E_m = 29{,}000 + 0.2 \times 3674 \times \dfrac{90.87}{9.13} = 36{,}313$ ksi

Eq. (10.25): $h_2 = 10$ in. $0.3 h_2 = 3$ in. > 2.02;   use $r_s = 3$ in.

$$\lambda_c = \frac{KL}{\pi r_s} \sqrt{\frac{F_{my}}{E_m}} = \frac{0.85 \times 15 \times 12}{3\pi} \sqrt{\frac{73.9}{36{,}313}} = 0.732 \qquad < 1.5$$

$$F_{cr} = 0.658^{\lambda_c^2} F_y = 0.658^{0.732^2} \times 73.9 = 59.04 \text{ ksi}$$

$$\phi_c P_n = 0.85 \times 9.13 \times 59.04 = 458 \text{ kips}$$

The design capacity is 458 kips.

10.5 Composite Beam Design Examples

Example 10.6

Design a simply supported composite beam by Sec. I3 of AISCS.

> *Given*:   Beam spacing: 10 ft
> Span: 40 ft 0 in.
> Slab thickness: $t_s = 5$ in.
> $f'_c = 3$ ksi $n = 9$ $F_y = 36$ ksi
> $\frac{3}{4}$-in. ϕ headed studs
> Superimposed dead load (partitions, fireproofing); 10 lb/ft^2
> Live load: 50 lb/ft^2 (office occupancy)
> Unshored construction
> Maximum live-load deflection: $\dfrac{\text{span}}{360}$
> Total depth/span $> \frac{1}{24}$
> Cost of studs: \$1/stud
> Cost of steel: \$0.50/lb
> Formwork provides lateral bracing during construction.

Solution   Dead load:

$$\text{Concrete:} \quad \frac{5}{12} \times 150 \text{ lb/ft}^3 = 62.5 \text{ lb/ft}^2$$

$$\text{Superimposed load} = \frac{10 \text{ lb/ft}^2}{72.5 \text{ lb/ft}^2}$$

Live load:

Live-load reduction (ASCE7 Load Code):

$$L = L_0 \left(0.25 + \frac{15}{\sqrt{A_I}} \right) \qquad L_0 = 50 \text{ lb/ft}^2$$

$$A_I = \text{influence area} = 2 A_T \text{ for beams}$$

$$A_T = \text{tributary area} = 40 \times 10 = 400 \text{ ft}^2$$

$$L_n = 50 \left(0.25 + \frac{15}{\sqrt{2 \times 400}} \right) = 39 \text{ lb/ft}^2$$

Required construction load:

Dead load: concrete + steel beam

$$w_{CD} = 10 \text{ ft} \times \frac{62.5}{1000} + 0.075 \text{ (estimate of beam weight)} = 0.700 \text{ kip/ft}$$

Live load: formwork, equipment, workers, 20 lb/ft^2

$$w_{CL} = 10 \times \frac{20}{1000} = 0.2 \text{ kip/ft}$$

Required load:

$$w_{Cu} = 1.2w_{CD} + 1.6w_{CL} = 1.2 \times 0.700 + 1.6 \times 0.2 = 1.16 \text{ kips/ft}$$

Required moment for construction load:

$$M_{Cu} = \frac{w_{Cu}L^2}{8} = \frac{1.16 \times 40^2 \times 12}{8} = 2784 \text{ kip-in.}$$

$$(Z_x)_{req} = \frac{2784}{0.9 \times 36} = 85.9 \text{ in}^3$$

Required minimum beam depth:

$$\frac{\text{depth}}{\text{span}} = \frac{d+5}{480} = \frac{1}{24} \qquad d \geq 15 \text{ in.}$$

For construction load need a W21 × 44, $Z_x = 95.4$ in^3

Required design load:

$$w_D = 10 \times \frac{72.5}{1000} + 0.075 = 0.80 \text{ kip/ft} \qquad w_L = 10 \times \frac{39}{1000} = 0.39 \text{ kip/ft}$$

$$w_u = 1.2 \times 0.80 + 1.6 \times 0.39 = 1.58 \text{ kips/ft} \qquad M_u = \frac{1.58 \times 40^2 \times 12}{8} = 3801 \text{ kip-in.}$$

Try the W21 × 44 as a composite beam.

$$\text{Effective width } b_e = \min\left(\frac{L}{4}, s\right) = \min\left(\frac{40 \times 12}{4}, 10 \times 12\right) = 120 \text{ in.}$$

From spreadsheet output: $M_p = M_n = 6817$ kip-in. $\phi_b M_n = 0.9 \times 6817 = 5794, > 3801$ kip-in. OK; construction load governs design.

Spreadsheet data

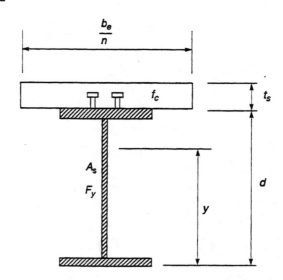

Doubly-symmetric wide-flange composite beam (W21 × 44).

$$A_s = 13 \text{ in}^2 \qquad t_s = 5 \text{ in.}$$
$$d = 20.66 \text{ in.} \qquad b_e = 120 \text{ in.}$$
$$I_s = 843 \text{ in}^4$$
$$f'_c = 3 \text{ ksi}$$
$$F_y = 36 \text{ ksi} \qquad M_u = 3801 \text{ kip-in.}$$
$$n = 9$$
$$b_f = 6.5 \text{ in.} \qquad t_f = 0.45 \text{ in.} \qquad t_w = 0.35 \text{ in.}$$

Elastic properties:

$$y = 21.081 \text{ in.} \qquad I_t = 2.772 \cdot 10^3 \text{ in}^4$$

Plastic moment:

$$M_{pn} = 6.817 \times 10^3 \text{ kip-in.}$$

End of spreadsheet data

Check live load deflection:

$$w_L = 0.39 \text{ kip/ft} \qquad I_t = 2772 \text{ in}^4$$

$$\Delta = \frac{5wL^4}{384EI} = \frac{5 \times (0.39/12) \times (40 \times 12)^4}{384 \times 29{,}000 \times 2772} = 0.28 \text{ in.} \qquad < \frac{\text{span}}{360} = \frac{480}{360} = 1.33 \text{ in.} \quad \text{OK}$$

Use W21 × 44. Required number of $\frac{3}{4}$-in.-diameter studs:

Area of one stud: $\qquad A_{sc} = \dfrac{\pi \times 0.75^2}{4} = 0.442 \text{ in}^2$

Capacity of one stud: $\quad Q_n = 0.5 A_{sc} \sqrt{f'_c E_c}$

$$E_c = 150^{1.5}\sqrt{3} = 3182 \text{ ksi} \qquad Q_n = 0.5 \times 0.442 \times \sqrt{3 \times 3182} = 21.6 \text{ kips}$$
$$0.85 f'_c b_e t_s = 0.85 \times 3 \times 120 \times 5 = 1530 \text{ kips}$$
$$A_s F_y = 13 \times 36 = 468 \text{ kips} \qquad < 1530 \text{ kips}$$
$$n = \frac{468}{21.6} = 22$$

Use 24 studs per half-span. Stud spacing, s: use two studs per location on flange.

$$s = \frac{20 \times 12}{24} = 10 \text{ in.}$$

From AISCS I5.6

Minimum allowable spacing: $6 \times 0.75 = 4.5$ in.   < 10 in.   OK
Maximum allowable spacing: $8 \times 5 = 40$ in.   > 10 in.   OK

Cost:

$$\begin{array}{ll}
\text{Beam:} & 40 \times 44 \times 0.5 = \$880 \\
\text{Studs:} & 48 \times 1 \qquad\quad = \underline{48} \\
& \qquad\qquad\qquad\quad \$928
\end{array}$$

Example 10.7

Design a two-span composite beam using Sec. I3 of AISCS. It is assumed that live load will always be present simultaneously on both spans.

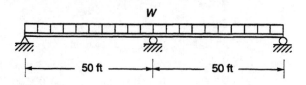

Given: Slab thickness: 6 in.
$f'_c = 3$ ksi $n = 9$ $F_y = 50$ ksi
Beam spacing: 10 ft 0 in.
Superimposed dead load: 15 lb/ft²
Live load: 100 lb/ft²
Construction load: 25 lb/ft²
Unshored construction
$\frac{3}{4}$-in.-diameter stud shear connectors

Solution *Design by LRFD:* Plastic design may be used for this method. The plastic mechanism is shown below. The moment in the interior of the beam is

$$M_x = \frac{wLx}{2} - \frac{wx^2}{2} - \frac{M_p x}{L} \tag{10.26}$$

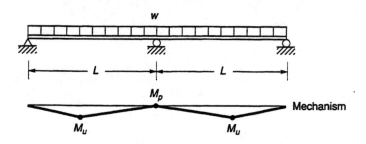

Mechanism

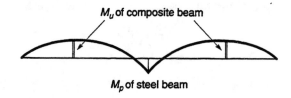

M_u of composite beam

M_p of steel beam

The moment is maximum when

$$\frac{dM_x}{dx} = 0 = \frac{wL}{2} - wx - \frac{M_p}{L}$$

from which

$$x = \frac{L}{2} - \frac{M_p}{wL} \tag{10.27}$$

Substitution of x from Eq. (10.27) into Eq. (10.26), taking note of the fact that $M_x = M_u$, the plastic moment of the composite section, leads to

$$M_u = \frac{w}{2}\left(\frac{L}{2} - \frac{M_p}{wL}\right)^2$$

Calculate the loads. Load to be supported by the steel beam alone during construction:

Slab: $\frac{6}{12} \times 150 = 75$ lb/ft^2

Estimated weight of beam: 100 lb/ft

$$w_{cD} = \frac{75 \times 10 + 100}{1000} = 0.85 \text{ kip/ft} \qquad \text{construction dead load}$$

$$w_{cL} = \frac{25 \times 10}{1000} = 0.25 \text{ kip/ft} \qquad \text{construction live load}$$

$$w_{cu} = 1.2 \times 0.85 + 1.6 \times 0.25 = 1.42 \text{ kips/ft}$$

Required moment for construction loading:

$$M_{pu} = \frac{w_{cu}L^2}{11.66} = \frac{1.42 \times 50^2 \times 12}{11.66} = 3654 \text{ kip-in.} \text{ [Eq. (8.1)]}$$

$$(Z_x)_{req} = \frac{3654}{0.9 \times 50} = 81.2 \text{ in}^2$$

Need a W21 × 44 beam for construction loads, $Z_x = 95.4$ in^3 OK

Load to be supported by the composite beam:

$$w_D = \frac{10 \times (75 + 15) + 100}{1000} = 1.0 \text{ kip/ft}$$

$$w_L = \frac{10 \times 100}{1000} = 1.0 \text{ kip/ft}$$

Design load:

$$w = 1.2 \times 1.0 + 1.6 \times 1.0 = 2.8 \text{ kips/ft (with load factors)}$$

Try a W21 × 44.

Plastic moment of the steel section:

$$M_p = Z_x F_y = 95.4 \times 50 = 4770 \text{ kip-in.}$$

Plastic moment of the composite section:

$$A_s F_y = 13 \times 50 = 650 \text{ kips}$$

$$b_e = \text{spacing} = 10 \text{ ft} = 120 \text{ in.}$$

$$0.85 f'_c b_e t_s = 0.85 \times 3 \times 120 \times 6 = 1836 \text{ kips}$$

Since $A_sF_y < 0.85 f'_c b_e t_s$, the plastic neutral axis is in the slab. From Eqs. (10.7) and (10.8),

$$a = \frac{F_y A_s}{0.85 f'_c b_e} = \frac{50 \times 13}{0.85 \times 3 \times 120} = 2.12 \text{ in.} \qquad < 6 \text{ in.} = t_s \quad \text{OK}$$

$$M_u = \frac{F_y A_s d}{2} + 0.85 f'_c b_e a \left(t_s - \frac{a}{2} \right) = 9924 \text{ kip-in.}$$

Design condition:

$$\phi M_u \geqslant M_u \text{ computed for factored loads} \qquad \text{[Eq. (10.28)]}$$
$$\phi = 0.85, \text{ resistance factor for composite beams}$$
$$0.85 M_u = 0.85 \times 9924 = 8436 \text{ kip-in.}$$

Design moment [Eq. (10.28)]:

$$M_u = \frac{w}{2}\left(\frac{L}{2} - \frac{M_p}{wL}\right)^2 = \frac{2.8}{2}\left(\frac{50}{2} - \frac{4770/12}{2.8 \times 50}\right)^2 \times 12$$
$$= 8250 \text{ kip-in.} < 8436 \qquad \text{OK}$$

Use a W21 × 44.

Number of shear connectors for one-half of positive moment region:

$$n = \frac{A_s F_y}{q_u} = \frac{650}{21.6} = 30 \quad \text{use 30}$$

Total number of shear connectors:

$$4 \times 30 = 120$$

It is left as an exercise for the student to determine the elastic deflection of the beam under unfactored live loads. It is acceptable to perform the deflection analysis by assuming the entire beam has the transformed moment of inertia I_t.

PROBLEMS

10.1. Design a composite beam.

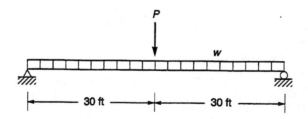

Given: P = 50 kips dead load, 50 kips live load
Beam spacing = 10 ft
Slab thickness = 6 in.
$f'_c = 3$ ksi, $n = 9$, normal-weight concrete
Live load = 100 lb/ft²
Dead load = concrete slab + beam weight

$F_y = 36$ ksi or 50 ksi

$\frac{7}{8}$-in.-diameter studs

Unshored construction

10.2. Design a composite beam.

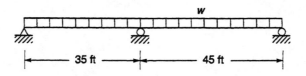

Given: $F_y = 36$ ksi

$f'_c = 3$ ksi, normal-weight concrete

$w_D = 0.86$ kip/ft

Left span = 0.41 kip/ft $\left.\begin{array}{l}\\ \\\end{array}\right\}$ reduced from 50 lb/ft^2 unfactored loads
Right span = 0.38 kip/ft

Beam spacing = 10 ft

Slab thickness: 4.5 in.

$\frac{5}{8}$-in.-diameter studs

11

Special Topics
in Beam Design

11.1 Introduction

Chapters 3, 7, and 8 included beam and plate girder design problems for which specification coverage is adequate, including the usual problems arising from lack of lateral support of rolled shapes having an axis of symmetry in the plane of the loads. Chapter 10 covered composite steel and concrete construction. We now consider special problems that arise in beam design as a result of lack of lateral support, in combination with loads that are not in a plane of symmetry, such as is usually the case when the channel, angle, or other unsymmetrical section is used.* Special attention is given to the problem of combined bending and torsion, and a simplified procedure is presented for W beam shapes.

11.2 Torsion

If torsion in a structural member is a major part of the load system, either a cylindrical or box tube should be used if possible. The cylindrical tube utilizes material in the most effective way possible for resistance to torsion. Box shapes are a close second.

---

* Refer to Section 3.1 for a review of support conditions required to permit use of simple bending theory in design.

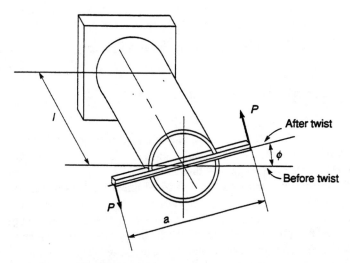

Figure 11.1   Hollow cylinder in pure torsion.

The tube in Fig. 11.1 is loaded in pure torsion, as is the drive shaft of an automobile or the propeller shaft of a ship under its primary load. In Fig. 11.1 the torsional moment, M_t, is equal to Pa and the end twists through a total angle ϕ. For a tubular member under uniform pure torsion the angle of twist per unit length is constant:

$$\theta = \frac{\phi}{l} \tag{11.1}$$

The stress f_v in a torqued tube is "pure shear," as is indicated in Fig. 11.2. In a thin-walled tube the shear stress can be assumed constant through the wall thickness t, and each unit distance around the circumference exerts a tangential force equal to tf_v. The twisting moment about the central axis of the cylinder, at O, of each unit length of tangential shear force is $tf_v r$, where r is the mean radius of the cylinder. Summing up the contributions of each unit length of circumference, the total torsional moment is equal to $tf_v r$ multiplied by the circumferential length; hence

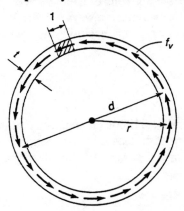

Figure 11.2   Section through a hollow cylinder in pure torsion.

$$M_t = 2\pi t f_v r^2 \tag{11.2}$$

The mean radius of the tube encompasses an area equal to

$$A_o = \pi r^2$$

Hence an alternative expression to Eq. (11.2) can be written

$$M_t = 2A_o t f_v \tag{11.3}$$

The form of Eq. (11.3) is useful in that it applies to square and rectangular box tubes as well as cylindrical. If the wall thickness of a box section varies, Eq. (11.3) still applies, but t should be taken as the thickness of the thinnest plate segment, as this will be the most highly stressed and will determine the allowable torsional moment.

When a box section is twisted, plane sections remain plane—or very nearly so—after twist, and their contribution to the torsional resistance is in proportion to their distance from the center of twist. When an open section, such as a wide-flange shape, is twisted, elements not centered on the axis of twist *tilt* or *warp*, as well as twist about their own axes. An element not centered on the center of twist, such as the flange of a W shape, can add appreciably to overall torsional resistance only if there is a resultant shear force in its midplane, as shown in Fig. 11.4(b). Such a shear force can develop only if the flange bends about its minor axis, which occurs if torsion is nonuniform along the length of the beam.

In the closed box section, warping of each rectangular element is prevented by continuity with adjacent elements. Thus the primary shear stress provides a stress resultant in the midplane of each element. Thus closed or box members are many times more rigid than open sections of the same general dimensions and weight per unit length. The torsional rigidity of a member is measured by the torsion constant J of the cross section, just as the bending rigidity is measured by the moment of inertia I. For a member under uniform torsion the general relationship between torsional moment and angle of twist per unit length is

$$M_t = JG\theta \tag{11.4}$$

For a circular cross section, solid or hollow, J is equal to the polar moment of inertia; for any noncircular section it is always *less* than the polar moment of inertia. For a closed box section of *any* shape, enclosing only one internal cell,

$$J = \frac{4A_0^2}{\displaystyle\sum_{i=1}^{n} (s_i/t_i)} \tag{11.5}$$

The denominator in Eq. (11.5) is the summation of the length/thickness ratios of all the n component parts of the tube around the periphery of the cross section. Thus for a thin-walled hollow cylinder, $A_o = \pi d^2/4$ and $\Sigma(s_i/t_i) = \pi d/t$, and for the circular pipe or tube

$$J = \frac{\pi d^3 t}{4} \tag{11.6}$$

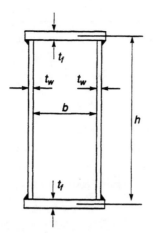

Figure 11.3 Box-section nomenclature.

For the box beam shown in Fig. 11.3, $\Sigma(s_i/t_i) = 2(h/t_w + b/t_f)$, and by Eq. (11.5),

$$J = \frac{2b^2h^2}{(h/t_w) + (b/t_f)} \qquad (11.7)$$

The torsion constant of a solid rectangular bar section, several times wider than its thickness, is approximately

$$J = \frac{1}{3}bt^3 \qquad (11.8)$$

The torsion constant of open (i.e., nontubular) structural shapes, such as the wide-flange beam or angle, is simply approximated by summing Eq. (11.8) for the various component rectangular parts. More accurately, making corrections for fillets and flange edge effects, AISCM lists J for standard shapes.

The maximum shear stress in a structural shape of open section under uniform torsion is approximately

$$f_v = \frac{M_t t}{J} \qquad (11.9)$$

In Eq. (11.9) the maximum shear stress is obviously located where the thickness t is greatest.

When flange warping is unrestrained, the distribution of shear stress in a wide-flange shape under uniform torsion is as shown in Fig. 11.4(a). When warping is completely restrained, the shear stress is nearly constant through the flange thickness, as shown in Fig. 11.4(b). The flanges are then stressed in shear as they would be in a box girder.

Warping restraint may be obtained locally by adding longitudinal stiffeners as shown in Fig. 11.4(c), combined with lateral stiffeners at the ends of the longitudinal segments.

At sections away from a restrained location, the stress distribution is a combination of that shown in Fig. 11.4(a) and (b), approaching that of Fig. 11.4(a) as the distance

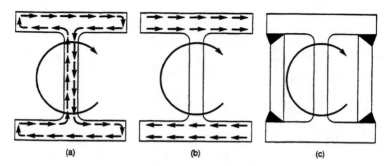

Figure 11.4 Torsional shear stress distributions at unrestrained and restrained locations in a wide-flange beam.

from the restrained location increases. A *warping constant C_w* is also tabulated in AISCM. Another torsion constant, the *torsion bending constant, a*, gives a rough approximation of the distance along a beam away from a restrained location that will permit the effect of restraint to be dissipated and the condition of Fig. 11.4(a) to be approached. The torsion bending constant may be calculated readily from the listed values of J and C_w in AISCM. Values of a as well as all other torsion coefficients are tabulated for all rolled shapes in Ref. 11.1 and in Part 1 of AISCM.

$$a = 1.61\sqrt{\frac{C_w}{J}} \qquad (11.10)$$

The dimensionless parameter l/a will be used in the next section in a simplified presentation of combined bending and torsion with comparative design examples involving both box and open sections.

11.3 Combined Bending and Torsion

In the structural design of buildings it is desirable and usually possible to avoid the complications of torsion by proper location of members. But if torsion cannot be avoided in a laterally loaded beam, the problem involves *combined bending and torsion*, and the rectangular box section should be used if feasible. AISCM provides dimensions and properties of available pipe and both square and rectangular box tubing.

In designing a box beam for combined bending and torsion, the preliminary selection may be made for bending moment alone, with a subsequent check on the combined shear stress due to both bending and torsion. Usually, the preliminary selection will be adequate. Where local concentration of torque load is introduced, care should be taken to guard against local distortion of the cross section. External stiffeners or internal diaphragms may be needed. If the framing is conducive to the use of a triangular closed box section, cross-sectional distortion is automatically eliminated.

For very short and stubby members, a W section may be suitable in combined bending and torsion. If the length l of a cantilever beam is less than $0.5a$ [Eq. (11.10)], the individual flanges may be assumed to take all the torsional moment as if they were individual cantilever beams, loaded in opposite directions. The cantilever beam shown in Fig. 11.5(a) is to be designed for a cantilever bending moment equal to Pl

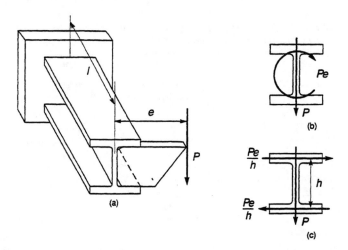

Figure 11.5 Very short W beam section assumed to resist torsion by flange shear forces.

combined with a torsional moment equal to Pe. These may be considered separately for the loads shown in Fig. 11.5(b) and, for the relatively short beam with l/a less than 0.5, as shown in Fig. 11.5(c). If the ratio $C_n = S_x/S_y$ can be estimated (see the discussion in Section 3.8 on biaxial bending), a preliminary estimate of the required section modulus can be made. The section modulus of an individual flange of a W section is approximately $S_y/2$. Thus for bending and torsional moments,

$$M_x = Pl \quad \text{and} \quad M_f = \frac{Pel}{h}$$

and the required section modulus is

$$S_x = \frac{M_x + 2C_nM_f}{\phi F_y} \tag{11.11}$$

For cantilever loads producing bending about the weak bending axis a more direct selection may be made, the required section modulus about yy being

$$S_y = \frac{M_y + 2M_f}{\phi F_y} \tag{11.12}$$

Simplified formulas for the stress in very long W beams can be written; but unless the maximum permissible angle of twist is very large, a tubular member will be desirable. Tables 11.1 and 11.2 tabulate simple approximate formulas for maximum flange moment and maximum angle of twist for the six cases of combined bending and torsion that are most commonly encountered. The formulas include simple approximations for short and long beams and interpolation formulas for intermediate-length beams. The principal advantage of these formulas over the exact ones is the fact that they may be used for direct design checks without reference to hyperbolic or exponential functions. Formulas for the maximum shear stress due to torsion are not tabulated.

For short beams in the minimum l/a category, the shear stress will be mostly of the type shown in Fig. 11.4(b). Neglecting the shear stress of the type in Fig. 11.4(a),

TABLE 11.1   Approximate Formulas for Flange Moment and Total Twist
Angle: Three Concentrated Load Cases, 1, 2, and 3.

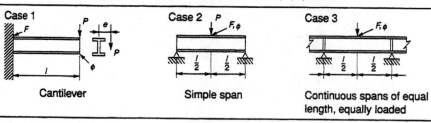

Case 1	Case 2	Case 3
Cantilever	Simple span	Continuous spans of equal length, equally loaded

Maximum lateral bending moment in each flange at location F:

l/a less than 0.5:	l/a less than 1.0:	l/a less than 2.0:
$M_f = \dfrac{Pel}{h}$	$M_f = \dfrac{Pel}{4h}$	$M_f = \dfrac{Pel}{8h}$
l/a more than 2.0:	l/a more than 4.0:	l/a more than 8.0:
$M_f = \dfrac{Pea}{h}$	$M_f = \dfrac{Pea}{2h}$	$M_f = \dfrac{Pea}{2h}$

Maximum total angle of twist at end of a cantilever or a midspan of other beams, at location ϕ:

l/a less than 0.5:	l/a less than 1.0:	l/a less than 2.0:
$\phi_t = 0.32 \dfrac{Pea}{JG}\left(\dfrac{l}{a}\right)^3$	$\phi_t = 0.16 \dfrac{Pea}{JG}\left(\dfrac{l}{2a}\right)^3$	$\phi_t = 0.32 \dfrac{Pea}{JG}\left(\dfrac{l}{4a}\right)^3$
l/a more than 2.0:	l/a more than 4.0:	l/a more than 8.0:
$\phi_t = \dfrac{Pe}{JG}(l - a)$	$\phi_t = \dfrac{Pe}{2JG}\left(\dfrac{l}{2} - a\right)$	$\phi_t = \dfrac{Pe}{JG}\left(\dfrac{l}{4} - a\right)$

Interpolation formulas for bending moment in each flange at F and maximum total twist angle at location ϕ. *

Case 1. Cantilever: l/a more than 0.5 and less than 2.0:

$$M_{f=} \frac{Pea}{h}\left[0.05 + 0.94\left(\frac{l}{a}\right) - 0.24\left(\frac{l}{a}\right)^2\right], \; \phi_t = \frac{Pe}{JG}\left[-0.029 + 0.266\left(\frac{l}{a}\right)^2\right]$$

Case 2. Simple span: l/a more than 1.0 and less than 4.0:

$$M_{f=} \frac{Pea}{2h}\left[0.05 + 0.94\left(\frac{l}{2a}\right) - 0.24\left(\frac{l}{2a}\right)^2\right], \; \phi_t = \frac{Pea}{2JG}\left[-0.029 + 0.266\left(\frac{l}{2a}\right)^2\right]$$

Case 3. Continuous spans of equal length, equally loaded.
l/a more than 2.0 and less than 8.0:

$$M_{f=} \frac{Pea}{2h}\left[0.05 + 0.94\left(\frac{l}{4a}\right) - 0.24\left(\frac{l}{4a}\right)^2\right], \; \phi_t = \frac{Pea}{JG}\left[-0.029 + 0.266\left(\frac{l}{4a}\right)^2\right]$$

*Note: At the extreme range of application the interpolation formulas are more accurate
than the formulas given for locations outside this range, yielding values slightly less
than the others, which err slightly on the side of conservative design estimates.

the maximum shear stress may be assumed to be 1.5 times the average at the location
along the beam where shear due to torsional flange bending is greatest. To calculate
the flange shear due to torsional flange bending, each individual flange may be
assumed to act as a beam, loaded as shown in Fig. 11.5(c), with load magnitudes equal
to the total beam load times e/h. The load distributions and beam end conditions for

TABLE 11.2 Approximate Formulas for Flange Moment and Total Twist
Angle: Three Concentrated Load Cases, 4, 5, and 6.

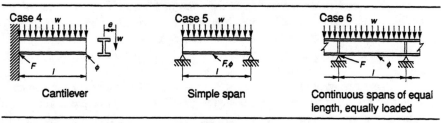

Case 4	Case 5	Case 6
Cantilever	Simple span	Continuous spans of equal length, equally loaded

Maximum lateral bending moment in each flange at location F:

l/a less than 0.5:	l/a less than 1.0:	l/a less than 2.0:
$M_f = \dfrac{wl^2e}{2h}$	$M_f = \dfrac{wl^2e}{8h}$	$M_f = \dfrac{wl^2e}{12h}$
l/a more than 3.0:	l/a more than 6.0:	l/a more than 8.0:
$M_f = \dfrac{wea}{h}\left(1-\dfrac{a}{l}\right)$	$M_f = \dfrac{wea^2}{h}$	$M_f = \dfrac{wlea}{h}\left(\dfrac{1}{2}-\dfrac{a}{l}\right)$

Maximum total angle of twist at end of a cantilever or a midspan of other beams, at location ϕ:

l/a less than 0.5:	l/a less than 1.0:	l/a less than 2.0:
$\phi_t = 0.114\dfrac{wlea}{JG}\left(\dfrac{l}{a}\right)^3$	$\phi_t = 0.094\dfrac{wlea}{JG}\left(\dfrac{l}{2a}\right)^3$	$\phi_t = 0.151\dfrac{wlea}{JG}\left(\dfrac{l}{4a}\right)^3$
l/a more than 3.0:	l/a more than 6.0:	l/a more than 8.0:
$\phi_t = \dfrac{wlea}{JG}\left(\dfrac{l}{2a}-1+\dfrac{a}{l}\right)$	$\phi_t = \dfrac{wlea}{JG}\left(\dfrac{l}{8a}-\dfrac{a}{l}\right)$	$\phi_t = \dfrac{wlea}{JG}\left(\dfrac{l}{8a}-\dfrac{1}{2}\right)$

Interpolation formulas for bending moment in each flange at F and maximum total twist
angle at location ϕ. *

Case 4. Cantilever: l/a more than 0.5 and less than 3.0:

$$M_f = \frac{wlea}{h}\left[0.041 + 0.423\frac{l}{a} - 0.068\left(\frac{l}{a}\right)^2\right]$$

$$\phi_t = \frac{wlea}{JG}\left[-0.023 + 0.029\frac{l}{a} + 0.086\left(\frac{l}{a}\right)^2\right]$$

Case 5. Simple span: l/a more than 1.0 and less than 6.0:

$$M_f = \frac{wlea}{h}\left[0.097 + 0.094\frac{l}{2a} - 0.0255\left(\frac{l}{2a}\right)^2\right]$$

$$\phi_t = \frac{wlea}{JG}\left[-0.032 + 0.062\frac{l}{2a} + 0.052\left(\frac{l}{2a}\right)^2\right]$$

Case 6. Continuous spans of equal length, equally loaded.
l/a more than 2.0 and less than 8.0:

$$M_f = \frac{wlea}{2h}\left[0.005 + 0.342\frac{l}{4a} - 0.078\left(\frac{l}{4a}\right)^2\right]$$

$$\phi_t = \frac{wlea}{2JG}\left[-0.029 + 0.266\left(\frac{l}{4a}\right)^2\right]$$

*Note: See footnote to Table 11.1.

the individual flanges in torsional bending will be identical with those shown in Tables 11.1 and 11.2 for bending of the complete beam. Likewise, the maximum beam moments and flange bending moments occur at identical locations, and the direct stresses in the flanges due to the two causes are additive. Thus Eqs. (11.11) and (11.12), as applied to the short cantilever, may also be applied to any of the other five loading and support conditions.

In the design of intermediate and long W beams for combined bending and torsion, the amount of torsion to be resisted is greatly dependent on the degree to which external resistance to twist exists where the load is applied. If such resistance is negligible or uncertain, the use of W shapes will be feasible only if the load eccentricities are very small.

Direct flange stresses should be checked for combined bending and torsion, and the maximum twist determined. Away from restrained ends the torsional shear stress, of the pattern shown in Fig. 11.4(a), may be calculated by Eq. (11.9), but it is bound to be small and of little design significance if the total twist angles are small.

Several design examples will now illustrate the foregoing procedures and indicate situations where W shapes are satisfactory in combined bending torsion, as well as other cases where the reverse is true.

Example 11.1

A 4-kip pull is applied at any angle, tangential to the circumference of a 20-in. diameter, as shown, at the top of a 20-in.-long W beam that is fixed at its base. Select beam size using A36 steel, $F_y = 36$ ksi. Obviously, the design may be based on the pull being directed so as to cause bending about the weak axis of the beam, as shown at the bottom of the accompanying drawing.

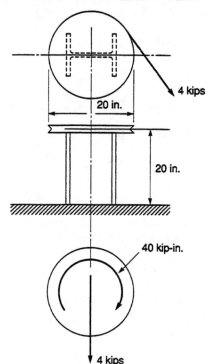

Solution   Refer to case 1, Table 11.1, and assume that l/a is less than 0.5—to be checked after selection. Try a W12 beam and assume that $h = 11.5$ in.

$$M_f = \frac{Pel}{h} = \frac{4 \times 10 \times 20}{11.5} = 69.6 \text{ kip-in.}$$

$$M_y = Pl = 4 \times 20 = 80.0 \text{ kip-in.}$$

By Eq. (11.12), the required section modulus S_y is determined:

$$S_y = \frac{80.0 + 2 \times 69.6}{0.9 \times 36} = 6.76 \text{ in}^3$$

Try W12 × 35:

$$S_y = 7.47 \text{ in}^3$$
$$J = 0.74 \text{ in}^4$$
$$C_w = 879 \text{ in}^6$$
$$h = 12.50 - 0.52 = 11.98 \text{ in.}$$

Check l/a by Eq. (11.10):

$$b_f = 6.56 \text{ in.}$$

$$a = 1.61 \sqrt{\frac{879}{0.74}} = 55.5 \text{ in.}$$

$$\frac{l}{a} = \frac{20}{55.5} = 0.36 \qquad < 0.50 \qquad \text{OK}$$

Check the direct stress in the flange:

$$f_b = \frac{80}{7.47} + \frac{2 \times 4 \times 10 \times 20}{11.98 \times 7.47} = 28.6 \text{ ksi} \qquad < 0.9 \times 36 = 32.4 \text{ ksi} \quad \text{OK}$$

Check the maximum shear stress:

$$A_f = 6.56 \times 0.52 = 3.41 \text{ in}^2$$

$$V_{max} = \frac{4}{2} + \frac{4 \times 10}{11.98} = 5.34 \text{ kips}$$

$$f_v = \frac{1.5 \times 5.34}{3.41} = 2.35 \text{ ksi} \qquad < 0.9 \times 0.6 \times 36 = 19.44 \text{ ksi}$$

Note that in spite of the very short length of the beam and the orientation that makes the direct flange shear forces due to torsion and bending directly additive, the maximum shear stress is not significant.

Example 11.2

Redesign for Example 11.1, changing W section to a pipe section. Ignore torsion in preliminary selection and assume that $F_y = 36$ ksi.

Solution   The section modulus required is

$$S = \frac{80}{0.9 \times 36} = 2.47 \text{ in}^3$$

Try a 4-in. standard pipe (refer to AISCM):

$$S = 3.21 \text{ in}^3 \qquad \text{O.D.} = 4.5 \text{ in.}$$
$$A = 3.17 \text{ in}^2 \qquad \text{mean diameter: } d_m = 4.5 - 0.237 = 4.26 \text{ in.}$$
$$t = 0.237 \text{ in.}$$

Check the shear stress due to torsion and bending. (The shear shape factor for a circular tube in bending is 1.33.)

Due to bending:

$$f_v = \frac{1.33 \times 4}{3.17} = 1.68 \text{ ksi}$$

Due to torsion:

$$f_v = \frac{M_t}{2tA_o} = \frac{4 \times 10}{2 \times 0.237 \times \pi \times 2.13^2} = 5.92 \text{ ksi} \qquad [\text{Eq. (11.3)}]$$

Due to both bending and torsion:

$$f_v = 1.68 + 5.92 = 7.60 \text{ ksi} \qquad < 19.44 \quad \text{OK}$$

Direct stress due to bending:

$$f_b = \frac{80}{3.21} = 24.92 \text{ ksi}$$

Reduced allowable direct stress because of shear stress [Eq. (11.13)]*:

$$F_{rt} = \left[1 - \left(\frac{5.92}{32.4}\right)^2\right] 32.4 = 31.32 \text{ ksi} \qquad > 24.92 \quad \text{OK}$$

Examples 11.1 and 11.2 showed that either a W or pipe section could serve satisfactorily in combined bending and torsion for a short cantilever member. The required W section weighed three times more than the pipe. In the design of relatively long members in combined bending and torsion, such as might be required for a highway direction sign attached to a single vertical member, the pipe is the only suitable member because of the unduly large torsional deflections that would result from use of a W section.

Examples 11.3 and 11.4 illustrate the combined bending and torsion problem in the design of a continuous spandrel beam supporting an exterior wall. It is assumed that the framing situation does not permit reduction or elimination of the torsion component by means of a laterally contiguous floor slab or framed beams.

Example 11.3

Continuous 24-ft spans of a spandrel beam carry a wall dead load of 600 lb/ft, 4 in. from the center of the beam. Use A36 steel and a W beam. Wall support brackets will not be considered as adding to the beam section.

---

* *Note regarding combined shear and direct stress:* AISCS does not provide any limitation on direct stress combined with shear, except in the case of connection design. To keep the maximum principal stress less than $0.9F_y$, the allowable direct stress, the solution of the quadratic principal stress formula may be avoided for the simple case of a single direct stress component by the determination of a reduced direct stress (F_{rt}) as follows:

$$F_{rt} = \left[1 - \left(\frac{f_v}{0.9F_y}\right)^2\right] 0.9F_y \qquad (11.13)$$

If $f_v/0.9F_y$ is less than 0.2, the effect of shear stress on the allowable stress may be ignored. The case in Example 11.2 is borderline.

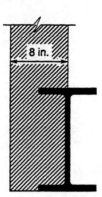

Solution   A preliminary trial selection will be chosen by designing for bending alone, but at a greatly reduced design stress, say, one-third of $0.9F_y$, or 11 ksi. Because of lack of lateral support, the provisions of AISCS, Sec. A5.1, for reduced end moment in the continuous beam cannot be applied. Design moment, assuming a beam weight of 50 lb/ft, is calculated:

$$w_u = 1.4 \times (0.6 + 0.05) = 0.91 \text{ kip/ft}$$

$$M_x = \frac{0.91 \times 24^2 \times 12}{12} = 524.2 \text{ kip-in.}$$

Required section modulus for trial beam at 11.0 ksi,

$$S_x = \frac{524.2}{11.0} = 47.7 \text{ in}^3$$

Refer to AISCM and choose a W14 × 38 as a trial, for which

$$S_x = 54.6 \text{ in}^3$$
$$S_y = 7.88 \text{ in}^3$$
$$J = 0.80 \text{ in}^4$$
$$C_w = 1230.0 \text{ in}^6$$
$$h = 14.10 - 0.51 = 13.59 \text{ in.}$$

By Eq. (11.10),

$$a = 1.61 \sqrt{\frac{1230}{0.80}} = 63.1 \text{ in.}$$

$$\frac{l}{a} = \frac{288}{63.1} = 4.56$$

(More than 2.0, less than 8.0; hence use interpolation formulas of Table 11.2, case 6.) Using the interpolation formula, the maximum moment in a flange due to torsional bending is found:

$$M_f = \frac{(0.91)(288)(4.0)(63.1)}{(12)(13.59)} \left[0.005 + 0.342 \left(\frac{4.56}{4} \right) - 0.078 \left(\frac{4.56}{4} \right)^2 \right] = 119.1 \text{ kip-in.}$$

The stress due to torsional bending alone is

$$f_{bt} = \frac{119.1 \times 2}{7.88} = 30.2 \text{ ksi}$$

too great, obviously, for the torsion bending stress alone. To make a better selection, assume $M_f = 120$ kip-in., as in this case, $S_x/S_y = 54.6/7.88 = 6.93$. Then, by Eq. (11.11),

$$\text{req'd } S_x = \frac{524.2 + 2 \times 6.96 \times 120}{32.4} = 67.7 \text{ in}^3$$

requiring, by AISCM, a W21 × 44. But noting that S_x/S_y would be more than 10, the requirement for S_x would be escalated. It is obviously desirable to stay with a wider and less deep cross section. Noting, also, that the next group of sections heavier than the W14 × 30 trial selection have $S_x/S_y = 5.5$ for the median of the group (AISCM), the required S_x would be reduced to

$$\frac{524.2 + 2 \times 5.5 \times 120}{32.4} = 56.9 \text{ in}^3$$

Try a W14 × 43, for which $S_x = 62.7 \text{ in}^3$.

Other needed properties of the W14 × 43 are

$$S_y = 11.3 \text{ in}^3$$
$$J = 1.05 \text{ in}^4$$
$$C_w = 1950.0 \text{ in}^6$$
$$h = 13.66 - 0.53 = 13.13 \text{ in.}$$
$$a = 1.61\sqrt{\frac{1950}{1.05}} = 69.4 \text{ in.}$$
$$\frac{l}{a} = \frac{288}{69.4} = 4.15 \text{ in.}$$

Again, use interpolation formulas from Table 11.2, case 6, for flange moment:

$$M_f = \frac{(0.91)(288)(4)(69.4)}{(12)(13.13)}\left[0.005 + 0.342\left(\frac{4.15}{4}\right) - 0.078\left(\frac{4.15}{4}\right)^2\right] = 127.3 \text{ kip-in.}$$

Stress due to bending moment M_x:

$$f_{bx} = \frac{0.91 \times 24^2 \times 12}{12 \times 62.7} = 8.4 \text{ ksi}$$

Stress due to torsional flange bending:

$$f_{bt} = \frac{127.3 \times 2}{11.3} = 22.5 \text{ ksi}$$

Total direct stress due to combined bending and torsion:

$$f_b = 8.4 + 22.5 = 30.9 \text{ ksi} \qquad < 34.2 \quad \text{OK}$$

provided that $f_b = 8.4$ ksi is less than the flexural design strength determined from Sec. F1 of AISCS.

Assume, conservatively, that $C_b = 1.0$ and $L_b = 24$ ft. For the W14 × 38 the following data are taken from AISCM Load Factor Design Selection Table: $L_r = 20.0$ ft $< L_b$; hence AISCS Eq. (F1-13) applies.

$$M_n = \frac{C_b S_x X_1 \sqrt{2}}{L_b/r_y}\sqrt{1 + \frac{X_1^2 X_2}{2(L_b/r_y)^2}}$$

$$\phi F_{cr} = \frac{0.9 M_n}{S_{cr}} = \frac{0.9 X_1 \sqrt{2}}{L_b/r_y}\sqrt{1 + \frac{X_1^2 X_2}{2(L_b/r_y)^2}}$$

From Part 1 of AISCM:

$X_1 = 2190$ ksi $X_2 = 0.00685$ ksi^{-2} $r_y = 1.55$ in.

$$\frac{L_b}{r_y} = \frac{24 \times 12}{1.55} = 186$$

$$\phi F_{cr} = \frac{0.9 \times 2190 \times \sqrt{2}}{186}\sqrt{1 + \frac{2190^2 \times 0.00685}{2 \times 186^2}} = 18.22 \text{ ksi}   > 8.4 \text{ ksi   OK}$$

Check maximum twist at center by interpolation formula from Table 11.2, case 6:

$$\phi_t = \frac{(0.91)(288)(4)(69.4)}{(12)(2)(1.05)(11,200)}\left[-0.029 + 0.266\left(\frac{4.15}{4}\right)^2\right] = 0.067 \text{ rad}$$

Thus, if the masonry wall lacked internal restraint (if placed to a height of 50 in. before setting up of mortar), a point 50 in. above the top of the beam (assuming the beam to be about half loaded) would tend to deflect outward:

$$\frac{1}{2} \times 0.067 \times 50 = 1.67 \text{ in.}$$

Since the deflection would increase gradually as the wall was placed, it might be partially offset by progressive correction as bricks or blocks were placed. Thus, for eccentricities of *only an inch or two*, the use of W sections for spandrel beams, designed as above for combined bending and torsion, would be feasible without excessive twist.

Example 11.4

Alternative design using a box (rectangular tube) section for the same support and load conditions as Example 11.3. See the sketch.

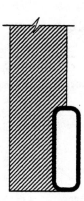

Solution   Initial design will be for full bending moment at $\phi F_y = 0.9 \times 36 = 32.4$ ksi, neglecting torsion. Assume that the weight of the member is 0.03 kip/ft.

$$M_x = \frac{0.88 \times 24^2 \times 12}{12} = 506.9 \text{ kip-in.}$$

The required section modulus is

$$S_x = \frac{506.9}{32.4} = 15.5 \text{ in}^3$$

Refer to the AISCM, for properties of rectangular structural tubing.

Try TS12 × 4 × 0.250; weight = 25.82 lb/ft:

$$S_x = 21.1 \text{ in}^3$$
$$t = 0.25 \text{ in.}$$
$$h = 12.0 - 0.25 = 11.75 \text{ in.}$$
$$b = 4.0 - 0.25 = 3.75 \text{ in.}$$
$$A_0 = 3.75 \times 11.75 = 44.1 \text{ in}^2$$

Check the shear stress due to combined bending and torsion.
Due to beam bending:

$$f_{vb} = \frac{0.88 \times 12}{0.5 \times 12} = 1.76 \text{ ksi}$$

Due to torsion:

$$M_t = \frac{wel}{2} = \frac{0.88 \times 4 \times 288}{12 \times 2} = 42.2 \text{ kip-in.}$$

$$f_{vt} = \frac{42.2}{2 \times 44.1 \times 0.25} = 1.92 \text{ ksi}$$

$$f_v = 1.76 + 1.92 = 3.68 \qquad < 0.9 \times 0.6 \times 36 = 19.44$$

From AISCS, Eq. (F1-14) for symmetrical box sections:

$$M_{cr} = \frac{57,000 C_b \sqrt{JA}}{L_b/r_y}$$

$$C_b = 1.0 \qquad A = 7.59 \text{ in}^2 \qquad r_y = 1.71 \text{ in.} \qquad S_x = 21.1 \text{ in}^3$$

The torsion constant J is calculated by Eq. (11.7):

$$J = \frac{2b^2h^2}{h/t_w + b/t_f} = \frac{2 \times 3.75^2 \times 11.75^2}{(11.75 + 3.75)/0.25} = 62.7 \text{ in}^4$$

Contrast this with $J = 1.05$ in^4 for the W14 × 43 beam selection in Example 11.3!

$$\phi_b F_{cr} = \frac{\phi_b M_{cr}}{S_x} = \frac{0.9 \times 57,000 \times 1.0 \times \sqrt{62.7 \times 7.59}}{[(24 \times 12)]/1.71 \times 21.1} = 314 \text{ ksi} \qquad > \phi_b F_y = 32.4 \text{ ksi}$$

Hence

$$\phi_b F_n = 32.4 \text{ ksi} \qquad > \frac{576}{21.1} \quad \text{OK}$$

The *average* torsional moment between one end and the center of the span is $wel/4$ and

$$\phi_t = (\theta_{av})\left(\frac{1}{2}\right) = \frac{wel^2}{8JG}$$

$$= \frac{0.88 \times 4 \times 288^2}{12 \times 8 \times 62.7 \times 11,200} = 0.004 \text{ rad}$$

to be contrasted with ϕ_t of 0.067 rad for the W-beam selection of Example 11.3.

Examples 11.3 and 11.4 have demonstrated the strength and stiffness advantages of the closed box section in comparison with the open W section in resisting torsion. However, the determination of torsional strength and stiffness was based on the extremely conservative assumption that the masonry wall does not participate in providing torsional resistance. The torsion problem may be minimized by making the steel member composite with the concrete wall by use of the principles enunciated in Chapter 10. Similarly, in the absence of composite construction, the torsion problem may be minimized by providing continuity with contiguous framing members in such a way that they absorb much or most of the torsional moment. Examples 11.5A and 11.5B will illustrate.

Example 11.5

A. A W24 × 55 beam, 18 ft in length, carrying a total live load of 45 kips, is framed into two supporting beams of 34 ft span at their midlength. A bracket, also at midspan, as shown in sketch (a), carries a live load of 45 kips, 12 in. eccentric to the web center. In sketch (b), the W24 × 55 beam is framed to the supporting beams by bolted clip angles. The bolts have a gage of $2\frac{1}{2}$ in. and will be assumed to be centered 3 in. from the center of the web.

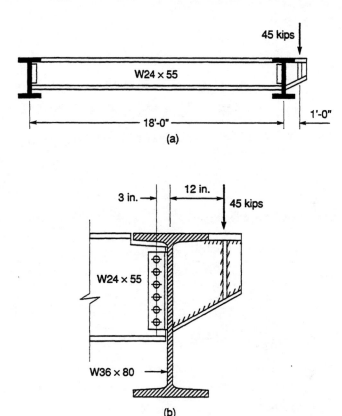

(a)

(b)

    Solution   The bolted clip angles cannot be counted on to transmit more than a few percent of full beam capacity moment. Hence the 34-ft beam will be designed for direct stress due to bending moment plus warping stress induced by the central applied torque of

$$M_t = 1.6 \times (45 \times 12 - 22.5 \times 3) = 756 \text{ kip-in.}$$

The required section modulus, neglecting torsion, can be used as a trial lower limit beam size:

Assuming a dead load of 0.12 kip/ft:

$$M_b = \frac{1.6 \times 67.5 \times 34 \times 12}{4} + \frac{1.2 \times 0.12 \times 34^2 \times 12}{8} = 11{,}265 \text{ kip-in.}$$

Assuming that $F_y = 36$ ksi, the required section modulus is

$$S_x = \frac{11{,}265}{0.9 \times 36} = 348 \text{ in}^3$$

Required beam size, W33 × 118, $S_x = 359$ in^3.

After several trials, designed to include torsion, a W36 × 280 was found to be required. Calculations follow.

Section properties:

$$S_x = 1030 \text{ in}^3$$
$$S_y = 144 \text{ in}^3$$
$$J = 52.6 \text{ in}^4 \qquad h = 36.52 - 1.57 = 34.95 \text{ in.}$$
$$a = 134 \text{ in.} \qquad \frac{L}{a} = \frac{34 \times 12}{134} = 3.04$$

Formulas for case 2 of Table 11.1 apply ($1 < L/a < 4$). Using the interpolation formula, the flange moment due to torsion is

$$M_f = \frac{756 \times 134}{2 \times 34.95}\left[0.05 + \frac{0.94 \times 34 \times 12}{2 \times 134} - 0.24\left(\frac{34 \times 12}{2 \times 134}\right)^2 \right] = 1340 \text{ kip-in.}$$

section modulus of one flange = $\dfrac{144}{2} = 72$ in^3

warping direct stress $f_w = \dfrac{1340}{72} = 18.62$ ksi

Bending moment:

$$M_b = \left(\frac{1.6 \times 67.5 \times 34}{4} + \frac{1.2 \times 0.28 \times 34^2}{8} \right) \times 12 = 11{,}599 \text{ in.-kip}$$

$$f_b = \frac{11{,}599}{1030} = 11.26 \text{ ksi}$$

Combined direct stress due to bending and torsion:

$$f_b + f_w = 11.26 + 18.62 = 29.88 \text{ ksi} \qquad < 32.4 \quad \text{OK}$$

Check unsupported length requirement in Sec. F2.2 of AISCS.
Unsupported length = $17 \times 12 = 204$ in.

$$L_r = 53.0 \text{ ft} \quad \text{and} \quad \phi M_r = 2010 \text{ kip-ft}$$

$$\phi F_{cr} = \frac{2010 \times 12}{1030} = 23.42 \text{ ksi} \qquad > 11.26 \text{ ksi} \quad \text{OK}$$

B. If the clip angle connection of Example 11.5A is replaced by a fully continuous welded detail as shown, the moment applied by the bracket ($12 \times 45 \times 1.6 = 864$ kip-in.) will be resisted proportionally by the bending stiffness of the 18-ft, W24 × 55 beam and by the torsional resistance of the 34-ft beam. The bending stiffness of the beam, assuming the far end to be hinged, is equal to $3EI/L$ (see Fig. 8.1). The torsional stiffness of the 34-ft beam equals Pe/ϕ_t and is obtained from the equation for ϕ_t in Table 11.1, case 2. As a trial beam selection, we use a W33 × 118, previously shown in Example 11.5A to be adequate if torsion is entirely neglected.

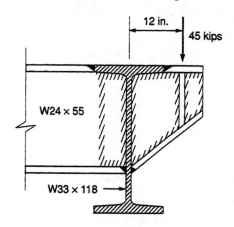

Solution   Section properties of W33 × 118:

$$S_x = 359 \text{ in}^3 \qquad a = 154 \text{ in.}$$
$$S_y = 32.6 \text{ in}^3 \qquad h = 32.86 - 0.740 = 32.12 \text{ in.}$$
$$J = 5.30 \text{ in}^4$$

Bending stiffness of W24 × 55 beam at bracket:

$$\frac{3EI}{L} = \frac{3 \times 29,000 \times 1350}{18 \times 12} = 544 \times 10^3 \text{ kip-in./rad}$$

Torsional stiffness at center of W33 × 118 beam (Table 11.1, case 2):

$$\frac{Pe}{\phi_t} = \frac{2 \times 5.30 \times 11,200}{154\left[-0.029 + 0.266\left(\dfrac{34 \times 12}{2 \times 154}\right)^2\right]} = 1.76 \times 10^3 \text{ kip-in./rad}$$

At this point it can be seen that the torsional stiffness is negligible compared to the beam bending stiffness and that torsion could be neglected entirely. However, for the sake of completeness, the torsional warping direct stress will be calculated:

Torsional moment resisted by the W33 × 118 beam:

$$M_t = \frac{1.76}{544 + 1.76} \times 864 = 2.78 \text{ kip-in.}$$

Flange bending moment due to torsion:

$$M_f = \frac{2.78 \times 154}{2 \times 32.12}\left[0.05 + 0.94\left(\frac{34 \times 12}{2 \times 154}\right) - 0.24\left(\frac{34 \times 12}{2 \times 154}\right)^2\right] = 5.83 \text{ kip-in.}$$

Flange direct stress due to torsional warping restraint:

$$f_w = \frac{5.83}{16.3} = 0.36 \text{ ksi}$$

Direct stress in flange due to bending:

$$f_b = \frac{11,265}{359} = 31.38 \text{ ksi}$$

Combined direct stress in flange:

$$f_b + f_w = 31.38 + 0.36 = 31.74 \text{ ksi} \qquad < 32.4 \quad \text{OK}$$

It is seen that the torsional warping stress has been reduced in the revised design from a major factor to nearly negligible proportions and the 34-ft beam weight is reduced from 280 lb/ft to 118 lb/ft. Further examples of moment distribution principles applied to torsional problems will be found in Refs. 11.1 and 11.2. Reference 11.1 also provides the more exact equations on which Tables 11.1 and 11.2 are based. Both references also provide graphical charts which permit rapid calculation. However, as pointed out earlier, torsion-producing loads should be avoided entirely if at all possible.

11.4 Biaxial Bending and Lateral-Torsional Buckling

Properties for structural sections as listed in AISCM are in terms of the xx and yy axes and, except for the angle and Z^* sections, the xx and yy axes are also the *principal* axes of the cross sections, as will always be the case if one or both of the two axes are axes of symmetry. As explained in Chapter 3, if lateral support is provided, either continuously or at the locations of applied concentrated load, a beam of any shape may be designed on the basis of simple bending theory. If lateral support is not provided, the possibility of lateral-torsional buckling about the weakest principal axis is always present, and the design procedure to be recommended here requires the determination of the orientation of the principal axes, in cases where they are not already known, and the calculation of the principal moments of inertia, designated herein as I_1 and I_2. After these are determined, and the effect of lateral-torsional buckling in reducing the allowable stress in bending about the strong bending axis is estimated, the procedure parallels that of Chapter 3.

In the general case, to determine the principal moments of inertia, it will be necessary to calculate the product of inertia, I_{xy}. The readers should refresh their acquaintance with this parameter by reference to their text on strength of materials. It will be recalled that

$$I_x = \int y^2 \, dA \qquad I_y = \int x^2 \, dA \qquad I_{xy} = \int xy \, dA$$

Note that I_x and I_y are always positive quantities, but that I_{xy} may be either positive or negative, and that for the same section this will depend on the arbitrary way in which the positive directions of x and y are chosen. If, as is so common, the structural shape is made of rectangular component parts, the contribution of any one rectangle to I_{xy} may be determined by the *parallel-axis theorem*,

$$I_{xy} = I_{xyo} + Ax_o y_o \tag{11.14}$$

If the x and y axes include one that is an axis of symmetry, they are then principal axes, and I_{xyo} is zero. Thus, for a rectangular component, I_{xyo} is different from zero only if the sides are tilted at an angle to the x and y axes, as illustrated in Fig. 11.6, which is explanatory of Eq. (11.15):

_____

* No longer listed in AISCM.

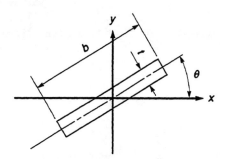

Figure 11.6

$$I_{xyo} = \frac{b^3 t - b t^3}{12} \sin\theta \cos\theta \tag{11.15}$$

If b is relatively large in relation to t, the term bt^3 may be omitted from Eq. (11.15) with but little error. It should be noted that when θ is either zero or 90°, I_{xyo} is zero. Another useful relationship is the fact that, regardless of the orientation of x and y, the sum of I_x and I_y is a constant. Thus also,

$$I_x + I_y = I_1 + I_2 \tag{11.16}$$

Equation (11.16) is useful in the determination of the principal moments of inertia of an angle, using the information listed in the AISCM, as illustrated by Example 11.6.

Example 11.6

Determine the principal moments of inertia of an L6 × 4 × $\frac{1}{2}$.

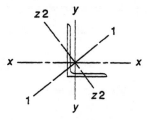

Solution   Referring to AISCM,

$$I_x = 17.4 \text{ in}^4 \qquad I_y = 6.27 \text{in}^4$$

and the minimum radius of gyration about the principal axis zz is listed as 0.870 in. The area A is given as 4.75 in², and from the relationship $I = Ar^2$, the minimum moment of inertia, I_2, is

$$I_2 = 4.75 \times 0.87^2 = 3.60 \text{ in}^4$$

Equation (11.16) now provides the calculation of I_1:

$$I_1 = 17.4 + 6.27 - 3.6 = 20.1 \text{ in}^4$$

The complete problem of calculating section properties in the more general case for which no handbook information is available will be illustrated in Example 11.7. If the orientation of the principal axes is not known, it may be determined by Eq. (11.17), in which θ is the angle between the x axis and the principal axes. Only one of the angles need be calculated since they are 90° apart:

$$\tan 2\theta = \frac{2I_{xy}}{I_y - I_x}$$ (11.17)

The magnitudes of the principal moments of inertia are given by

$$I_1, I_2 = \frac{I_x + I_y}{2} \pm \sqrt{\left(\frac{I_x - I_y}{2}\right)^2 + I_{xy}^2}$$ (11.18)

Stresses may now be determined by resolving the loads into components in the principal planes and superposing the stresses as calculated by the ordinary beam stress formula applied successively to the bending moments about each of the two principal axes, as in the case treated in Chapter 3 [Eq. (3.16)], which was applicable when the handbook x and y axes were also principal axes. A scale layout will be helpful to determine distances to extreme fibers in the stress calculations.

However, if there is no lateral support, as is always the case if applied loads cause stress about both principal axes, there will be a reduction in the design stress required for bending about the strong axis because of the lateral-torsional buckling effect. The AISCS interaction procedure, also explained and illustrated in Chapter 3 [Eq. (3.17)], may then be applied, provided that the reduced design stress for bending about the strong axis is determined, as will be discussed later.

Example 11.7

Determine the section properties about the x and y axes, and about the principal axes, for the shape shown on the following page. As a close approximation in the calculations it is assumed that the section is made up of two rectangular parts with breadths of 8 and 10 in., respectively.

Solution The area is

$$A = 0.75(8.0 + 10.0) = 13.5 \text{ in}^2$$

Locate the neutral axes. Use point A, the centroid of the 10-in. segment, as a reference origin:

$$\bar{x}_A = \frac{6.0 \times 3.46}{13.5} = 1.54 \text{ in.}$$

$$\bar{y}_A = \frac{6.0 \times 7.0}{13.5} = 3.11 \text{ in.}$$

Having located the centroid, this becomes the origin of x and y distances to be used in calculation of the moments of inertia, I_x, I_y, and I_{xy}.

Determine I_x:

$$6.0 \times 3.89^2 + \frac{6.0 \times 4.0^2}{12} = 98.79$$

$$7.5 \times 3.11^2 + \frac{7.5 \times 10.0^2}{12} = 135.06$$

$$I_x = \overline{233.85} \text{ in}^4$$

Determine I_y:

$$6.0 \times 1.92^2 + \frac{6.0 \times 6.92^2}{12} = 46.06$$

$$7.5 \times 1.54^2 + \frac{7.5 \times 0.75^{2*}}{12} = 18.14$$

$$I_y = \overline{64.20} \text{ in}^4$$

* This term could have been omitted with but little error.

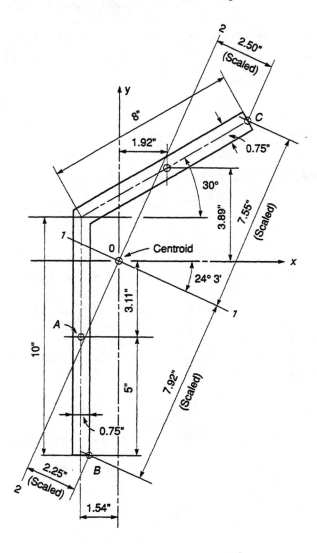

Determine I_{xy}: Calculate I_{xyo} of the 8×0.75 in. segment, by Eq. (11.15), for $\theta = 30°$,

$$\sin\theta = 0.500 \qquad \cos\theta = 0.866$$

Equation (11.15), for calculation purposes, may be more conveniently expressed as

$$I_{xyo} = \frac{bt(b^2 - t^2)}{12}\sin\theta\cos\theta$$

The t^2 term will be included for sake of completeness:

$$I_{xyo} = \frac{6.0(8.0^2 - 0.75^2)}{12}(0.5 \times 0.866) = 13.73 \text{ in}^4$$

Thus for the complete section, by Eq. (11.14),

$$I_{xy} = (7.5)(-3.11)(-1.54) + (6.0)(+3.89)(+1.92) + 13.73 = 94.47 \text{ in}^4$$

Orientation of the principal axes is obtained by use of Eq. (11.17),

$$\tan 2\theta = \frac{2(+94.47)}{64.20 - 233.85} = -1.114$$

Therefore, $2\theta = -48°5'$; hence $\theta = -24°3'$ and

$$\sin \theta = -0.4075 \qquad \cos \theta = 0.9132$$

The principal moments of inertia are now calculated by Eq. (11.18):

$$I_1, I_2 = \frac{233.85 + 64.20}{2} \pm \sqrt{\left(\frac{233.85 - 64.20}{2}\right)^2 + 94.47^2}$$

$$I_1 = 275.99 \text{ in}^4$$

$$I_2 = 22.07 \text{ in}^4$$

Suppose, now, that lateral support is provided for the section of Example 11.7, either continuously, if the load is continuous, or by rods attached at all load locations, as shown in Fig. 11.7. With such support, bending is forced to be about the xx axis and the stress may be calculated by the usual Mc/I formula. The load-carrying capacity is increased by such lateral support, and, if the use of supports is optional, the cost of supports can be weighed against the cost of a heavier member that would be required if such supports were lacking. The force P_H in the support may be calculated by

$$P_H = \frac{P_V I_{xy}}{I_x} \qquad\qquad (11.19)$$

The design of a laterally supported unsymmetrical section is illustrated next.

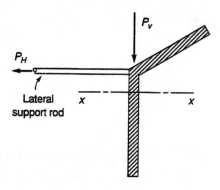

Figure 11.7

Example 11.8

A member having the properties and cross section of Example 11.7 has a span of 18 ft and is loaded vertically at the third points. If the design stress is $0.9 \times 36 = 32.4$ ksi (A36 steel), what load can the member safely carry if lateral supports are provided at the load locations as shown in the sketch?

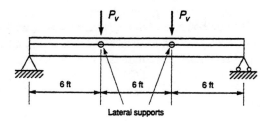

Solution   Calculate the bending moment:

$$M_x = 72P_V \text{ kip-in.}$$

Referring to Example 11.7, the maximum stress in tension, 8.11 in. below the neutral axis, will determine the load capacity:

$$f_b = 32.4 = \frac{72P_V \times 8.11}{233.85}$$

Solving,

$$P_V = 12.98 \text{ kips}$$

In trying to bend laterally, about its weak axis, the member would obviously tend to deflect to the right. Thus the stress in the lateral support rods would be tension in the amount

$$P_H = \frac{12.98 \times 94.47}{233.85} = 5.24 \text{ kips}$$

If there is no lateral support, the design stress in bending about the strong axis will probably need to be reduced to provide safety against lateral-torsional buckling. The AISCS covers only specific cross sections. In the present case a conservative estimate of the critical moment that will cause lateral-torsional buckling is provided by

$$M_{cr} = \frac{\pi}{l}\sqrt{JGEI_2} \tag{11.20}$$

For steel, $\sqrt{GE} = 18{,}000$, and

$$M_{cr} = \frac{18{,}000\pi}{l}\sqrt{JI_2}$$

Although overconservative for bent beams, it is convenient in situations not covered by the specifications to convert the beam-buckling problem into an equivalent column problem and thus permit direct use of tables of column design stresses.

The buckling stress in compression is calculated:

$$f_{cr} = \frac{18{,}000\pi c_c}{lI_1}\sqrt{JI_2} \tag{11.21}$$

where c_c is the distance from the 1-1 principal axis to the extreme fiber in compression.

Table 8, Appendix A, of AISCS, gives column design buckling stresses, assuming elastic behavior. Thus, if the beam buckling stress by Eq. (11.21) is calculated, one can enter Table 8 with the corresponding stress and read out the "equivalent" column slenderness ratio, Kl/r. This can be used to determine a safe beam buckling stress by use of the design stress tables for columns, as given in AISCM. The result is overconservative because Eqs. (11.20) and (11.21) neglect the bending contribution to torsional resistance. The equivalent column procedure is now illustrated in Example 11.9.

Example 11.9

Same as Example 11.8, but without any lateral support.

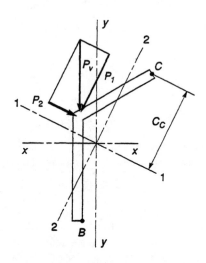

Solution   Resolve P_v into principal-plane components P_1 and P_2:

$$P_1 = P_v \cos\theta = 0.9132P_v$$
$$P_2 = P_v \sin\theta = 0.4075P_v$$

Stress at B = 32.4 ksi (trial):

$$32.4 = \frac{0.9132P_v \times 6 \times 12 \times 7.92}{275.99} + \frac{0.4075P_v \times 6 \times 12 \times 2.25}{22.07}$$

$$32.4 = 1.887P_v + 2.991P_v = 4.878P_v$$

$$P_v = 6.64 \text{ kips}$$

Check the selection by the AISCS interaction formula procedure (refer to Chapter 3, Sec. 3.8). Referring to Example 11.7, the scaled distance from the 1–1 principal axis to the location of average maximum compression stress (C), distance c_c, is 7.55 in. The torsion constant, J, by Eq. (11.8), is

$$J = \frac{(10+8)\,0.75^3}{3} = 2.53 \text{ in}^4$$

By Eq. (11.21),

$$f_{cr} = \frac{18,000 \times 3.1416 \times 7.55}{216 \times 275.99}\sqrt{2.53 \times 22.07} = 53.5 \text{ ksi}$$

Referring to AISCS, Table 8, Appendix A, the equivalent

$$\frac{Kl}{r} = 73.1$$

Note: For remarks on load and support locations in Examples 11.8 and 11.9, refer to Section 11.5.

Now enter Column Design Stress Table 3-36, AISCS, and obtain

$$\phi_c F_{cr} = 23.26 \text{ ksi}$$

for a column with $Kl/r = 73.2$. This will be a safe maximum stress for the beam buckling condition of this problem. It is obvious that the value of $P_v = 6.64$, based on a maximum stress of 32.4 ksi for bending about both axes, is too great. Try $P_v = 5$ kips and calculate bending stresses f_{b1} and f_{b2} separately:

$$f_{b1} = \frac{0.9132 \times 5 \times 72 \times 7.55}{275.99} = 8.99 \text{ ksi}$$

$$= \text{maximum compression stress for bending about 2–2 axis}$$

$$f_{b2} = \frac{0.4075 \times 5 \times 72 \times 2.50}{22.07} = 16.62 \text{ ksi}$$

$$= \text{maximum tension stress for bending about 1–1 axis}$$

Applying the interaction formula,

$$\frac{8.99}{23.26} + \frac{16.62}{32.4} = 0.90 \qquad < 1.0 \quad \text{OK}$$

11.5 Shear Center

If certain structural sections, such as the channel and angle, are loaded through their centroidal axis without any torsional support or torsional restraint at the load points, they will twist. The design problem is then one of combined bending and torsion, as covered in Section 11.3, a complication that may be avoided if the member can be loaded and supported through its *shear-center axis*. In the case of the channel the three-dimensional free-body equilibrium diagram sketched in Fig. 11.8(a) illustrates a short segment, Δx in length, cut from a beam shown loaded through the shear-center axis, so as to avoid twist. Since the loads are parallel to a principal axis of the cross section, the member will be in simple bending and without twist. The torsional couple $\Delta V_f h$ is held in equilibrium by the opposed torsional couple $we_o\Delta x$, as illustrated. Referring to Fig. 11.8(b), the distance from the middle plane of the web to the shear-center axis is

$$e_o = \frac{x_o h^2}{4r_x^2} \tag{11.22}$$

where r_x is the radius of gyration about the xx axis. Equation (11.22) applies to channels with nonparallel flange faces as well as parallel.

If a channel supports beams that frame into it, the arrangement in Fig. 11.9(a) is preferred over that of Fig. 11.9(b). The channel may be assumed to be without twist and loaded through its shear-center axis, in either case, provided that the supported beams are designed for span L measured in each case from the shear-center axis. In addition, in Fig. 11.9(a) connecting bolts should be located as near as possible to a vertical line passing through the shear-center axis, whereas in Fig. 11.9(b) the connection should preferably be designed for an eccentricity of load equal to the distance from the shear-center axis to the bolt line. The channel as a spandrel beam loaded through the shear center is treated in Example 11.11.

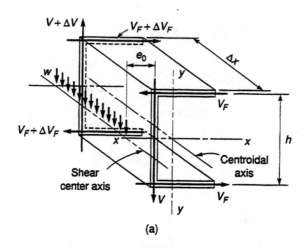

(a)

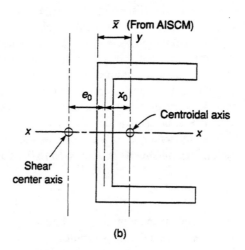

(b)

Figure 11.8   Shear center of channel.

Example 11.10

Select a channel as a 22-ft simple span spandrel beam for a dead load of 2 kips/ft, including weight of channel, and locate 8-in. wall so as to eliminate torsion of beam.

Solution   The maximum bending moment is

$$M_x = \frac{1.4 \times 2 \times 22^2 \times 12}{8} = 2033 \text{ kip-in.}$$

The required section modulus, assuming that $\phi F_y = 32.4$ ksi, is

$$S_x = \frac{2033}{32.4} = 62.7 \text{ in}^3$$

Referring to the beam selection tables, AISCM, MC18 × 51.9 with $S_x = 69.7$ is OK, as is the lighter weight W21 × 44, but if it is desired to hide the beam by the concrete wall, the channel offers advantages in the elimination of the combined bending and torsion problem, taking advantage of the shear-center location, as shown in the sketch.

During construction, at least temporary lateral support would be needed. If no permanent lateral support is provided, it would be desirable to provide shear connectors to make the wall and channel composite. From AISCM, the properties of the MC18 × 51.9 are

$$S_x = 69.7 \text{ in}^3 \qquad r_x = 6.41 \text{ in.} \qquad \bar{x} = 0.858$$

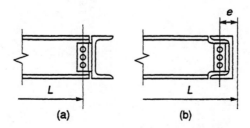

Figure 11.9   Alternative framing arrangements for channel beam.

In Eq. (11.22), $x_0 = \bar{x} - (t_w/2) = 0.858 - 0.300 = 0.558$ in. and $h = d - t_f = 18.0 - 0.625 = 17.375$ in. By Eq. (11.22),

$$e_o = \frac{0.558 \times 17.375^2}{4 \times 6.41^2} = 1.02 \text{ in.}$$

suggesting the relative channel and wall location as shown in the sketch.

Another situation conducive to use of a channel beam may occur if loads must be suspended by hanger rods, as shown in Fig. 11.10. These may be located at the shear center and again the combined bending and torsion problem is avoided. Twist must be prevented at end supports.

Formulas for the shear-center location of other shapes and procedures for determining the location for shapes having no axis of symmetry are given in Ref. 11.3. When a section consists of only two rectangular component parts, the shear center is at the intersection of their two middle planes, as shown for the angle and tee section in Fig. 11.11. Referring back to Examples 11.8 and 11.9, it should be noted that loads and supports were both introduced at the shear-center axis, thus eliminating torsion from the problem.

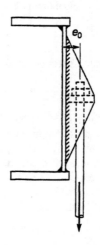

Figure 11.10   Channel girder supporting hanger loads.

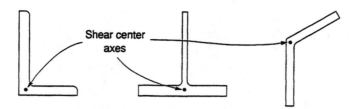

Figure 11.11   Shear-center locations for two-part sections.

Example 11.11

Calculate the maximum concentrated design force that can be applied to a 3-m-long cantilever angle member. The force acts through the shear center of the member. Neglect the weight of the member when determining the force. Use the provisions of the Specification for Load and Resistance Factor Design of Single Angle Members in Sec. 6 of AISCM.

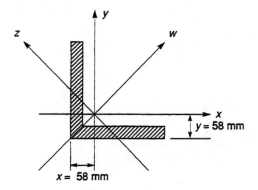

Given: Cross section: L 203 mm × 203 mm × 19 mm (L 8 in. × 8 in. × $\frac{3}{4}$ in.)

$L = 3$ m
$b = 203$ mm
$t = 19$ mm
$A = 7380$ mm^2
$I_x = 28.9 \ 10^6$ mm^4
$I_y = I_x$
$x = 57.8$ mm
$y = x$
$r_z = 40.1$ mm
$I_z = A r_z^2$
$I_w = I_x + I_y - I_z$
$I_z = 1.187 \times 10^7$ mm^4
$I_w = 4.593 \times 10^7$ mm^4
$F_y = 250$ MPa
$E = 200{,}000$ MPa

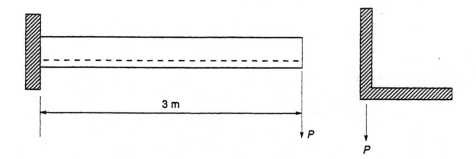

Solution   Use Sec. 5.1 in AISCS Single-Angle Specification. Limit state: local buckling (Sec. 5.1.1)

$$\frac{b}{t} = 10.684 \qquad > 0.382 \sqrt{\frac{E}{F_y}} = 10.805 \qquad \begin{array}{l} \text{no reduction of moment capacity because of} \\ \text{local buckling} \end{array}$$

Geometry and loading of cross section:

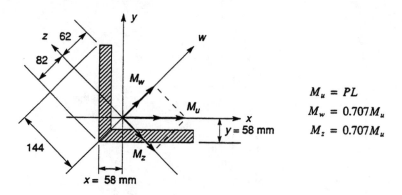

$$M_u = PL$$
$$M_w = 0.707 M_u$$
$$M_z = 0.707 M_u$$

Major axis bending (about w axis), (AISCS, Sec. 5.3.1a):

$C_b = 1.75$ moment varies from zero at the free end to maximum at the fixed end

$$M_{ob} = \frac{C_b \cdot 0.46 b^2 t^2 E}{L} \quad \text{(elastic buckling moment)} = 798.369 \text{ kN·'m}$$

$$M_{yw} = F_y \frac{I_w}{144} = 79.745 \text{ kN·'m} \qquad < M_{ob} = 798.369 \text{ kN·'m}$$

Hence Eq. (5.13) applies for the inelastic buckling strength.

$$M_{nw} = \left(1.58 - 0.83 \sqrt{\frac{M_{yw}}{M_{ob}}}\right) M_{yw} = 105.078 \text{ kN·'m} \qquad > 1.25 M_{yw} = 99.681 \text{ kN·'m}$$

Therefore,

$$M_{nw} = 99.681 \text{ kN·m.}$$

Minor axis bending (about z axis), (AISCS, Sec. 5.3.1b):

$$M_{yz} = \frac{F_y I_z}{82} \quad \text{(yield moment about } z \text{ axis)} = 36.18 \text{ kN·'m}$$

$$M_{nz} = 1.25 M_{yz} = 45.225 \text{ kN·'m}$$

Use interaction equation (6-1b):

$$\frac{M_w}{\phi M_{nw}} + \frac{M_z}{\phi M_{nz}} = \frac{M_u}{\sqrt{2}\phi}\left(\frac{1}{M_{uw}} + \frac{1}{M_{nz}}\right) = \frac{PL}{\sqrt{2}\phi}\left(\frac{1}{M_{nw}} + \frac{1}{M_{nz}}\right)$$

$$\phi = 0.9$$

$$P_u = \frac{\phi\sqrt{2}}{L}\frac{1}{1/M_{nz} + 1/M_{nw}} = 13.199 \text{ kN}$$

REFERENCES

11.1. *Torsional Analysis of Steel Members*, American Institute of Steel Construction, Chicago, 1983.

11.2. BRUCE G. JOHNSTON, "Design of W-Shapes for Combined Bending and Torsion,"*AISC Engineering Journal*, Vol. 19, No. 2, 1982, with discussions by C. G. Salmon, Vol. 19, No. 4, and by P. H. Lin, Vol. 20, No. 2, 1983.

11.3. R. L. BROCKENBROUGH and B. G. JOHNSTON, *The USS Steel Design Manual*, rev. ed., The United States Steel Corporation, Pittsburgh, PA, 1981.

PROBLEMS[*]

11.1. Compare (1) torsional stiffness as measured by the torsion constant *J* for the following sections, which have approximately the same cross-sectional area, and (2) the torsional moment, for which the maximum shear stress is equal to 19.44 ksi. In the case of the W section, calculate the torsion constant by Eq. (11.8) and compare with the handbook value.

(a) Pipe 8 std. ($A = 8.40$ in^2).

(b) Structural tube TS $6 \times 4 \times 0.50$ ($A = 8.36$ in^2).

(c) W8 $\times$ 28 ($A = 8.25$ in^2).

11.2. A vertical pipe supports a sign, as shown, which is to be designed for a wind force of 30 lb/ft^2. Neglecting the weight of the pipe and the sign, and neglecting the wind force on the pipe, select a size for the bending moment and check the maximum shear stress due to combined torsion and bending. Redesign if necessary.

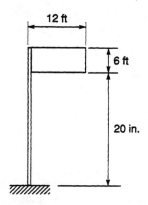

11.3. For the box section as shown in cross section, what is the torsional moment capacity for a maximum shear stress of $0.9 \times 0.6 \times 36 = 19.44$ ksi? At this stress, what would the total angle of twist for a member 28 ft long amount to?

---

[*] A36 steel unless otherwise specified.

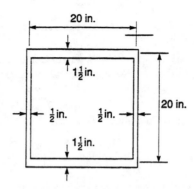

11.4. Using procedures applicable only to very short W sections, as described in connection with Fig. 11.5, determine the maximum normal and shear stress in the flange of the beam made up of three plates as shown.

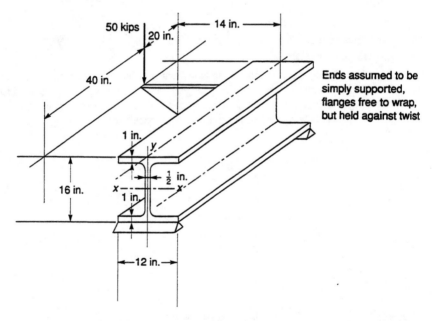

Ends assumed to be simply supported, flanges free to wrap, but held against twist

11.5. Similar to Example 11.3. Move the W section 1 in. laterally to reduce the eccentricity of load from 4 to 3 in. Design as a simple beam instead of continuous and reduce the span from 24 to 18 ft.

11.6. Redesign the situation in Problem 11.5 using a standard rectangular structural tube. Procedure is similar to that of Example 11.4.

11.7. Determine the principal moments of inertia of an $L5 \times 3\frac{1}{2} \times \frac{5}{8}$ by the procedure used in Example 11.6.

11.8. A zee bar, cross section as shown, is used as a simple beam, 21 ft between supports, and is loaded at the third points with loads of 2000 lb each. To prevent lateral movement and force bending to remain in the plane of the web, lateral ties are introduced at the third points as shown. Determine the maximum stress due to bending and the force in the lateral ties. In which direction is the tie force applied?

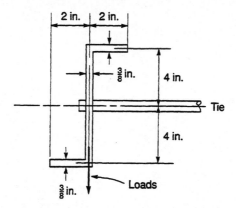

11.9. If the lateral ties are removed (refer to Problem 11.8), what is the maximum stress in the zee bar for the same loads that were applied in Problem 11.8?

11.10. Using a steel with $F_y = 50$ ksi, check the adequacy of the laterally unsupported zee section (refer to Problem 11.9) by use of the adaptation of the AISCM interaction formula used in Example 11.9, utilizing the equivalent column allowable stress procedure for bending about the minor principal axis.

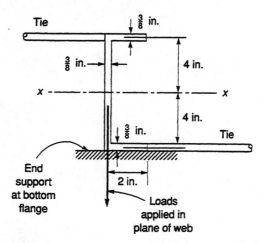

11.11. For a channel section with the same I_x and same area as the zee section of Problem 11.8, and with the same loads and span, ties must now be provided to prevent twisting, as shown, at the third points where the vertical loads are applied. For what force must the ties be designed and in what direction are the tie forces applied?

11.12. Design a welded channel girder of the type shown in Fig. 11.10 for a span of 42 ft to support third-point 60-kip factored loads $2\frac{1}{2}$ in. from the center line of the web with the channel proportioned so that the loads pass through the shear center to eliminate torsion. E70 electrodes and A36 steel may be assumed.

Index

A

AISC Manual (AISCM), 2
AISC Specification (AISCS), 2
AISC Specification, beam-column
     design criteria, 122–125
  beam design, 49, 55–63
  bolted and riveted connection
     design, 139–143
  column design criteria, 103
  intermediate stiffener design, 195
  plate girder flange design, 190
  plate girder web design, 188
  second-order effects, 228–320
  tension member design, 30
AISI Specification for cold formed
     steel, 6
Allowable stress design, 9, 12
American Institute of Steel
     Construction, 2, 21
American Iron and Steel Institute, 6,
     21
Angle beam, 311–312
Angle struts, 96, 106–108, 115–116
Articulated beam design, 224–225
Axially loaded column, 86–92

B

Beam, biaxial bending, 65–65, 73–76
  classification of sections, 57–58
  compactness related limit states,
     57–58
  cross section nomenclature, 47–48
  definition, 45, 55
  deflection limitation, 64–65
  design criteria, 49, 55–63
  design for lateral-torsional
     buckling, 60–63
  elastic bending theory, 50–52
  elastic flexure stresses, 51–52
  elastic shear stresses, 52–53
  fatigue design, 65, 71–72
  flexural design, 55–63
  forces acting on it, 46
  hybrid definition, 55

Beam, biaxial bending *(cont.)*
  inelastic bending theory, 53–55
  lateral buckling, 49, 60–63
  lateral restraint, 49
  lateral support requirements,
      63–64
  limit states, 56
  load tables, 68
  plastic moment, 54
  shear design, 63
  simple, 45–46
  singly-symmetric cross section, 62,
      76–81
  support details, 66–67
Beam-column, biaxial bending,
      132–134
  definition, 119–121
  design by interaction equation,
      122–125
  design criteria, 122–125
  equivalent axial compression load,
      125
  load deflection curve, 121
  second-order effects, 124–125
Bearing plate, 67–68
Bearing stiffeners, 197–198
Bending and torsion, 286–300
Biaxial bending, 65–66, 73–76,
      300–307
  beam-columns, 132–134
Biaxial bending and lateral-torsional
      buckling, 300–307
Block shear rupture strength,
      147–148
Box beams, 59
Box columns, 95, 105
Bracket connection, 146–148,
      158–159
Building codes, 7
Butt-splice design, 144–146, 157

C

Cable, 24, 25–26
Calibration, 16
Cold-formed steel specification, 7
Column, base plate design, 101
  basic strength, 87–92
  composite, 273–275
  cross section types, 94–99
  definition, 86
  design criteria, 103
  effective length, 88–89, 92–94

initial imperfection, 90
  slenderness ratio, 89
  width-thickness ratios, 99–101
Column Research Council, 89
Columns with lacing, battens or
      perforated cover plates, 98–99
Compact cross section, 57–58
Components of a structure, 3
Composite beam, effective width,
      270–272
  elastic behavior, 261–264
  plastic behavior, 264–270
  shear connector design, 272
  shoring, 270
Composite columns, 273–275
Composite construction, 259–261
Computer aided optimization,
      257–258
Computer aided technology, 249–258
Connection, definition of types,
      136–137
  eccentrically loaded, 159–166
  for building frames, 166–179
  friction type, 141
  in shear, 139
  plate-girder elements, 198–202
  prying action, 175–176, 178
  riveted or bolted, 139–148
  welded, 150–159
Connectors in combined shear and
      tension, 142
Continuous beam, elastic design,
      218–224
  plastic design, 237–238
Continuous frame, elastic design,
      225–234
  plastic design, 238–246

D

Deflection limitations of beams, 64–65
Design strength, in tension, 30

E

Eccentric connection, bolted, 160–162,
      164–165
Eccentric connection, welded,
      162–166
Economy of structures, 10–12
Effective length of columns, 88–89,
      92–94
Effective net area in tension, 31–32

Effective width of composite beams, 270–272
Eyebar, 24, 27–29

F

Factor of safety, 9, 12–13
Fatigue design, 32–33, 43
Fillet weld design, 153–155
Fillet weld geometry, 153
Fillet welded angle connections, 156–157
Flange local buckling limit state, 57–58
Flexible framing connection, 166
Flowchart for computer aided design system, 252
Flowchart of column design, 103
Flowchart programming, 250–251
Flowcharts for plate girder design, 188, 190, 195
Flowcharts for tension member design, 33–35
Frame second-order effect, 228–230
Friction-type connections, 141
Fully restrained connection, 136–137, 138

H

Hybrid beams, 55

I

Initial imperfection of column, 90
Interaction equations, 122–125

L

Lap joint design, 143–144, 157
Lateral buckling of beams, 49
Lateral restraint on beams, 49
Lateral support requirements, 64
Lateral-torsional buckling limit state, 60–63, 300–307
Leaner columns, 226–227
Limit of structural usefulness, 14
Limit states design, 10
Load and resistance factor design, 9
Load effects, 13
Load factors, 13, 16
Loads, ASCE load standard, 7, 22
Loads, dead load, 7

Loads, live load, 7
Local buckling of columns, 99–101
LRFD criteria, 13–17
LRFD Specification, 2

M

Material properties of steel, 4–6
Mechanical properties of steel, 4–6
Moment distribution method, 216–218, 220–223
Moment-resisting connections, 173–179

N

Non-compact cross section, 57–58

P

Partially restrained connection, 137, 138
Pinned connections, 148–150
Pipe columns, 95
Plastic design theory, 234–236
Plastic moment, 54
Plastic section modulus, 54
Plate girder, bearing stiffener, 197–198
    definition, 47, 55
    flange selection, 189–194
    interaction between bending and shear, 196–197
    intermediate stiffener design, 194–196
    tension-field action, 186
    types, 183–185
    vertical buckling strength, 187
    web selection, 185–188
Probalistic design, 14–15
Prying action, 175–176, 178

R

Reliability index, 15
Repeated loads, tension member design, 32–33, 43
Resistance, 14–16
Resistance factors, 13, 16
Riveted and bolted connections, 139–148
Rod, 24, 26–27

20

S

Safety of structures, 12–19
Seat angle connection, 167–173
Second-order effect on frame design, 228–320
Semi-rigid connections, 176–179
Shape factor, 54
Shear center, 307–312
Shear connector design, 272
Shear design of beams, 63
Shored construction, 270
Slender cross section, 57–58
Specification, definition, 2
Specification for structural joints, 142
Steel, common types and their properties, 5
  material properties, 4–6
  tensile strength values, 5
  yield stress values, 5
Steel Joist Institute, 7, 22
Stiffener design for plate girders, 194–201
Structural Stability Research Council Guide, 89

Structure, construction:
  details, 17–19
  planning, 17
  site exploration, 17

T

Tension member, common shapes, 29, 30
  definition, 2
  design for repeated loads, 32–33, 43
  effective net area, 31–32
  flowchart for design, 33–35
  limit states, 30
Torsion, 282–299
  approximate formulas, 288–289
  closed section, 283–285
  open section, 286–290
Torsion constant, 284–285

W

Web crippling, 67
Web local buckling limit state, 57–58
Welded connections, 150–159
Welded joint types, 151–153